Exploring Abiotic Stress in Plants: Mechanisms, Adaptations, and Mitigation Strategies

Exploring Abiotic Stress in Plants: Mechanisms, Adaptations, and Mitigation Strategies

Guest Editor

Tarek Alshaal

MDPI

Basel • Beijing • Wuhan • Barcelona • Belgrade • Novi Sad • Cluj • Manchester

Guest Editor
Tarek Alshaal
Institute of Applied Plant
Biology
University of Debrecen
Debrecen
Hungary

Editorial Office
MDPI AG
Grosspeteranlage 5
4052 Basel, Switzerland

This is a reprint of the Special Issue, published open access by the journal *International Journal of Molecular Sciences* (ISSN 1422-0067), freely accessible at: https://www.mdpi.com/journal/ijms/special_issues/22ZOM078BI.

For citation purposes, cite each article independently as indicated on the article page online and as indicated below:

Lastname, A.A.; Lastname, B.B. Article Title. *Journal Name* **Year**, *Volume Number*, Page Range.

ISBN 978-3-7258-7561-0 (Hbk)
ISBN 978-3-7258-7562-7 (PDF)
https://doi.org/10.3390/books978-3-7258-7562-7

Cover image courtesy of Tarek Alshaal

Contents

About the Editor

Tarek Alshaal

Tarek Alshaal is a senior research fellow at the Institute of Applied Plant Biology at the University of Debrecen, Hungary, and at the Department of Food Biotechnology of Széchenyi István University Mosonmagyaróvár, Hungary. His research focuses on soil fertility and plant nutrition, sustainable soil management, plant responses to abiotic stress, selenium biofortification, soil microbiology, and the development of biostimulants and biofertilizers for improving crop productivity and resilience. He has also contributed to research on marginal land restoration using bioenergy crops, including giant reed, and on green biorefinery approaches such as protein isolation from biomass plants and agricultural residues.

He has participated as an investigator and co-investigator in several national and international research projects related to soil health, environmental remediation, plant nutrition, and sustainable agricultural production systems. His academic activities include supervision of graduate students and mentoring young researchers in soil science and plant biotechnology. He has authored more than sixty peer-reviewed journal articles and over one hundred scientific contributions with more than two thousand citations. His work has been recognized through multiple scientific excellence and publication awards, as well as international research scholarships.

Tarek Alshaal serves as Academic Editor for *Indoor Air* and *BMC Plant Biology* and as a Topic Advisory Panel Member for the *International Journal of Molecular Sciences* and has contributed as Guest Editor to several international journals, including the *International Journal of Molecular Sciences, Plants,* and *Frontiers in Nanotechnology*.

Preface

This Reprint brings together selected contributions that advance our current understanding of plant responses to abiotic stress and their relevance for sustainable agriculture and environmental stability. Its subject centers on the mechanisms through which plants perceive, integrate, and respond to adverse environmental conditions, including drought, heat, salinity, heavy metal exposure, nutrient imbalance, and anthropogenic chemical stress. By presenting research that spans molecular regulation, physiological adaptation, metabolic reprogramming, and applied mitigation strategies, this Reprint reflects the multidisciplinary character of contemporary plant stress biology.

The aim of this Reprint is to provide readers with a coherent overview of recent progress in identifying key regulatory pathways and adaptive processes that support plant resilience under challenging environmental conditions. Particular attention is given to the integration of omics technologies, physiological analyses, and agronomic approaches that together contribute to the development of effective strategies for improving stress tolerance in crops and other plant systems. The collected studies illustrate how mechanistic knowledge can support practical solutions that enhance productivity and stability in both agricultural and natural ecosystems.

This Reprint was prepared in response to the growing need for integrated scientific perspectives on plant adaptation to environmental stress in the context of climate variability, soil degradation, and increasing pressure on global food production systems. It is intended for researchers, graduate students, plant breeders, agronomists, and professionals working in plant biology, biotechnology, and environmental sciences who are interested in recent advances and future directions in abiotic stress research.

Tarek Alshaal
Guest Editor

International Journal of *Molecular Sciences*

MDPI

Editorial

Special Issue: Exploring Abiotic Stress in Plants—Mechanisms, Adaptations, and Mitigation Strategies

Tarek Alshaal [1,2,3]

[1] Faculty of Agricultural and Food Sciences and Environmental Management, Institute of Applied Plant Biology, University of Debrecen, Böszörményi Str. 138, 4032 Debrecen, Hungary; alshaal.tarek@agr.unideb.hu or alshaal.tarek.ali.ahmed.ibrahim@sze.hu or tarek.ibrahim@agr.kfs.edu.eg

[2] Department of Food Biotechnology, Albert Kazmer Mosonmagyarovar Faculty, Széchenyi István University, Egyetem Sqr. 1, 9026 Győr, Hungary

[3] Soil and Water Department, Faculty of Agriculture, University of Kafrelsheikh, Kafr El-Sheikh 33516, Egypt

Abiotic stress remains one of the biggest obstacles globally to plant growth, productivity, and ecological stability. Understanding how plants sense, integrate, and respond to non-biological stress is a fundamental scientific and societal challenge due to the rapid intensification of the agricultural sector, soil degradation, climate change, and environmental pollution. The *International Journal of Molecular Sciences* Special Issue on "Exploring Abiotic Stress in Plants: Mechanisms, Adaptations, and Mitigation Strategies" offers a thorough and integrated forum for current developments in this rapidly developing topic.

Together, the 10 papers in this collection cover a wide range of topics addressing abiotic stress biology, including wounding, heat, salinity, drought, heavy metal toxicity, nutritional imbalance, and xenobiotic exposure. They cover subjects including whole-plant physiology, metabolomics, agronomic interventions, molecular regulation, and phosphoproteomics. Collectively, these investigations highlight the intricacy of stress reactions, while also pointing to viable paths toward reductions via genetic engineering, metabolic engineering, and creative agronomic techniques.

The goal of plant abiotic stress research is to fundamentally understand how sessile organisms preserve homeostasis in dynamic, frequently hostile environments. Plants use complex sensor systems, signaling cascades, and the adaptive reprogramming of metabolism and development since they are unable to avoid stress. This systems-level viewpoint is reflected in the Special Issue, which emphasizes the interconnection of metabolic plasticity, nutritional signaling, transcriptional regulation, redox biology, and epigenetic and post-translational alterations.

One of the most urgent environmental issues is heavy metal stress, especially in soils that have been affected by industrialization and intensive farming. Toxic levels of cadmium (Cd) and copper (Cu), which frequently co-occur in contaminated areas, interfere with photosynthesis, disturb cellular redox balance, and prevent root growth. The potential of folcisteine (NATCA) to promote root development and maize seed germination under combined copper–cadmium stress is examined by Dong et al. [contribution 1]. Their research highlights the critical function of redox buffering mechanisms in preserving cellular integrity under metal-induced oxidative stress by showing that ascorbate–glutathione (AsA–GSH) cycle regulation is essential for stress reduction. Folcisteine is a promising exogenous regulator that can improve early developmental resilience by maintaining glutathione homeostasis and increasing antioxidant enzyme activity.

Marques et al. [contribution 2] investigate phytochelatins as important molecular agents in cadmium mitigation, which supports this physiological viewpoint. By chelating

https://doi.org/10.3390/ijms27052182

heavy metals and facilitating their sequestration into vacuoles, phytochelatins—which are enzymatically generated from glutathione—reduce cytosolic toxicity. The authors describe genetic approaches for modifying phytochelatin pathways in order to enhance functional tolerance. A forward-looking review of phosphoproteomics in plants exposed to cadmium is provided by Marques et al. [contribution 3], who highlight how post-translational phosphorylation processes fine-tune stress signaling networks. Together, these two papers highlight the importance of combining biochemical and proteomic investigations for understanding stress adaptation by bridging traditional detoxifying systems and state-of-the-art omics techniques.

Other abiotic stimuli, like heat, salinity, and drought, challenge plants through ion imbalance, membrane instability, and water deficiency, while heavy metals mainly cause oxidative damage and metabolic interference, upsetting cellular homeostasis. Tang et al. [contribution 4] provide a thorough examination of maize's resistance to heat stress and drought, emphasizing hormonal interactions, transcriptional reprogramming, and physiological adaptations. Their observations support the idea that coupled stresses often result in non-additive effects, the authors calling for integrative research that takes outdoor conditions into account rather than relying solely on single-stressor lab models.

Similarly, Jia et al. [contribution 5] show that tomato tolerance to salt stress and drought is improved by overexpressing the transcription factor SlPLATZ17. This study serves as an example of how osmotic adjustment, antioxidant defense, and stress-responsive gene expression can all be coordinated through the focused manipulation of regulatory genes. These works connect molecular mechanisms to crop enhancement tactics, thereby advancing translational research.

Plant stress responses are further modulated by fertilization techniques and nutrient availability. Vukmirović et al. [contribution 6] examine the impact of phosphorus fertilization on the antioxidant responses of drought-stressed common beech and sessile oak provenances. Their findings highlight the significance of provenance selection and nutrient management in forestry and ecosystem resilience, showing genotype-specific relationships between oxidative metabolism and nutrient status. Optimizing the nutrient supply in agricultural systems can improve adaptive capacity and mitigate the effects of stress.

In order to understand the mechanisms behind phenolic compound increases in *Lonicera japonica* caused by nitrate nitrogen supply, Cao et al. [contribution 7] combine metabolomics and photosynthetic study. Phenolic chemicals have two functions: they are both signaling molecules and antioxidants. This work demonstrates how nutritional regimes influence metabolic plasticity and the generation of bioactive compounds by connecting nitrogen feeding to secondary metabolism and photosynthetic performance. These results are especially pertinent to high-value crops and medicinal plants, since stress-responsive metabolites are tightly linked to qualitative features.

A growing aspect of plant stress biology is anthropogenic chemical exposure, which extends the concept of abiotic stress beyond environmental extremes. According to Zhang et al. [contribution 8], nitrogen-doped carbon dots can reduce tomato pesticide toxicity by controlling the antioxidant system. This novel technique based on nanotechnology raises new concerns about environmental safety and long-term ecological effects, while demonstrating the potential of designed nanomaterials to reduce chemical stress. It also serves as an example of how interdisciplinary approaches, such as combining plant physiology with nanoscience, can produce useful mitigation techniques.

Even though it is frequently linked to biotic interactions, mechanical damage is a type of abiotic stress that causes significant metabolic changes. In their study of metabolic reactions in injured Hippeastrum bulb scales, Wiczkowski et al. [contribution 9] find dynamic changes in primary and secondary metabolism. The intersections of wound-induced

https://doi.org/10.3390/ijms27052182

metabolic remodeling with defense priming, oxidative signaling, and resource allocation highlight the common molecular mechanisms underlying various stress responses.

The rearrangement of primary carbon metabolism is at the heart of many of these investigations. Hu et al. [contribution 10] investigate the regulating function of phosphoenolpyruvate (PEP) and its associated metabolites in plant growth and development. PEP links glycolysis, the shikimate pathway, and biosynthetic fluxes, making it a crucial metabolic node. Changes in PEP availability and utilization under stress can affect secondary metabolite synthesis, energy balance, and general growth dynamics. This contribution offers a conceptual link between stress adaptation and metabolic management by placing PEP into a larger regulatory network.

A number of recurrent themes emerge throughout this Special Issue.

First, the unifying principle of abiotic stress tolerance is redox homeostasis. Plants constantly activate antioxidant systems to limit reactive oxygen species (ROS) in response to many stressors, including heavy metals, drought, salt, pesticide exposure, and wounding. Superoxide dismutase, catalase, peroxidases, the AsA–GSH cycle, and non-enzymatic antioxidants are frequently mentioned as key players. However, ROS serve as signaling molecules in addition to having harmful consequences. Stress recognition and subsequent transcriptional activation are supported by the delicate balance between ROS production and scavenging.

Second, adaptive plasticity is orchestrated by transcriptional and post-translational regulation. The rapid reprogramming of signaling networks is shown in phosphoproteomic studies, and gene expression landscapes are coordinated by transcription factors such SlPLATZ17. Plants can dynamically integrate external stimuli with their internal metabolic status because of these regulatory layers.

Third, metabolic flexibility is both the result of and a factor in stress adaptation. The re-allocation of resources toward survival is reflected in changes in phenolic biosynthesis, carbon allocation, nitrogen assimilation, and key intermediates like PEP. The significance of systems-level techniques is demonstrated through the integration of metabolomics with physiological measures, as shown in multiple contributions.

Fourth, mitigation techniques are becoming more complex. These consist of new nanomaterials, the genetic engineering of detoxifying and regulatory mechanisms, the exogenous administration of protective chemicals (such as folcisteine), and optimized fertilization regimens. Crucially, these tactics are based on mechanical knowledge rather than empirical experimentation alone.

The urgent need to ensure crop yield and ecosystem stability in the face of global environmental change forms the larger backdrop of this Special Issue. According to climate projections, heatwaves, extended droughts, and soil salinization are becoming more frequent. Heavy metal buildup and chemical residues in soils are concurrently caused by industrialization and intensive agriculture. Water scarcity and nutrient imbalances affect forest ecosystems, and high-value horticultural and medicinal crops need to be preserved under adverse conditions.

Molecular biology, biochemistry, physiology, ecology, agronomy, and biotechnology must all be integrated to address these problems. The strength of such an integration is demonstrated by the contributions in this volume. Additionally, these articles emphasize the significance of model and crop species diversity, ranging from medicinal plants and forest trees to maize and tomatoes. Breeding and management techniques require an understanding of the intrinsic species and genotype specificity of stress tolerance.

Crucially, this Special Issue demonstrates that reactions to abiotic stress are rarely isolated events. Sequential and combined stresses are the rule rather than the exception. Future studies must use multifactorial designs more often and make use of computational

https://doi.org/10.3390/ijms27052182

modeling, sophisticated phenotyping platforms, and high-throughput omics technologies. Predictive frameworks for resilience will be made possible by integrating transcriptome, proteomic, metabolomic, and phenomic data.

The goal of translational applications is still crucial. Real-world impact is determined by agronomic viability and field validation, although molecular discoveries serve as the basis for this. Translating mechanistic insights into resilient cultivars and sustainable management techniques will require cooperation between academic researchers, breeders, agronomists, and policymakers.

As this Special Issue's Guest Editor, I want to express my sincere gratitude to every one of the contributing authors for their inventive and hard work, as well as to the reviewers for their helpful criticism. The contributions' richness and breadth demonstrate how vibrant the field of abiotic stress research is. This collection aims to serve as a resource for experts and a gateway for researchers wishing to delve into the intricacies of plant stress biology by combining mechanistic insights and practical viewpoints.

The goal of turning this Special Issue into a printed book is to present a coherent story that places each study inside a single conceptual framework. The book charts paths for further research while capturing a point in time in a field that is developing quickly. It emphasizes how important findings in signaling, metabolism, and redox biology can guide workable answers to environmental and agricultural problems.

Ultimately, the more general objectives of food security, environmental sustainability, and climate resilience are inextricably linked to comprehending and reducing abiotic stress in plants. Terrestrial life and human sustenance are based on plants. Improving their resilience to environmental stress is a scientific challenge as well as a social obligation. The contributions gathered here are significant steps in the direction of that goal.

Acknowledgments: The Guest Editor sincerely thanks all authors who contributed their high-quality research and valuable efforts to this Special Issue of the *International Journal of Molecular Sciences*. This work is supported by the University of Debrecen Program for Scientific Publication.

Conflicts of Interest: The author declares no conflicts of interest.

List of Contributions:

1. Dong, L.; Zhao, M.; Hou, K.; Wei, J.; Liu, Z.; Wang, R.; Zhou, Y.; Gu, W. Does Folcisteine (NATCA) Play a Role in Facilitating Seed Germination, Root Development, and Elevating Root AsA-GSH Cycle Efficiency Under Combined Copper–Cadmium Stress in Maize? *Int. J. Mol. Sci.* **2025**, *26*, 11220. https://doi.org/10.3390/ijms262211220.
2. Marques, D.N.; Thiengo, C.C.; Azevedo, R.A. Phytochelatins and Cadmium Mitigation: Harnessing Genetic Avenues for Plant Functional Manipulation. *Int. J. Mol. Sci.* **2025**, *26*, 4767. https://doi.org/10.3390/ijms26104767.
3. Marques, D.N.; Piotto, F.A.; Azevedo, R.A. Phosphoproteomics: Advances in Research on Cadmium-Exposed Plants. *Int. J. Mol. Sci.* **2024**, *25*, 12431. https://doi.org/10.3390/ijms252212431.
4. Tang, H.; Zhang, L.; Xie, X.; Wang, Y.; Wang, T.; Liu, C. Resilience of Maize to Environmental Stress: Insights into Drought and Heat Tolerance. *Int. J. Mol. Sci.* **2025**, *26*, 5274. https://doi.org/10.3390/ijms26115274.
5. Jia, X.; Pang, X.; Xu, M.; Wang, C.; Wei, H.; Liu, J.; Zhang, H.; Li, D.; Xu, X.; Zhao, T. The Overexpression of SlPLATZ17 Can Increase the Tolerance of Tomatoes to Drought and Salt Stress. *Int. J. Mol. Sci.* **2025**, *26*, 1836. https://doi.org/10.3390/ijms26051836.
6. Vukmirović, A.; Škvorc, Ž.; Bogdan, S.; Krstonošić, D.; Bogdan, I.K.; Karažija, T.; Bačurin, M.; Brener, M.; Sever, K. The Role of Phosphorus Fertilization in Antioxidant Responses of Drought-Stressed Common Beech and Sessile Oak Provenances. *Int. J. Mol. Sci.* **2025**, *26*, 3053. https://doi.org/10.3390/ijms26073053.

7. Cao, Y.; Yang, Y.; Tan, Z.; Feng, X.; Tian, Z.; Liu, T.; Pan, Y.; Wang, M.; Su, X.; Liang, H.; et al. Metabolomics Combined with Photosynthetic Analysis Reveals Potential Mechanisms of Phenolic Compound Accumulation in Lonicera Japonica Induced by Nitrate Nitrogen Supply. *Int. J. Mol. Sci.* **2025**, *26*, 4464. https://doi.org/10.3390/ijms26094464.
8. Zhang, X.; Xin, Y.; Wang, H.; Dang, Y.; Wang, W.; Gao, Y.; Han, Y.; Kang, R.; Shi, Q.; Du, H. Nitrogen-Doped Carbon Dots Alleviate Pesticide Toxicity in Tomato by Regulating Antioxidant Systems. *Int. J. Mol. Sci.* **2025**, *26*, 9916. https://doi.org/10.3390/ijms26209916.
9. Wiczkowski, W.; Lahuta, L.B.; Szawara-Nowak, D.; Stałanowska, K.; Saniewski, M.; Marasek-Ciołakowska, A.; Góraj-Koniarska, J.; Horbowicz, M. The Metabolic Screening in Wounded Scales of Hippeastrum × Hybridum Hort. Bulbs. *Int. J. Mol. Sci.* **2025**, *26*, 9179. https://doi.org/10.3390/ijms26189179.
10. Hu, R.; Yu, H.; Deng, J.; Chen, S.; Yang, R.; Xie, H.; Tang, X.; Yu, Y.; Duan, Y.; Zhang, M.; et al. Phosphoenolpyruvate and Related Metabolic Pathways Contribute to the Regulation of Plant Growth and Development. *Int. J. Mol. Sci.* **2025**, *26*, 391. https://doi.org/10.3390/ijms26010391.

https://doi.org/10.3390/ijms27052182

International Journal of
Molecular Sciences

MDPI

Article

Does Folcisteine (NATCA) Play a Role in Facilitating Seed Germination, Root Development, and Elevating Root AsA-GSH Cycle Efficiency Under Combined Copper–Cadmium Stress in Maize?

Ling Dong, Meng Zhao, Kangbo Hou, Jingwen Wei, Ziwen Liu, Runze Wang, Yu Zhou * and Wanrong Gu *

College of Agriculture, Northeast Agricultural University, Harbin 150030, China; dongling@neau.edu.cn (L.D.); zmeng1116@163.com (M.Z.); s220301018@neau.edu.cn (K.H.); s2503022090@neau.edu.cn (J.W.); s2503022063@neau.edu.cn (Z.L.); s2503012109@neau.edu.cn (R.W.)

* Correspondence: zhouyu@neau.edu.cn (Y.Z.); wanronggu@neau.edu.cn (W.G.)

Abstract: Copper (Cu) and cadmium (Cd) are common co-occurring environmental pollutants inducing combined stress, which severely harms maize growth. Previous studies have confirmed the involvement of the ascorbate–glutathione (AsA-GSH) cycle in heavy metal stress tolerance, but the regulatory effect of NATCA on this cycle under Cu-Cd combined stress—especially during maize seed germination and root development—remains unelucidated. Exogenous folcisteine (NATCA, 3-acetylthiazolidine-4-carboxylic acid) can enhance plant tolerance to abiotic stress. This study investigated the role of NATCA, a novel plant growth regulator with antioxidant potential, in alleviating Cu-Cd combined stress in maize. Two maize (*Zea mays* L.) varieties—Jiuyuan 15 (Cu/Cd-tolerant) and Longfuyu 6 (Cu/Cd-intolerant)—were exposed to combined stress (80 mg·L^{-1} $CuSO_4$ + 100 mg·L^{-1} $CdCl_2$) with/without 20 mg·L^{-1} NATCA. Germination and hydroponic experiments were conducted to investigate NATCA's effects on seed germination, growth, root traits, photosynthetic characteristics, reactive oxygen species (ROS) metabolism, AsA-GSH cycle (ascorbate–glutathione cycle), and endogenous hormones under stress. The results showed that combined Cu/Cd stress inhibited seed germination (reduced vigor, rate, index), while NATCA significantly reversed these declines, increased tolerance index, lowered relative damage rate, and improved seed activity—with more pronounced effects on Longfuyu 6. Stress stunted seedling growth (reduced dry/fresh weight, water content; increased water deficit), whereas NATCA promoted growth (taller plants, less leaf chlorosis, more fibrous roots), enhanced dry matter accumulation, and improved water metabolism. Stress impaired root morphology (shorter length, smaller surface area/volume, fewer tips) and absorption capacity; NATCA improved root traits, stress tolerance, and vitality. Stress weakened ROS scavenging, but NATCA elevated antioxidant enzyme activity and non-enzymatic antioxidant content, strengthened AsA-GSH cycle-mediated ROS clearance, mitigated stress damage, and maintained intracellular ROS balance in maize seedling root. These findings not only reveal a new regulatory role of NATCA in enhancing heavy metal stress tolerance via the AsA-GSH cycle but also provide a potential eco-friendly strategy for improving maize production in heavy metal-contaminated soils.

Keywords: maize; folcisteine; seed germination; root development; AsA-GSH cycle; copper–cadmium stress

1. Introduction

Soil heavy metal pollution, a complex and insidious environmental issue, presents a formidable challenge precisely because it defies direct assessment and intuitive perception by humans. Unlike more overt forms of pollution, such as visible waste dumps or oil slicks on water surfaces, heavy metal contaminants within the soil matrix remain concealed from our immediate senses. This inability to readily detect or gauge their presence through straightforward observational means has led to soil heavy metal pollution emerging as an elusive hidden danger that lurks silently beneath our feet, yet has the potential to wreak havoc on various ecological and human systems [1]. Characterized by concealment, latency, cumulativeness, and irreversibility, heavy metal contamination of soil is often overlooked despite its severe detrimental effects [2]. The damage inflicted on soil ecosystems by heavy metals is irreversible, and remediation of contaminated soil remains technically challenging and economically costly. Atmospheric deposition constitutes a primary source of regional soil heavy metal pollution [3]. Additionally, extensive discharge of industrial wastewater into farmland as irrigation water contributes significantly to such pollution [4,5]; notably, excessive application of phosphorus fertilizers can elevate soil cadmium (Cd) content. Concurrently, Cd pollution arising from atmospheric deposition and agricultural irrigation is on the rise [6]. The global concern over heavy metal pollution in agricultural soils is exemplified by China's 2014 survey, which revealed exceedance rates of 16.1% for soil and 19.4% for arable land, with Cu and Cd notably impairing crop yield, quality, and human health via the food chain [7–9]. A 2018 report revealed moderate heavy metal pollution in southern Songnen Plain, Heilongjiang, with Cu and Cd being predominant, evidenced by a concentration of 42.8 $mg \cdot kg^{-1}$ in a carex scoparia swamp and 0.77% of samples exhibiting moderate Cd pollution, respectively [10,11]. Thus, investigating methods to mitigate heavy metal harm to crops and their underlying response mechanisms holds significant importance.

The phytotoxicity of both essential Cu and non-essential Cd manifests in suppressed seed germination and growth through the disruption of metabolic processes, including limited starch fixation, nutrient supply, and decreased proteolytic enzyme activity [12]. The suppressive effect of Cu and Cd on seed germination potential, rate, and index is primarily mediated by disrupting metabolic homeostasis—affecting soluble sugar and trace element metabolism—and by inhibiting the activities of α-amylase and acid phosphatase [13]. Phytotoxicity assays revealed that Cu^{2+} and Cd stress significantly inhibit seed germination and early seedling growth in white clover and wheat, adversely affecting key parameters such as germination rate, potential, and root morphology [14,15]. Research has found that plants treated with Cd^{2+} exhibit symptoms such as short and brown roots, reduced lateral roots, slow stem growth, yellowing, curling, and spotted leaves, short stature, and yellowing of leaves [16]. By inhibiting the activity of starch-hydrolyzing enzymes like α-amylase, high-concentration Cd treatment disrupts the mobilization of energy reserves, leading to insufficient metabolic processes for seed germination [17–20]. Cu and Cd can bind to the catalytic sites of proteolytic enzymes, thereby disrupting protein hydrolysis in seeds. This impairment leads to insufficient nitrogen supply to the embryo, ultimately hindering seed germination [21]. Additionally, Cu and Cd treatments disrupt the physiological balance of seedlings, resulting in reduced carbon and nitrogen metabolism, decreased chlorophyll content, damaged cell membrane structure, and subsequent cell fluid leakage [22]. As Cu and Cd accumulate in plants, the extent of damage intensifies, manifesting as stunted growth, abnormal leaf coloration, and overall inhibited seedling development [23].

Root system morphology, critical for Cu and Cd uptake and xylem transport following initial metal perception, plays a definitive role in regulating the metal's accumulation within the plant [24–26]. Exposure to Cu and Cd significantly inhibits seedling root growth, leading to reduced length, number, and volume, as well as yellowing and curling, which consequently impairs water and nutrient uptake [27]. Previous studies indicate that Cu^{2+} concentration is positively correlated with the root inhibition rate in grasses, demonstrating a clear dose-dependent inhibition of growth [28]. Most plant species exhibit higher concentrations of Cu and Cd in roots than in stems, with limited translocation of these metals to the xylem [29]. The subcellular accumulation of Cd in roots, primarily within the symplast and cell walls, exhibits a dynamic distribution that is highly dependent on cultivation conditions [30]. Over 81% of Cd in rice roots is compartmentalized within the soluble fraction and cell wall, owing to the presence of Cd-complexing ligands such as organic acids, amino acids, and phytochelatins that facilitate its accumulation [31]. The Cu and Cd ions adsorbed on maize roots, primarily precipitated in the epidermis and outer layers, are partitioned into exchangeable, complex-bound, and precipitated fractions, which in turn dictate their specific biological impacts [32]. Most Cu^{2+} and Cd^{2+} bind to the cell wall, leading to cross-linking of pectin in the middle layer and inhibiting cell elongation and growth, thereby suppressing root length [33–37]. Functioning as a key regulatory mechanism, the AsA-GSH cycle enhances plant tolerance by mitigating oxidative damage induced by diverse abiotic stresses [38]. Research indicates that enhancing the activity of key enzymes and the pool sizes of AsA and GSH within the AsA-GSH cycle effectively reduces reactive oxygen species (ROS) levels, thereby boosting plant stress tolerance [39]. In plant cells, ascorbate peroxidase (APX) scavenges H_2O_2, generating monodehydroascorbic acid (MDHA); MDHA further converts to dehydroascorbic acid (DHA). Monodehydroascorbate reductase (MDHAR) facilitates ascorbic acid (AsA) regeneration and MDHA scavenging, while dehydroascorbate reductase (DHAR) primarily reacts with DHA to regenerate AsA [40–42]. Glutathione (GSH) can reduce oxidized glutathione (GSSG) back to GSH; AsA reduces free metal ions and maintains antioxidant enzyme activity, and GSH acts as an antioxidant [43–48]. Previous studies on plant leaves have found that stress initially induces phytotoxicity in leaves, characterized by a suppression of key antioxidant enzymes (APX, GR, MDHAR, DHAR), yet subsequently triggers an adaptive response marked by proline accumulation and the recovery of APX and GR activities [49,50].

Folcisteine (NATCA), also known as 3-acetylthiazolidine-4-carboxylic acid or N-acetylthioproline, is a derivative of organic amino acids. The molecular formula of NATCA is $C_6H_9NO_3S$, with a molecular weight of 175.2056. Folcisteine is an intermediate drug used to treat myocardial ischemia. NATCA is used in agriculture as a plant growth regulator. Alleviating heavy metal-induced damage to plants via exogenous substances has become a current research focus. Specifically, the application of certain exogenous substances can alter the speciation of heavy metals, mitigate their inhibitory effects on plant growth, stimulate the expression of stress-resistant genes, reduce their toxicity by regulating bioavailability, and ultimately promote plant growth [51]. Using maize varieties of differential heavy metal sensitivity, this study aimed to investigate how exogenous NATCA modulates seed germination, root growth, and the root AsA-GSH cycle under Cu-Cd co-stress, offering new methodologies and a theoretical basis for improving maize tolerance.

2. Results

2.1. Evaluation of Germination and Copper–Cadmium Tolerance Parameters During Stress

As shown in Table 1, NATCA treatment enhanced the germination potential, rate, and index of both maize varieties compared to CK, whereas CuCd stress caused a decline in these parameters. After NATCA treatment of Jiuyuan 15 and Longfuyu 6, compared with CK, the germination vigor of the two varieties increased by 5.79% and 25.93%, the germination rate increased by 1.39% and 2.81%, and the germination index increased by 17.23% and 18.23%, respectively. After CuCd treatment, the germination potential of Jiuyuan 15 and Longfuyu

6 decreased by 23.91% and 61.72%, respectively, compared to the CK, the germination rate decreased by 20.83% and 23.24%, respectively, and the germination index decreased by 31.20% and 46.85%, respectively. The decline in germination vigor, germination rate, and germination index of Longfuyu 6 is higher than that of Jiuyuan 15, indicating that Jiuyuan 15 has a higher ability to resist CuCd than Longfuyu 6. Compared with CuCd, after NATCA + CuCd treatment, the germination potential of Jiuyuan 15 and Longfuyu increased by 25.93% and 25.71%, the germination rate increased by 2.81% and 20.17%, the germination index increased by 18.23% and 31.15%, the relative damage rate decreased by 75.52% and 76.22%, and the CuCd tolerance index increased by 31.19% and 69.85%, respectively. The CuCd tolerance index of Longfuyu 6 treated with NATCA + CuCd was 69.87% higher than that of Jiuyuan 15, indicating that NATCA has a more significant effect on the germination of Longfuyu 6 seeds (Table 1).

Table 1. Alleviating effects of NATCA on seed germination under copper–cadmium combined stress by modulating physiological responses in maize.

Variety	Treatment	Germination Vigor (%)	Germination Rate (%)	Germination Index	Relative Damage Degree	Tolerance Index
Jiuyuan 15	Ck	92.00 ± 0.12 a	96.00 ± 0.16 a	26.44 ± 0.12 b		
	NATCA	97.33 ± 0.06 a	97.33 ± 0.67 a	31.00 ± 5.67 a		
	CuCd	70.00 ± 0.02 c	76.00 ± 0.02 b	18.19 ± 2.33 d	0.20 ± 0.34	0.69 ± 0.01
	NATCA + CuCd	88.00 ± 0.23 b	91.33 ± 0.31 a	23.86 ± 2.31 c	0.05 ± 0.36	0.90 ± 0.02
Longfuyu 6	Ck	54.00 ± 0.24 b	94.67 ± 1.76 ab	15.78 ± 4.18 b		
	NATCA	68.00 ± 0.02 a	97.33 ± 0.91 a	18.66 ± 9.01 a		
	CuCd	20.67 ± 0.18 d	72.67 ± 0.55 c	8.39 ± 5.46 c	0.23 ± 3.23	0.53 ± 0.04
	NATCA + CuCd	35.33 ± 0.03 c	90.67 ± 0.13 b	14.27 ± 2.88 b	0.05 ± 0.02	0.90 ± 0.02

Notes: Data are expressed as mean ± standard deviation, where NATCA denotes 1/2 Hoagland nutrient solution containing 20 $mg \cdot L^{-1}$ NATCA; CuCd denotes 1/2 Hoagland nutrient solution containing 80 $mg \cdot L^{-1}$ $CdCl_2$ and 100 $mg \cdot L^{-1}$ $CuSO_4$; and NATCA + CuCd denotes 1/2 Hoagland nutrient solution containing 20 $mg \cdot L^{-1}$ NATCA, 80 $mg \cdot L^{-1}$ $CdCl_2$, and 100 $mg \cdot L^{-1}$ $CuSO_4$, with different letters in the same column indicating a significant difference at the 5% level.

2.2. Growth Phenotype of Maize Seedlings

From Figure 1, it can be clearly observed that the growth of maize seedlings treated with CuCd is inhibited, resulting in stunted growth, yellowing and wilting of leaves, white stripes at the edge of leaf tips, slow and short root growth, less rooting, and yellowing and darkening of roots. Among the two varieties, Longfuyu 6 showed more significant damage after being subjected to CuCd stress. Compared with CuCd treatment, NATCA + CuCd treatment reduced the area of leaf chlorosis and improved root growth in both varieties, with an increase in the number and length of fibrous roots. Compared with CK, NATCA treatment resulted in the best growth of maize seedlings, with high plant height, wide leaves, and well-developed root systems (Figure 1).

2.3. Dry and Fresh Weight, Moisture Content, and Natural Saturation Deficit

As indicated in Table 2, NATCA-treated maize seedlings exhibited the highest fresh weight, dry weight, and moisture content, along with the lowest natural saturation deficit, whereas the CuCd-treated group showed the opposite trend across all parameters. The effects of different treatments on the fresh weight, dry weight, and natural saturation deficit of Jiuyuan 15 and Longfuyu 6 maize seedlings are summarized in Table 2. Compared to the CK group, NATCA treatment alone increased fresh and dry weights in both varieties, whereas CuCd stress significantly reduced fresh weight (by 43.25% and 36.83%, respectively) and dry weight, and markedly increased the natural saturation deficit (by over

120%). The application of NATCA + CuCd mitigated the stress-induced damage, leading to increases in fresh and dry weights and a reduction in the natural saturation deficit by more than 54% compared to the CuCd treatment alone (Table 2).

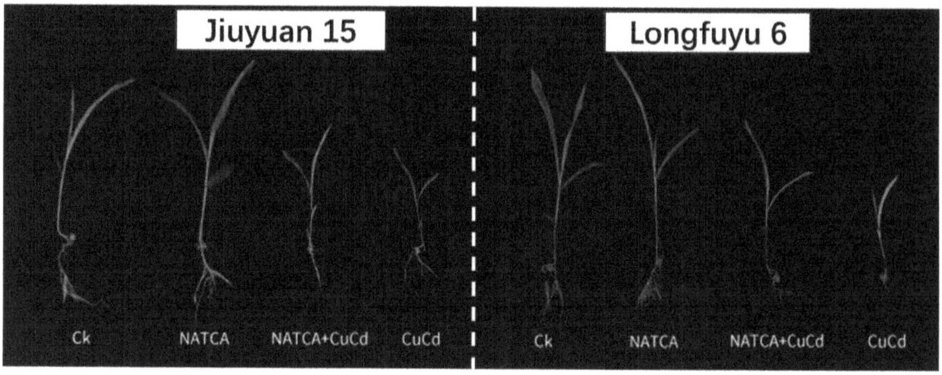

Figure 1. Ameliorative effect of NATCA on the toxicity of copper–cadmium combined stress in maize seedlings.

Table 2. Response of dry matter and water content to NATCA under copper–cadmium combined stress on the fifth day in maize seedlings.

Cultivars	Treatment	Fresh Weight (g·plant^{-1})	Dry Weight (g·plant^{-1})	Water Content (%)	Natural Saturation Deficit (%)
Jiuyuan 15	CK	6.15 ± 0.06 b	0.51 ± 0.01 b	91.78 ± 0.09 a	2.38 ± 0.01 d
	NATCA	7.17 ± 0.15 a	0.56 ± 0.01 a	92.24 ± 0.01 a	3.95 ± 0.01 c
	CuCd	3.49 ± 0.03 d	0.50 ± 0.04 b	85.76 ± 0.05 b	28.34 ± 0.01 a
	NATCA + CuCd	4.36 ± 0.11 c	0.57 ± 0.01 a	86.91 ± 0.01 b	7.70 ± 0.02 b
Longfuyu 6	CK	5.24 ± 0.03 b	0.61 ± 0.04 b	87.48 ± 0.05 b	7.28 ± 0.02 c
	NATCA	6.04 ± 0.04 a	0.66 ± 0.07 a	90.00 ± 0.03 a	12.71 ± 0.06 b
	CuCd	3.31 ± 0.01 d	0.46 ± 0.01 d	84.03 ± 0.02 c	26.24 ± 0.05 a
	NATCA + CuCd	3.66 ± 0.01 c	0.53 ± 0.04 c	87.39 ± 0.07 b	13.19 ± 0.06 b

Notes: Data are expressed as mean ± standard deviation, where NATCA denotes 1/2 Hoagland nutrient solution containing 20 mg·L^{-1} NATCA; CuCd denotes 1/2 Hoagland nutrient solution containing 80 mg·L^{-1} $CdCl_2$ and 100 mg·L^{-1} $CuSO_4$; and NATCA + CuCd denotes 1/2 Hoagland nutrient solution containing 20 mg·L^{-1} NATCA, 80 mg·L^{-1} $CdCl_2$, and 100 mg·L^{-1} $CuSO_4$, with different letters in the same column indicating a significant difference at the 5% level.

2.4. Root Characteristic Parameters

As summarized in Table 3, CuCd stress significantly inhibited maize root growth, as evidenced by reductions in root length, diameter, surface area, volume, and tip number, whereas NATCA treatment promoted these morphological parameters. Root length increased under NATCA (by up to 71.79% in Jiuyuan 15) but decreased under CuCd stress; NATCA + CuCd treatment resulted in a net increase compared to CK (by 56.60% and 17.66%). Average diameter saw a moderate increase with NATCA and a decrease with CuCd, but was recovered and even promoted beyond CK levels with NATCA + CuCd, especially in Longfuyu 6 (+16.83% vs. CuCd). Root surface area and volume followed a similar pattern, with NATCA + CuCd treatment effectively reversing the losses caused by CuCd stress. Total root tip number was the most responsive parameter, dramatically enhanced by NATCA (by 246.94% in Jiuyuan 15), severely reduced by CuCd, and remarkably restored by NATCA + CuCd, with increases exceeding 159% over the CuCd treatment (Table 3).

Table 3. Alleviating effect of NATCA on copper–cadmium stress by modifying root architectural traits at day 5 in maize.

Cultivars	Treatment	Length (cm)	Average Diameter (cm)	Surface Area (cm^2)	Volume (cm^3)	Root Tips
Jiuyuan 15	Ck	8.69 ± 0.53 c	1.09 ± 0.06 a	3.03 ± 0.04 c	0.08 ± 0.05 ab	170.00 ± 6.08 a
	NATCA	14.93 ± 0.90 a	1.11 ± 0.03 a	4.60 ± 0.02 a	0.11 ± 0.02 a	49.00 ± 8.50 c
	CuCd	7.28 ± 0.28 c	0.98 ± 0.07 a	2.49 ± 0.18 d	0.07 ± 0.02 b	28.67 ± 4.48 c
	NATCA + CuCd	11.41 ± 0.54 b	1.05 ± 0.03 a	3.78 ± 0.28 b	0.10 ± 0.08 ab	78.00 ± 6.08 b
Longfuyu 6	Ck	16.07 ± 1.02 ab	1.00 ± 0.09 bc	5.46 ± 0.29 a	0.14 ± 0.05 b	214.00 ± 18.33 a
	NATCA	17.33 ± 0.93 a	1.16 ± 0.04 a	5.61 ± 0.23 a	0.16 ± 0.04 a	101.00 ± 9.29 b
	CuCd	11.61 ± 0.61 c	0.95 ± 0.08 c	4.24 ± 0.03 b	0.12 ± 0.04 b	37.00 ± 2.31 c
	NATCA + CuCd	13.66 ± 0.45 bc	1.11 ± 0.02 ab	4.70 ± 0.17 b	0.13 ± 0.04 b	96.00 ± 4.36 b

Notes: Data are expressed as mean ± standard deviation, where NATCA denotes 1/2 Hoagland nutrient solution containing 20 mg·L^{-1} NATCA; CuCd denotes 1/2 Hoagland nutrient solution containing 80 mg·L^{-1} $CdCl_2$ and 100 mg·L^{-1} $CuSO_4$; and NATCA + CuCd denotes 1/2 Hoagland nutrient solution containing 20 mg·L^{-1} NATCA, 80 mg·L^{-1} $CdCl_2$, and 100 mg·L^{-1} $CuSO_4$, with different letters in the same column indicating a significant difference at the 5% level.

2.5. Activities of Key Enzymes in the Root Ascorbate–Glutathione (AsA-GSH) Cycle in Maize Root

As the primary site perceiving CuCd stress, maize roots exhibited altered AsA-GSH cycle enzyme activities (Figure 2), while APX activity increased under stress compared to CK, GR, DHAR, and MDHAR decreased, and these suppressed activities were restored by NATCA supplementation. As shown in Figure 2, CuCd stress induced a complex time-dependent response in the root antioxidant enzyme activities of both maize varieties. While CuCd treatment initially increased APX activity but markedly decreased GR, DHAR, and MDHAR activities over 1–5 days compared to CK, the application of NATCA + CuCd not only further enhanced APX activity but also dramatically reversed the suppression of the other three enzymes. The restorative effect of NATCA was particularly strong on GR activity, which saw increases of over 160% in Jiuyuan 15 and 120% in Longfuyu 6 by day 5 compared to the CuCd treatment (Figure 2).

2.6. Contents of Ascorbic Acid (AsA) and Glutathione (GSH) in Maize Root

AsA and GSH, as crucial reducing substances in plants, play a key role in preventing membrane lipid peroxidation. As shown in Figure 3, compared with the CK treatment, CuCd stress led to a significant time-dependent decrease in the contents of AsA and GSH in the roots of both maize varieties. However, the addition of NATCA (NATCA + CuCd treatment) effectively reversed this trend, resulting in a substantial increase in the contents of AsA and GSH compared to the CuCd treatment alone. The increase in GSH content was particularly remarkable, with an elevation ranging from 106% to 760% over 1–5 days, indicating that NATCA alleviates oxidative damage by significantly enhancing the ASA-GSH cycle. Overall, compared with the CuCd treatment, the application of NATCA + CuCd increased the contents of AsA and GSH in maize roots. For instance, compared with the CuCd treatment, the AsA content in the roots of Jiuyuan 15 seedlings treated with NATCA + CuCd for 1–5 days increased by 33.49%, 47.38%, 97.9%, 125.05%, and 162.5%, respectively. Similarly, the AsA content in Longfuyu 6 increased by 55.05%, 94.89%, 156.72%, 194.87%, and 162.8%, respectively. Meanwhile, the GSH content in Jiuyuan 15 increased by 106.26%, 201.03%, 317.5%, 547.76%, and 596.59%, respectively, while that in Longfuyu 6 increased by 121.28%, 285.52%, 382.35%, 686.22%, and 760.9%, respectively.

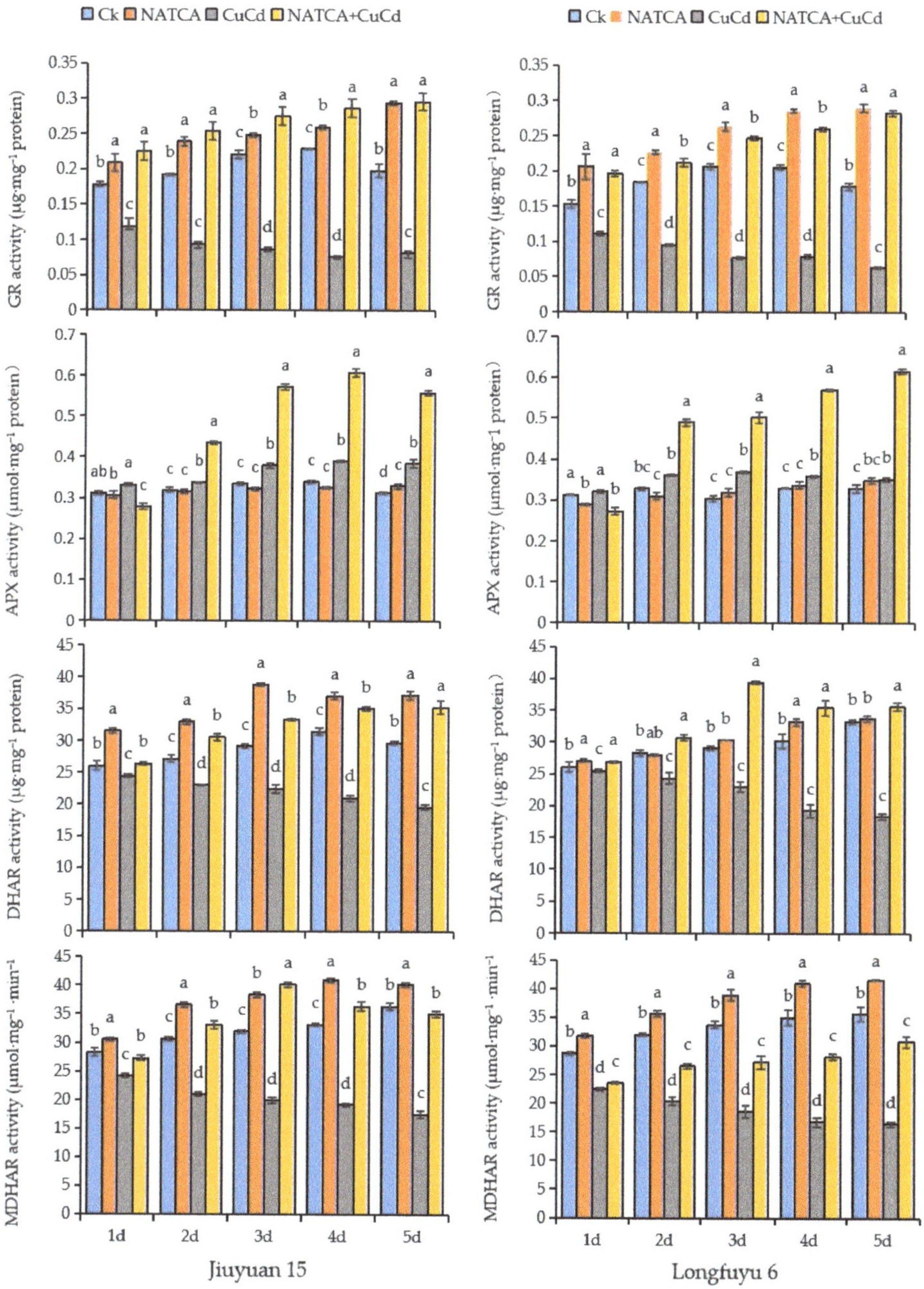

Figure 2. Influence of NATCA on the operation of the AsA-GSH cycle in roots under copper–cadmium combined stress. Values are mean ± SD (n = 3). Significant differences ($p < 0.05$) as determined by Duncan's multiple range test are indicated by different letters within a column for the following treatments: NATCA, which represents 1/2 Hoagland nutrient solution containing 20 $mg \cdot L^{-1}$ NATCA; CuCd, which represents 1/2 Hoagland nutrient solution containing 80 $mg \cdot L^{-1}$ $CdCl_2$ and 100 $mg \cdot L^{-1}$ $CuSO_4$; and NATCA + CuCd, which represents 1/2 Hoagland nutrient solution containing 20 $mg \cdot L^{-1}$ NATCA, 80 $mg \cdot L^{-1}$ $CdCl_2$, and 100 $mg \cdot L^{-1}$ $CuSO_4$.

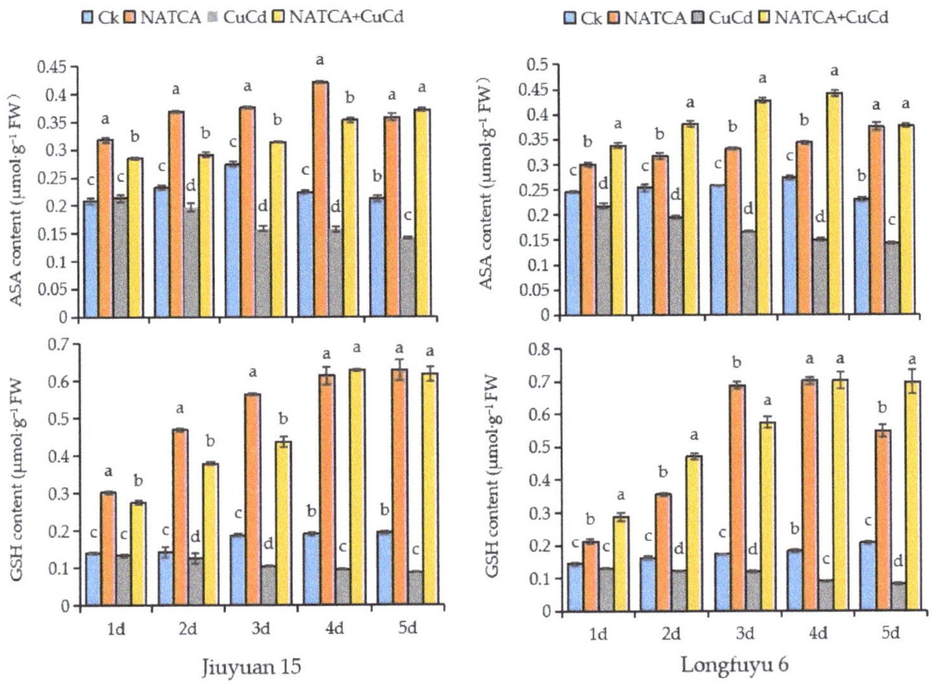

Figure 3. Effects of NATCA on the content of AsA and GSH in roots under combined copper—cadmium stress. Significant differences ($p < 0.05$) as determined by Duncan's multiple range test are indicated by different letters within a column for the following treatments: NATCA, which represents 1/2 Hoagland nutrient solution containing 20 mg·L^{-1} NATCA; CuCd, which represents 1/2 Hoagland nutrient solution containing 80 mg·L^{-1} $CdCl_2$ and 100 mg·L^{-1} $CuSO_4$; and NATCA + CuCd, which represents 1/2 Hoagland nutrient solution containing 20 mg·L^{-1} NATCA, 80 mg·L^{-1} $CdCl_2$, and 100 mg·L^{-1} $CuSO_4$.

In contrast, compared with the CK treatment, CuCd stress resulted in a decrease in the contents of AsA and GSH in maize roots. Specifically, in Jiuyuan 15, the AsA content showed no significant difference on the first day after CuCd treatment compared with CK, but decreased by 15.09%, 42.22%, 29.81%, and 33.23% on days 2–5, respectively. In Longfuyu 6, the AsA content decreased by 11.38%, 23.1%, 35.53%, 45.18%, and 37.68% over 1–5 days, respectively. The GSH content in Jiuyuan 15 decreased by 4.76%, 12.09%, 44.23%, 49.3%, and 54.61% over 1–5 days, respectively, while that in Longfuyu 6 decreased by 9.51%, 24.54%, 30.81%, 50.92%, and 60.87%, respectively. These results demonstrate that, compared with CK, CuCd stress caused a significant time-dependent reduction in the contents of AsA and GSH in the roots of both maize varieties, with the maximum decrease in GSH content exceeding 60%. In contrast, the addition of NATCA (NATCA + CuCd treatment) markedly elevated the levels of these two antioxidants. Particularly in the later stages of treatment, the GSH content showed a dramatic increase, by up to 547.76% to 760.9%, indicating that NATCA effectively reversed the inhibitory effect induced by CuCd stress (Figure 3).

3. Discussion

The germination process, which marks a plant's first encounter with heavy metal stress, is characterized by parameters such as vigor, rate, and index that collectively reflect seed vitality [52]. CuCd stress reduces seed activity and inhibits germination, as evidenced by decreased germination potential, rate, and index alongside increased relative damage rate in maize—consistent with prior findings that combined CuCd stress lowers these parameters in five wolf grass species [53,54]. Seed germination in Phoenix tail cockscomb was subject to concentration-dependent inhibition by Cu^{2+} [55,56]. This experimental study found that the length of the embryonic root decreased, and embryonic roots exhibited wilting after CuCd treatment. Previous studies have demonstrated that high concentrations of Cu^{2+} markedly suppress the germination and early seedling growth of adzuki beans [57]. In the germination experiment, it was also found that the embryonic roots of seeds treated with CuCd turned yellow and wilted, possibly due to the embryonic roots breaking through the seed coat first, resulting in the accumulation of more CuCd ions and causing greater damage to the embryonic roots. This study demonstrated that NATCA treatment significantly enhanced the germination potential, rate, and index of maize seedlings, suggesting its efficacy in breaking seed dormancy and improving overall seed vigor. According to literature analysis, it is possible that NATCA reversed the synthesis of ABA, promoted the production of ethylene, and disrupted the primary and secondary dormancy of seeds [58,59]. NATCA, a plant growth regulator, enhances seed germination and alleviates dormancy by modulating key enzymes and proteins through signal transduction and hormonal crosstalk. This study further confirms that NATCA treatment significantly reduces seed damage and improves CuCd tolerance in maize. This finding demonstrates that NATCA mitigates the phytotoxic impact of Cu-Cd co-exposure on germination, thereby enhancing seed tolerance throughout the germination process [59].

After being stimulated by Cu and Cd, maize will activate self-protection mechanisms to reduce or avoid the entry of Cu and Cd into the plant body as much as possible in order to better adapt to the environment. At the cellular level, the cell wall is a structure that was exposed to Cu and Cd stress earlier [60]. Due to the anionic nature of D-galacturonic acid in cell wall pectins, which exhibit a strong exchange capacity for divalent and trivalent cations such as Cu^{2+} and Cd^{2+}, these metal ions are effectively immobilized within the cell wall, thereby preventing their entry into the cell [61]. The tolerance and defense response of maize to Cd^{2+} stress is associated with significant alterations in cell wall polysaccharide components and their ratios to lignin and cellulose, suggesting a key role for cell wall remodeling in metal sequestration [62]. Beyond functioning as a primary barrier for toxic metal accumulation, the cell wall also dynamically adjusts its structure in response to cadmium stress as an adaptive mechanism. A key adaptive change is the increase in low-methylated pectin, which enhances the binding of divalent/trivalent metal ions, thereby boosting the cell wall's accumulation capacity and reducing their translocation to the protoplast [63–66]. Maize roots employ the secretion of low-molecular-weight organic acids (LMWOAs) as a strategy to mitigate the phytotoxic effects induced by heavy metal stress. The low molecular weight organic acids secreted by plant roots can alter the migration of cadmium ions in soil after forming soluble complexes [67]. The LMWOA secreted by plant roots, such as oxalic acid, malic acid, succinic acid, and citric acid, helps to form soluble complexes and chelates, and alters the mobility of heavy metals in the soil rhizosphere [68]. The immobilization of heavy metals in the soil significantly reduces the translocation of Cd, Ni, and Zn to various plant parts [69]. Metals can chelate with ligands outside of cells, such as through the chelation of organic acids flowing from roots, as well as metallothioneins (MT) and phytochelatins (PC).

Cu is an essential trace element for plants, while Cd is non-essential; both can induce plant growth stress when their environmental concentrations exceed a certain threshold. Cu-Cd combined stress inhibits maize seedling growth (e.g., stunted growth, wilting, shorter height, yellow leaves) and exerts similar effects on tomatoes, such as shortened, slender stems and yellow leaves [70]. Research shows Cd stress significantly reduces the plant height, stem thickness, fresh weight, and dry weight of thin-skinned cantaloupe [71]. This study indicates that the dry and fresh weight decreases after CuCd treatment, which inhibits the accumulation of dry matter in maize seedlings. This conclusion is consistent with previous research results. Higher Cd concentrations reduce wheat seedling height, slow leaf extension, and significantly decrease fresh weight, dry weight, and water content [72]. Research has shown that as the concentration of copper and cadmium increases, the plant height and individual weight of Ogi tend to decrease [73]. This study showed that NATCA application enhanced maize growth, resulting in thicker stems, wider and longer leaves, and increased dry and fresh weight. These findings indicate that NATCA can promote maize seedling growth and boost biomass. The study also showed that applying NATCA to maize seedlings under CuCd stress improved leaf yellowing, restored normal growth, and increased plant height, dry weight, and fresh weight. These results indicate that NATCA can alleviate the inhibitory effect of CuCd stress on maize seedling growth. The water status in plants is one of the important indicators for studying plant stress physiology. Natural saturation deficiency reflects the internal water shortage in plants. This study indicates that under CuCd stress, plant water content decreases while natural saturation deficit increases, suggesting maize seedlings suffer from water shortage. Most abiotic stresses (e.g., salt, drought, heavy metal stress) disrupt plant water balance, which is linked to the regulation of aquaporin subtype expression and activity [74]. Plant water absorption is closely related to water channel proteins, and cadmium accumulation can cause oxidative damage to lipids and proteins and reduce antioxidant enzyme activity, thereby affecting plant water status [75].

To survive under heavy metal stress, plants evolve morphological structures that avoid or tolerate heavy metals, with one key form being limiting root absorption of heavy metals [76]. This study shows that CuCd stress damages maize seedling roots—causing curling, yellowing, shortening, and fewer lateral roots—and research has found that such combined stress hinders root growth, with higher Cu and Cd content in roots than in aboveground parts. Under combined Cu-Cd stress, root accumulation of Cu and Cd is higher than under single Cu or Cd stress [77]. Cd stress can cause damage to root growth, inhibit root elongation, and significantly reduce root growth [78]. The root is the maize organ that directly contacts soil; as soil Cu and Cd concentrations gradually increase, various root indicators show a downward trend, and Cu-Cd stress can damage chromosomes and nucleoli—with stronger copper toxicity to nucleoli leading to a more potent inhibitory effect on root growth [79,80]. The inhibition of root elongation by Cu and Cd may be due to the interference of heavy metals on cell division, including inducing chromosomal aberrations and mitotic abnormalities [81]. The phenotypic changes in roots treated with Cu may be related to auxin. Under Cu stress, auxin activity is higher in the meristematic and elongation zones of roots. Excessive auxin accumulation in the meristematic and elongation zones may reduce cell division and elongation, causing primary root growth arrest [82]. Cd stress reduces root growth, followed by a significant decrease in cell viability, an increase in PAL and POD activity and H_2O_2 levels. This may affect the production and further polymerization of phenylpropane derivatives, as well as increase the production of lignin and its H and S units. Therefore, cadmium-induced root growth inhibition may be due to the excessive production of mono lignin, which polymerizes to form lignin, hardening the cell wall and limiting plant growth [83]. This study showed that NATCA treatment increased maize seedlings' root length, average diameter, surface area, volume, and total root tip number, indicating NATCA can promote root growth. Previous studies found that 2-(5-bromo-2-hydroxy-phenyl)-thiazolidine-4-carboxylic acid (BHTC) induces plant resistance to bacterial, oomycetous, and fungal pathogens, and low-dose BHTC promotes Arabidopsis root growth [84]. This study found that NATCA alleviates CuCd-induced root damage, restoring curled and yellow roots to normal white and firm states, and increasing lateral root number and root length. This study also found that in CuCd treatment, the growth direction of roots changed to horizontal and closely adhered to the hydroponic floating plate above the water surface as much as possible. After adding NATCA, the root system resumed downward growth, the degree of decay also decreased, and the number of primary roots increased.

Copper and cadmium stress indirectly induce oxidative stress in plants by generating ROS. The level of AsA GSH cycle is usually positively correlated with the expression of related enzyme genes, that is, high expression of related enzyme genes promotes the progression of AsA GSH cycle, increases its circulation level, and better maintains the homeostasis of ROS in plant cells. The ascorbic acid glutathione cycle is the main pathway for antioxidant defense, mainly detoxifying ROS in plant cells [85]. This study suggests that CuCd stress reduces root GR activity—possibly because DHAR, which participates in converting DHA to AsA, is involved in GR generation, leading to consistent changes in root GR and DHAR activities—and that the antioxidant system, beyond its detoxification function, may also be a sensitive target of Cd toxicity in plants. Low Cd levels increase APX activity in maize leaves—an early adaptive response of APX to Cd stress that may serve as a prospective biomarker for soil Cd in maize seedlings, and high AsA and GSH contents can protect cells from Cu- and Cd-induced oxidative damage [86]. This study indicates that under CuCd stress, the levels of AsA and GSH in maize seedlings decrease. Previous studies found that under Cu stress, AsA and GSH contents in wheat roots decreased compared to the control group, which may be due to tissue-absorbed copper ions reaching a threshold, leading to AsA and GSH accumulation in tissues [87]. In plants, antioxidant reactions mainly rely on the utilization and recycling of GSH and AsA—GSH not only has a primary antioxidant function but also serves as a substrate for regenerating other essential antioxidants (e.g., AsA)—and in the AsA-GSH cycle, the two oxidized metabolites (GSSG and DHA) are recycled by NADPH/GR and GSH/DHAR, respectively [88,89]. In this study, the roots were more affected by oxidative stress caused by CuCd, and NATCA had the most significant effect on the roots. In this study, only the physiological responses of maize to CuCd stress after NATCA treatment were observed, such as alleviated oxidative damage and improved growth indicators. It is speculated that this may involve the regulation of hormones such as ABA or ethylene, but this mechanism has not been directly verified through experiments such as determining endogenous hormone contents or analyzing the expression of genes related to hormone synthesis/signaling transduction. Therefore, the specific association between NATCA and plant hormone pathways still needs to be further clarified in subsequent studies through targeted determination and intervention experiments. One limitation of this study is the lack of data on copper and cadmium accumulation. Although this study focused on internal physiological responses, future research should determine if NATCA affects metal uptake beyond its observed role in mitigating oxidative stress, as this will provide a more complete understanding of the interaction between this compound and heavy metal stress in maize seedlings.

4. Materials and Methods

4.1. Experimental Materials

Based on preliminary experiments, Jiuyuan 15 (tolerant to CuCd stress) and Longfuyu 6 (intolerant to CuCd stress) were selected as materials, with seeds provided by Gansu Zhongyu Seed Industry Co., Ltd. (Zhangye, China) and Jilin Rongsheng Seed Industry Co., Ltd. (Changchun, China), respectively. Half leaf extract (NATCA, Folcisteine) was purchased from Zhengzhou Zhengshi Chemical Products Co., Ltd. (Zhengzhou, China), CAS number: 5025-82-1, molecular formula $C_6H_9NO_3S$, molecular weight 175.2056. Copper stress was treated with copper sulfate pentahydrate, molecular formula $CuSO_4.5H_2O$, molecular weight: 249.685, CAS number: 7758-99-8, provided by Tianjin Third Chemical Plant Co., Ltd. (Tianjin, China). Cadmium stress was treated with anhydrous cadmium chloride, molecular formula $CdCl_2$, molecular weight: 183.32, CAS number: 10108-64-2, provided by Shenyang Xinguang Chemical Plant (Shenyang, China).

4.2. Growth Condition of Seed Germination

Select 50 healthy and plump seeds from each of the two varieties, disinfect them with 10% sodium hypochlorite for 10 min, and clean them thoroughly. Put it into a culture dish with filter paper and divide it into four treatments: (1) distilled water (CK); (2) 20 $mg{\cdot}L^{-1}$ NATCA; (3) 80 $mg{\cdot}L^{-1}$ $CdCl_2$ + 100 $mg{\cdot}L^{-1}$ $CuSO_4$ (CuCd); (4) 20 $mg{\cdot}L^{-1}$ NATCA + 80 $mg{\cdot}L^{-1}$ $CdCl_2$ + 100 $mg{\cdot}L^{-1}$ $CuSO_4$ (NATCA + CuCd), setting up 3 replicates. Place the seeds in a constant temperature incubator for dark cultivation at 26 °C and 96% humidity. To ensure concentration, change the filter paper once a day. The number of germinated seeds was counted at fixed daily intervals, with germination defined as the stage when the lengths of the radicle and plumule reached half the length of the seed. CuCd tolerance index (%) = (mean of a certain growth index in the stress treatment group ÷ mean of the same growth index in the control group) × 100.

4.3. Growth Condition of Seedling Hydroponic Cultivation

Healthy and plump seeds were selected for this experiment, disinfected with 10% sodium hypochlorite solution, washed and soaked for 10 h, evenly spread into germination boxes, and transferred to a constant temperature incubator (26 ± 1 °C). After the seeds germinate and the embryo grows to 1.5–2 cm, select germinated seeds with consistent growth and plant them in a non-light culture pot containing 20 L of 1/2 Hoagland nutrient solution (pH = 6.0 ± 0.1). Change the nutrient solution once in 3D, ventilate the air pump at regular intervals (45 $min{\cdot}h^{-1}$), expose to light for 12 h, and maintain a temperature of 28 ± 1 °C. When the seedlings grow to three leaves and one center, select seedlings with consistent growth for treatment. Each group has 60 seedlings, divided into four groups.

The experiment comprised four treatments: the control (CK) with 1/2 Hoagland nutrient solution; the NATCA treatment with 20 $mg{\cdot}L^{-1}$ NATCA added to the solution [90]; the copper–cadmium stress (CuCd) treatment with 80 $mg{\cdot}L^{-1}$ $CdCl_2$ and 100 $mg{\cdot}L^{-1}$ $CuSO_4$ [91,92]; and the combined NATCA + CuCd treatment with 20 $mg{\cdot}L^{-1}$ NATCA, 80 $mg{\cdot}L^{-1}$ $CdCl_2$, and 100 $mg{\cdot}L^{-1}$ $CuSO_4$. Add 80 $mg{\cdot}L^{-1}$ $CdCl_2$ and 100 $mg{\cdot}L^{-1}$ $CuSO_4$ to Hoagland nutrient solution in maize three-leaf stage. The initial pH was adjusted to 6.0 ± 0.1, and measurements taken daily confirmed that the pH remained within this range (5.9–6.1) throughout the experimental period. The seedlings that require CuCd treatment should be added to the copper–cadmium treatment solution twice, with an interval of 12 h, to provide a buffer time for the seedlings. Seedlings that require NATCA treatment should be treated with copper–cadmium compound for 12 h before adding NATCA. Adjust pH once a day, with a photoperiod of 12/12 (day/night), temperature of (28 ± 1) °C/(25 ± 1) °C day/night, light intensity of around 400 $\mu mol{\cdot}m^{-2}{\cdot}s^{-1}$, and relative humidity of 60% to 70%. The 12th hour after adding various treatments is taken as the 1st day of sampling.

4.4. Determination of Seedling Growth Phenotypic

On the 5th day of treatment, select three maize seedlings with consistent growth in each treatment, clean them thoroughly, absorb the water with filter paper, and weigh their fresh weight. Kill at 105 °C for 30 min in an oven, and dry at 80 °C until constant weight. Weigh the dry weight separately and calculate the moisture content.

On the 5th day of treatment, three maize seedlings with consistent growth were selected for each treatment, and their leaf parts were taken. After weighing its fresh weight, soak the leaves in distilled water for 6 h, remove them, absorb the water with filter paper, and weigh them. Soak in distilled water for another 30 min, remove and dry the leaves, weigh them until they reach saturation weight to obtain the saturated fresh weight, then perform withering and drying to measure the dry weight, and calculate the water saturation deficit.

4.5. Determination of Root Characteristic Parameters

On the 5th day after treatment, the roots of maize seedlings were cut and washed. Three plants were taken from each treatment and washed with distilled water. The root morphology was scanned and analyzed using EPSON1680 root scanner (Winrhizo2005a root analysis software).

4.6. Determination of GR Activities

Take maize seedling roots (0.5 g) with consistent growth from each treatment at regular intervals every day for 1–5 days after treatment, and separate the roots. Glutathione reductase (GR) can catalyze the reduction in oxidized glutathione (GSSG) by nicotinamide adenine dinucleotide phosphate (NADPH) to regenerate reduced glutathione (GSH), while continuously consuming NADPH. The activity of GR can be determined by measuring the rate of decrease in absorbance at 340 nm due to the consumption of NADPH (with a molar extinction coefficient of 6.22 mM^{-1} cm^{-1}). The reaction mixture (1 mL) contains 100 mM sodium phosphate buffer (pH 7.8), 0.5 mM GSSG, 50 mL of extract, and 0.1 mM NADPH. One unit of GR activity is defined as the amount of enzyme that oxidizes 1 nmol of NADPH per minute per milligram of protein under the conditions of 25 °C and pH 8.0. Cut each sample into pieces, add PBS (pH = 7.4), grind to a homogenate at 2–8 °C, and leave for 20 min (3000 r/min) to obtain the supernatant. The GR activity was measured using the glutathione reductase activity detection kit provided by Nanjing Jiancheng Bioengineering Institute (Nanjing, China, http://www.njjcbio.com/) and assessed on 5 January 2022. Measure the absorbance using a spectrophotometer (340 nm) within 15 min after adding the stop solution.

4.7. Determination of APX Activities

Take maize seedling roots with consistent growth from each treatment at regular intervals every day for 1–5 days after treatment. The activity of APX was determined by monitoring the oxidation rate of ascorbate at 290 nm (with a molar extinction coefficient of 2.8 mM^{-1} cm^{-1}). The reaction mixture (3 mL) contained 1.5 mL of 0.1 M potassium phosphate buffer (pH 6.8), 0.5 mL of 6 mM ascorbate, 0.5 mL of 12 mM H_2O_2, and 0.5 mL of enzyme extract.

4.8. Determination of DHAR Activities

The preliminary preparation work is the same as the steps for GR activity determination. DHAR activity was detected using the dehydroascorbate reductase (DHAR) activity detection kit provided by Nanjing Jiancheng Bioengineering Institute (http://www.njjcbio.com/) and assessed on 5 January 2022, and the absorbance was measured using a spectrophotometer. DHAR activity was calculated by measuring the change in absorbance at 340 nm over 4 min due to DHA oxidation (with a molar extinction coefficient of 14 mM^{-1} cm^{-1}). The reaction mixture comprised 90 mM K-phosphate buffer (pH 7.0), 1 mM EDTA, 5.0 mM GSH, and enzyme extract. The reaction was initiated by the addition of 0.2 mM dehydroascorbic acid.

4.9. Determination of MDHAR Activities

The preliminary preparation work is the same as the steps for GR activity determination. The MDHAR activity was detected using the MDHAR activity detection kit from Nanjing Jiancheng Bioengineering Research Institute (http://www.njjcbio.com/) and assessed on 5 January 2022, and the absorbance was measured using a spectrophotometer. MDHAR activity was calculated by measuring the change in absorbance at 340 nm over 4 min due to NADH oxidation (with a molar extinction coefficient of 6.2 mM^{-1} cm^{-1}). The reaction mixture (1 mL) contained 90 mM K-phosphate buffer (pH 7.0), 0.0125% Triton X-100, 0.2 mM NADH, 2.5 mM ascorbate, 0.25 units of ascorbate oxidase, and enzyme extract. One unit of MDHAR activity was defined as the amount of enzyme that oxidizes 1 μmol of NADH per minute per milligram of protein.

4.10. Determination of AsA Contents

Take maize seedling roots with consistent growth from each treatment at regular intervals every day for 1–5 days after treatment. Grind 0.5 g of the sample with 5% hydrochloric acid, centrifuge 15,000× *g* for 20 min, and take the supernatant as the crude extract. Then add PBS and water to the supernatant (supernatant: 150 Mm PBS:water = 1:1:1). Vortex the mixture and let it stand at room temperature for 1 min. Add 10% TCA (Trichloroacetic Acid), 45% H_3PO_4, and 3% $FeCl_3$ for vortexing. Maintain at 37 °C for 1 h and measure using a spectrophotometer.

4.11. Determination of GSH Contents

Prepare the crude extract as shown in the AsA content determination, take the supernatant to measure the total content of GSH and GSSG, add the crude enzyme solution to the reaction solution, and then perform colorimetric analysis. Measure the content of GSSG again, add the crude enzyme solution diluted 50 times to the pyrimidine water bath for colorimetric analysis, and subtract the two to obtain the GSH content.

4.12. Data Analysis

The data were expressed by the measured mean value, analyzed by SPSS19.0 (IBM SPSS Statistics, 2010), and compared by Duncan's new complex difference method ($\alpha = 0.05$), and origin 8 is used for drawing.

5. Conclusions

Exogenous NATCA enhances maize seed germination and alleviates root growth inhibition by improving root traits under combined Cu/Cd stress. Additionally, NATCA regulates root development via modulating the AsA-GSH cycle, thus protecting root cell structure.

Author Contributions: Conceptualization, L.D., Y.Z. and W.G.; methodology, M.Z., K.H. and J.W.; software, L.D., J.W. and R.W.; validation, Z.L., M.Z., R.W. and J.W.; formal analysis, L.D., M.Z. and Z.L.; investigation, K.H., R.W., Y.Z. and W.G.; resources, L.D. and Y.Z.; data curation, L.D., M.Z. and K.H.; writing—original draft preparation, L.D. and M.Z.; writing—review and editing, L.D., M.Z. and W.G.; visualization, Y.Z. and W.G.; supervision, Y.Z. and W.G.; project administration, Y.Z. and W.G.; funding acquisition, Y.Z. All authors have read and agreed to the published version of the manuscript.

Funding: This research was supported by the Natural Science Foundation Project of Heilongjiang Province (YQ2022C009), the Backbone of Young Talent Scholar Project (20XG23) and the Collaborative Innovation and Promotion System for Green and Organic Agriculture in Heilongjiang Province (2025).

Institutional Review Board Statement: Not applicable.

Informed Consent Statement: Not applicable.

Data Availability Statement: The original contributions presented in this study are included in the article. Further inquiries can be directed to the corresponding authors.

Conflicts of Interest: The authors declare no conflicts of interest.

References

1. Eijsackers, H.; Reinecke, A.; Reinecke, S.; Maboeta, M. Heavy Metal Threats to Plants and Soil Life in Southern Africa: Present Knowledge and Consequences for Ecological Risk Assessment. *Rev. Environ. Contam. Toxicol.* **2019**, *249*, 29–70. [CrossRef] [PubMed]
2. You, Y.; Wu, C.N.; Zhong, W.J.; Hou, Z.L.; Liu, Y.; Du, S.T.; Jin, C.Y. Effects of dicyandiamide on cadmium accumulation in pakchoi under instant soluble nitrogen fertilizers. *Chin. J. Appl. Ecol.* **2020**, *31*, 3093–3100. [CrossRef]
3. Hu, W.; Wang, H.; Dong, L.; Huang, B.; Borggaard, O.K.; Bruun Hansen, H.C.; He, Y.; Holm, P.E. Source Identification of Heavy Metals in Peri-Urban Agricultural Soils of Southeast China: An Integrated Approach. *Environ. Pollut.* **2018**, *237*, 650–661. [CrossRef]
4. Zhao, R.F.; Cheng, B.; Hua, X.Z.; Wang, S.; Wang, Z.; Lyu, W. Evaluation and Distribution Characteristics of Heavy Metal Pollution in Soil of Xinzhou Irrigation District. *N. Hortic.* **2021**, *6*, 81–88. [CrossRef]
5. Yan, M. Evaluation and Analysis of Heavy Metal Pollution in Metal Mining Areas. *Chin. Resour. Compr. Util.* **2021**, *39*, 148–150. [CrossRef]
6. Xu, H.Z.; Wu, D.T.; Li, G.S.; Wu, L.T.; Ye, C.F.; Guo, B.; Ma, J.W.; Ye, Z.Q.; Liu, D. Input and output balance of cadmium (Cd) in cultivated land with moderate pollution in Songyang County. *J. Zhejiang AF Univ.* **2021**, *38*, 1231–1237.
7. Aladesanmi, O.T.; Oroboade, J.G.; Osisiogu, C.P.; Osewole, A.O. Bioaccumulation Factor of Selected Heavy Metals in *Zea mays*. *J. Health Pollut.* **2019**, *9*, 191207. [CrossRef]
8. Chen, N.C.; Zheng, Y.J.; He, X.F.; Li, X.F.; Zhang, X.X. Analysis of the Report on the national general survey of soil contamination. *J. Agro-Environ. Sci.* **2017**, *36*, 1689–1692. [CrossRef]
9. Winter, T.R.; Borkowski, L.; Zeier, J.; Rostás, M. Heavy Metal Stress Can Prime for Herbivore—Induced Plant Volatile Emission. *Plant Cell Environ.* **2012**, *35*, 1287–1298. [CrossRef]
10. Zhang, H.; Ma, X.P.; Su, H.; Sui, H.J.; Shi, X.L.; Li, X.Y.; Yang, H.; Zheng, Z.Z. Soil heavy metal background values and pollution degree in southern Songnen Plain of Heilongjiang Province. *Agric. Res. Arid. Areas* **2018**, *36*, 230–236. [CrossRef]
11. Li, H.X.; Zhao, W.J.; Wu, Y.F.; Liu, Z.H.; Wang, Q.B.; Fu, D.F.; Sun, X.X. Heavy Metal Contents and Pollution Assessment of the Soil of Natural Wetlands in Heilongjiang Sanjiang Natural Reserve. *Wetl. Sci. Manag.* **2018**, *14*, 38–41. [CrossRef]
12. Seneviratne, M.; Rajakaruna, N.; Rizwan, M.; Madawala, H.M.S.P.; Ok, Y.S.; Vithanage, M. Heavy Metal-Induced Oxidative Stress on Seed Germination and Seedling Development: A Critical Review. *Environ. Geochem. Health* **2019**, *41*, 1813–1831, Erratum in *Environ. Geochem. Health* **2019**, *41*, 1635. [CrossRef] [PubMed]
13. Mihoub, A.; Chaoui, A.; El Ferjani, E. Changements Biochimiques Induits Par Le Cadmium et Le Cuivre Au Cours de La Germination Des Graines de Petit Pois (*Pisum sativum* L.). *Comptes Rendus Biol.* **2004**, *328*, 33–41. [CrossRef] [PubMed]
14. Zhang, H.S.; Wang, Z.Y.; Zhang, Z.F.; Tian, H.; Xiong, J.B.; Lu, J.Y.; Liu, Y. Study on Effects of Copper Stress on Seed Germination and Root Growth of White Clover. *Seed* **2020**, *39*, 13–16. [CrossRef]
15. Pang, G.F.; Meng, Q.D.; Zhao, H.Y.; Qi, M.Y.; Lin, H.; Hu, J. Effects of Cd, Pb and Their Combined Stresses on Seed Germination and Seedling Root Growth of Wheat. *Mod. Agric. Sci. Technol.* **2020**, *3*, 21–22+24.
16. Schützendübel, A.; Schwanz, P.; Teichmann, T.; Gross, K.; Langenfeld-Heyser, R.; Godbold, D.L.; Polle, A. Cadmium-Induced Changes in Antioxidative Systems, Hydrogen Peroxide Content, and Differentiation in Scots Pine Roots. *Plant Physiol.* **2001**, *127*, 887–898. [CrossRef]
17. Karmous, I.; Khadija, J.; Chaoui, A.; El Ferjani, E. Proteolytic Activities in Phaseolus Vulgaris Cotyledons under Copper Stress. *Physiol. Mol. Biol. Plants* **2012**, *18*, 337–343. [CrossRef]
18. Siddiqui, M.M.; Abbasi, B.H.; Ahmad, N.; Ali, M.; Mahmood, T. Toxic Effects of Heavy Metals (Cd, Cr and Pb) on Seed Germination and Growth and DPPH-Scavenging Activity in *Brassica rapa* Var. *turnip*. *Toxicol. Ind. Health* **2014**, *30*, 238–249. [CrossRef]
19. Pokorska-Niewiada, K.; Rajkowska-Myśliwiec, M.; Protasowicki, M. Acute Lethal Toxicity of Heavy Metals to the Seeds of Plants of High Importance to Humans. *Bull. Environ. Contam. Toxicol.* **2018**, *101*, 222–228. [CrossRef]
20. Staszak, A.M.; Małecka, A.; Ciereszko, I.; Ratajczak, E. Differences in Stress Defence Mechanisms in Germinating Seeds of *Pinus sylvestris* Exposed to Various Lead Chemical Forms. *PLoS ONE* **2020**, *15*, e0238448. [CrossRef]
21. Xie, L.; Hao, P.; Cheng, Y.; Ahmed, I.M.; Cao, F. Effect of Combined Application of Lead, Cadmium, Chromium and Copper on Grain, Leaf and Stem Heavy Metal Contents at Different Growth Stages in Rice. *Ecotoxicol. Environ. Saf.* **2018**, *162*, 71–76. [CrossRef] [PubMed]
22. Huang, R.Z.; Li, Y.P.; Jiang, Y.J.; Jia, C.H.; Jiang, S.M.; Yian, X.P.; Qin, Z.X.; Luo, J. Effect of Cadmium and Lead Combined Stress on Growth of Mulberry Saplings and Contents of Heavy Metal in Mulberry leaf. *Acta Sericol. Sin.* **2018**, *44*, 665–671. [CrossRef]
23. Ghosh, S.; Sethy, S. Effect of Heavy Metals on Germination of Seeds. *J. Nat. Sci. Biol. Med.* **2013**, *4*, 272. [CrossRef] [PubMed]
24. Berkelaar, E.; Hale, B. The Relationship between Root Morphology and Cadmium Accumulation in Seedlings of Two Durum Wheat Cultivars. *Can. J. Bot.* **2000**, *78*, 381–387. [CrossRef]
25. Farrell, R.E.; McArthur, D.F.E.; Van Rees, K.C.J. Net Cd^{2+} Flux at the Root Surface of Durum Wheat (*Triticum turgidum* L. var. *durum*) Cultivars in Relation to Cultivar Differences in Cd Accumulation. *Can. J. Plant Sci.* **2005**, *85*, 103–107. [CrossRef]
26. Xia, S.; Deng, R.; Zhang, Z.; Liu, C.; Shi, G. Variations in the Accumulation and Translocation of Cadmium among Pak Choi Cultivars as Related to Root Morphology. *Environ. Sci. Pollut. Res.* **2016**, *23*, 9832–9842. [CrossRef]

27. Yu, H.; Guo, J.; Li, Q.; Zhang, X.; Huang, H.; Huang, F.; Yang, A.; Li, T. Characteristics of Cadmium Immobilization in the Cell Wall of Root in a Cadmium-Safe Rice Line (*Oryza sativa* L.). *Chemosphere* **2020**, *241*, 125095. [CrossRef]
28. Yu, Z.H.; Li, H.F.; Zhu, W.Q. The Inhibitory Effects of Cu^{2+} Stress on Seed Germination, Seedling Growth and Root Elongation for Forage Grasses. *Environ. Impact Assess.* **2011**, *33*, 19–23. [CrossRef]
29. Lux, A.; Martinka, M.; Vaculik, M.; White, P.J. Root Responses to Cadmium in the Rhizosphere: A Review. *J. Exp. Bot.* **2011**, *62*, 21–37. [CrossRef]
30. Hoy, K.S.; Uppal, J.S.; Le, X.C. Editorial: Effect of root anatomy and apoplastic barrier development on cadmium uptake in rice. *J. Environ. Sci.* **2019**, *79*, 361–363. [CrossRef]
31. Yu, H.; Wang, K.; Huang, H.; Zhang, X.; Li, T. The Regulatory Role of Root in Cadmium Accumulation in a High Cadmium-Accumulating Rice Line (*Oryza sativa* L.). *Environ. Sci. Pollut. Res.* **2021**, *28*, 25432–25441. [CrossRef] [PubMed]
32. Dong, G.; Xu, R.K. Acute Toxicity of Cu^{2+}/Cd^{2+} to Maize as Related to Chemical Forms of the Metals on Whole Plant Roots. *J. Ecol. Rural Environ.* **2021**, *37*, 387–393. [CrossRef]
33. Zhang, C.Q. Study on Rhizosphere Characteristics and Root Responses of *Malva sinensis* Cavan. Under Cadium Stress. Master's Thesis, Sichuan Agricultural University, Ya'an, China, 2017.
34. Peng, G.Y.; Hu, L.; Huang, C.; Yang, K.; Wan, W.; Huang, C.G. Transcriptome Analysis of Response to Heavy Metal Copper Stress in *Setcreasea purpurea* Root Tissue. *Biotechnol. Bull.* **2022**, *38*, 83–94. [CrossRef]
35. Sang, J.S. Copper Speciation in the Rhizosphere of Apple and Its Effect on Nutrient Uptake and Distribution in Young Trees. Master's Thesis, Shandong Agricultural University, Tai'an, China, 2022.
36. Li, W.; Guo, S.X.; Xing, L.J.; Zhai, Y.L.; Li, W.B.; Li, M.; Liu, R.J. Synergistic utilization of arbuscular mycorrhizal fungi and *Festuca elata* for remediating cadmium-contaminated soils. *Mycosystema* **2021**, *40*, 2785–2799. [CrossRef]
37. Singh, S.; Singh, A.; Srivastava, P.K.; Prasad, S.M. Cadmium Toxicity and Its Amelioration by Kinetin in Tomato Seedlings Vis-à-Vis Ascorbate-Glutathione Cycle. *J. Photochem. Photobiol. B* **2018**, *178*, 76–84. [CrossRef]
38. Hasanuzzaman, M.; Bhuyan, M.H.M.B.; Anee, T.I.; Parvin, K.; Nahar, K.; Mahmud, J.A.; Fujita, M. Regulation of Ascorbate-Glutathione Pathway in Mitigating Oxidative Damage in Plants under Abiotic Stress. *Antioxidants* **2019**, *8*, 384. [CrossRef]
39. Gill, S.S.; Tuteja, N. Reactive Oxygen Species and Antioxidant Machinery in Abiotic Stress Tolerance in Crop Plants. *Plant Physiol. Biochem.* **2010**, *48*, 909–930. [CrossRef]
40. Zhao, X.; Min, J.; He, J. Effect of Microwave Curing on Antimicrobial Activity of Chitosan Biguanidine Hydrochloride Treated Wool Fabrics. *J. Text. Inst.* **2011**, *102*, 801–807. [CrossRef]
41. Seifikalhor, M.; Aliniaeifard, S.; Bernard, F.; Seif, M.; Latifi, M.; Hassani, B.; Didaran, F.; Bosacchi, M.; Rezadoost, H.; Li, T. γ-Aminobutyric Acid Confers Cadmium Tolerance in Maize Plants by Concerted Regulation of Polyamine Metabolism and Antioxidant Defense Systems. *Sci. Rep.* **2020**, *10*, 3356. [CrossRef]
42. Małkowski, E.; Sitko, K.; Szopiński, M.; Gieroń, Ż.; Pogrzeba, M.; Kalaji, H.M.; Zieleźnik-Rusinowska, P. Hormesis in Plants: The Role of Oxidative Stress, Auxins and Photosynthesis in Corn Treated with Cd or Pb. *Int. J. Mol. Sci.* **2020**, *21*, 2099. [CrossRef]
43. Liu, Y.; Liu, L.; Qi, J.; Dang, P.; Xia, T. Cadmium Activates ZmMPK3-1 and ZmMPK6-1 via Induction of Reactive Oxygen Species in Maize Roots. *Biochem. Biophys. Res. Commun.* **2019**, *516*, 747–752. [CrossRef]
44. Fiala, R.; Repka, V.; Čiamporová, M.; Martinka, M.; Pavlovkin, J. The Effect of Cadmium-Nickel Interactions on Superoxide Production, Cell Viability and Membrane Potential (E_M) in Roots of Two Maize Cultivars. *Acta Biol. Hung.* **2015**, *66*, 192–204. [CrossRef]
45. Ding, S.; Ma, C.; Shi, W.; Liu, W.; Lu, Y.; Liu, Q.; Luo, Z.-B. Exogenous Glutathione Enhances Cadmium Accumulation and Alleviates Its Toxicity in Populus × Canescens. *Tree Physiol.* **2017**, *37*, 1697–1712. [CrossRef]
46. Zayneb, C.; Bassem, K.; Zeineb, K.; Grubb, C.D.; Noureddine, D.; Hafedh, M.; Amine, E. Physiological Responses of Fenugreek Seedlings and Plants Treated with Cadmium. *Environ. Sci. Pollut. Res.* **2015**, *22*, 10679–10689. [CrossRef]
47. Yetişsin, F.; Kurt, F. Gallic Acid (GA) Alleviating Copper (Cu) Toxicity in Maize (*Zea mays* L.) Seedlings. *Int. J. Phytoremediation* **2020**, *22*, 420–426. [CrossRef]
48. Roy, D.; Adhikari, S.; Adhikari, A.; Ghosh, S.; Azahar, I.; Basuli, D.; Hossain, Z. Impact of CuO Nanoparticles on Maize: Comparison with CuO Bulk Particles with Special Reference to Oxidative Stress Damages and Antioxidant Defense Status. *Chemosphere* **2022**, *287*, 131911. [CrossRef]
49. Yuan, M.; Xu, Y.C.; Niu, Y.Q.; Zhou, H.; An, Y.L.; Jin, Q.J.; Wang, Y.L. Effects of ethylene and NO on AsA-GSH in lotus under cadmium stress. *Chin. J. Appl. Ecol.* **2018**, *29*, 3433–3440. [CrossRef]
50. Lin, X.P.; Shi, M.T.; Chen, W.; Lu, T.T.; Ni, X.C. Effects of cadmium stress on active oxygen metabolism in leaves of balsam pear. *J. Fujian Agric. For. Univ.* **2010**, *39*, 143–146. [CrossRef]
51. Shi, G.Y.; Yu, Z.Q.; Shi, W.L. Research Progress on Mechanism and Application of Exogenous Substances in Phytoremediation of Heavy Metal Contaminated Soil. *Ecol. Environ. Sci.* **2021**, *30*, 655–666. [CrossRef]
52. Yao, F.N.; Wan, C.; Liu, J.P.; Xiu, W.C.; Zheng, X.F. Research on Biological Activity of Acetyl-Thiazolidine-4-Carboxylic Acid in Seedling Stage of Several Crops. *Mod. Agrochem.* **2021**, *20*, 62–64. [CrossRef]

53. Romero-Puertas, M.C.; Palma, J.M.; Gómez, M.; Del Río, L.A.; Sandalio, L.M. Cadmium Causes the Oxidative Modification of Proteins in Pea Plants. *Plant Cell Environ.* **2002**, *25*, 677–686. [CrossRef]
54. Jambunathan, N. Determination and Detection of Reactive Oxygen Species (ROS), Lipid Peroxidation, and Electrolyte Leakage in Plants. In *Plant Stress Tolerance*; Methods in Molecular Biology; Humana Press: Totowa, NJ, USA, 2010; Volume 639, pp. 291–297. [CrossRef]
55. Li, W.; Khan, M.A.; Yamaguchi, S.; Kamiya, Y. Effects of Heavy Metals on Seed Germination and Early Seedling Growth of Arabidopsis Thaliana. *Plant Growth Regul.* **2005**, *46*, 45–50. [CrossRef]
56. Deng, B.; Yang, K.; Zhang, Y.; Li, Z. Can Heavy Metal Pollution Defend Seed Germination against Heat Stress? Effect of Heavy Metals (Cu^{2+}, Cd^{2+} and Hg^{2+}) on Maize Seed Germination under High Temperature. *Environ. Pollut.* **2016**, *216*, 46–52. [CrossRef]
57. Xu, Z.M.; He, K.; Lin, T.; Chen, A.L.; Zhu, T.; Ding, G.C. Effect of Copper and Cadmium Combined Stress on Seed Germination Characteristics of Five Ornamental Grasses. *Shandong Agric. Sci.* **2019**, *51*, 59–64. [CrossRef]
58. Gao, Q.J.; Zhang, F.E.; Kong, D.Y.; Bao, W.M.; Zhang, W.W. Effects of Cu^{2+} and Pb^{2+} Stress on Seed Germination of Two Varieties of Cockscombs. *J. Northeast. For. Univ.* **2021**, *49*, 36–40. [CrossRef]
59. Li, H.Y.; Wu, Y.; Yang, R.; Wang, Z.Y.; Xiao, J.; Gao, C.H. Creening and Identification on Corn Varieties with Resistance to Heavy Metal Cd in Germination Stage. *Hortic. Seed* **2017**, 61–65. [CrossRef]
60. Liu, J.X.; Bai, Z.Z.; Wang, R.M.; Liu, L.Z.; Zhang, Z.H.; Wen, R.Y. Germination Characteristics and Accumulation Effects of Adzuki Bean under Heavy Metal Stress. *Crops* **2019**, *6*, 182–186. [CrossRef]
61. Khan, A.A. Cytokinins: Permissive Role in Seed Germination: With Other Plant Hormones, Cytokinins Regulate Germination and Dormancy by a Novel Mechanism. *Science* **1971**, *171*, 853–859. [CrossRef]
62. Song, S.Q.; Liu, J.; Yang, H.; Zhang, W.H.; Zhang, Q.; Gao, J.D. Research Progress in Seed Development, Dormancy and Germination Regulated by Cytokinin. *Chin. Bull. Bot.* **2021**, *56*, 218–231. [CrossRef]
63. Kollárová, K.; Kusá, Z.; Vatehová-Vivodová, Z.; Lišková, D. The Response of Maize Protoplasts to Cadmium Stress Mitigated by Silicon. *Ecotoxicol. Environ. Saf.* **2019**, *170*, 488–494. [CrossRef] [PubMed]
64. Asilian, E.; Ghasemi-Fasaei, R.; Ronaghi, A.; Sepehri, M.; Niazi, A. Chemical- and Microbial-Enhanced Phytoremediation of Cadmium-Contaminated Calcareous Soil by Maize. *Toxicol. Ind. Health* **2019**, *35*, 378–386. [CrossRef]
65. Vatehová, Z.; Malovíková, A.; Kollárová, K.; Kučerová, D.; Lišková, D. Impact of Cadmium Stress on Two Maize Hybrids. *Plant Physiol. Biochem.* **2016**, *108*, 90–98. [CrossRef]
66. Krzesłowska, M. The Cell Wall in Plant Cell Response to Trace Metals: Polysaccharide Remodeling and Its Role in Defense Strategy. *Acta Physiol. Plant.* **2011**, *33*, 35–51. [CrossRef]
67. Yang, C.; Qiu, W.; Chen, Z.; Chen, W.; Li, Y.; Zhu, J.; Rahman, S.U.; Han, Z.; Jiang, Y.; Yang, G.; et al. Phosphorus Influence Cd Phytoextraction in Populus Stems via Modulating Xylem Development, Cell Wall Cd Storage and Antioxidant Defense. *Chemosphere* **2020**, *242*, 125154. [CrossRef] [PubMed]
68. Yu, H.; Wu, Y.; Huang, H.; Zhan, J.; Wang, K.; Li, T. The Predominant Role of Pectin in Binding Cd in the Root Cell Wall of a High Cd Accumulating Rice Line (*Oryza sativa* L.). *Ecotoxicol. Environ. Saf.* **2020**, *206*, 111210. [CrossRef]
69. Vatehová-Vivodová, Z.; Kollárová, K.; Malovíková, A.; Lišková, D. Maize Shoot Cell Walls under Cadmium Stress. *Environ. Sci. Pollut. Res.* **2018**, *25*, 22318–22322. [CrossRef] [PubMed]
70. Fernández, R.; Fernández-Fuego, D.; Bertrand, A.; González, A. Strategies for Cd Accumulation in *Dittrichia viscosa* (L.) Greuter: Role of the Cell Wall, Non-Protein Thiols and Organic Acids. *Plant Physiol. Biochem.* **2014**, *78*, 63–70. [CrossRef]
71. Sun, J.; Cui, J.; Luo, C.; Gao, L.; Chen, Y.; Shen, Z. Contribution of Cell Walls, Nonprotein Thiols, and Organic Acids to Cadmium Resistance in Two Cabbage Varieties. *Arch. Environ. Contam. Toxicol.* **2013**, *64*, 243–252. [CrossRef]
72. Hu, J.; Wu, S.; Wu, F.; Leung, H.M.; Lin, X.; Wong, M.H. Arbuscular Mycorrhizal Fungi Enhance Both Absorption and Stabilization of Cd by Alfred Stonecrop (*Sedum alfredii* Hance) and Perennial Ryegrass (*Lolium perenne* L.) in a Cd-Contaminated Acidic Soil. *Chemosphere* **2013**, *93*, 1359–1365. [CrossRef]
73. Cui, X.M.; Hu, M.M.; Wang, J.Q.; Zhu, G.Y.P.; Lou, Y.H. The alleviating effect of humic acid on copper cadmium composite stress. In Proceedings of the Summary of the 20th Meeting of the Soil Environment Professional Committee of the Chinese Soil Society and the Seminar on Farmland Soil Pollution and Remediation, Hefei, China, 5 August 2018; p. 28.
74. Wang, Z.W.; Li, H.; Liang, S.H.; Chu, X.; Sun, X.W. Effects of Cadmium on Growth and Some Physiological of Oriental Melon Seedlings. *Acta Agric. Boreali-Sin.* **2020**, *35*, 81–88.
75. Zhang, L.H.; Li, P.J.; Li, X.M.; Meng, X.L.; Xu, C.B. Effects of cadmium stress on the growth and physiological characteristics of wheat seedlings. *Chin. J. Ecol.* **2005**, *24*, 458–460.
76. Ma, B.; Zhang, J.; Wei, Y.Y.; Xie, S.M. Effects of Lanthanum on the Growth and Physiological Characteristics of *Miscanthus sacchariflorus* under Single Stress of Copper and Cadmium. *J. Anhui Norm Univ. (Nat. Sci.)* **2020**, *43*, 467–474. [CrossRef]
77. Srivastava, A.K.; Penna, S.; Nguyen, D.V.; Tran, L.-S.P. Multifaceted Roles of Aquaporins as Molecular Conduits in Plant Responses to Abiotic Stresses. *Crit. Rev. Biotechnol.* **2014**, *36*, 389–398. [CrossRef] [PubMed]

78. Zhang, Y.; Wang, Z.; Chai, T.; Wen, Z.; Zhang, H. Indian Mustard Aquaporin Improves Drought and Heavy-Metal Resistance in Tobacco. *Mol. Biotechnol.* **2008**, *40*, 280–292. [CrossRef] [PubMed]
79. Li, G.; Kronzucker, H.J.; Shi, W. Root Developmental Adaptation to Fe Toxicity: Mechanisms and Management. *Plant Signal. Behav.* **2016**, *11*, e1117722. [CrossRef]
80. Bai, X.Y. Effects and Mechanisms of Nitric Oxide and Salicylic Acid on Alleviating Cd and Cu Stress in Ryegrass. Master's Thesis, Shandong Agricultural University, Tai'an, China, 2016.
81. Zhou, J.N.; Yang, C.; Song, Z.Y.; He, C.F.; He, J.H.; Huang, W.L.; Dang, Z. Influences of tetracycline and cadmium on rice roots: Growth and root exudates. *Acta Sci. Circumstantiae* **2021**, *41*, 1518–1528. [CrossRef]
82. Zheng, A.Z.; Song, W.Y. Effects of Cu on Seeds Germination and Seedling Early Growth of Cucumber and Corn. *North. Hortic.* **2009**, *9*, 7–10.
83. Agarwal, K.; Sharma, A.; Talukder, G. Copper Toxicity in Plant Cellular Systems. *Nucleus* **1987**, *30*, 131–158.
84. Jiang, W.; Liu, D.; Li, H. Effects of Cu^{2+} on Root Growth, Cell Division, and Nucleolus of *Helianthus annuus* L. *Sci. Total Environ.* **2000**, *256*, 59–65. [CrossRef]
85. Yuan, H.M.; Liu, W.C.; Jin, Y.; Lu, Y.T. Role of ROS and Auxin in Plant Response to Metal-Mediated Stress. *Plant Signal. Behav.* **2013**, *8*, e24671. [CrossRef]
86. Finger-Teixeira, A.; Lucio Ferrarese, M.D.L.; Ricardo Soares, A.; Da Silva, D.; Ferrarese-Filho, O. Cadmium-Induced Lignification Restricts Soybean Root Growth. *Ecotoxicol. Environ. Saf.* **2010**, *73*, 1959–1964. [CrossRef]
87. Rodriguez-Salus, M.; Bektas, Y.; Schroeder, M.; Knoth, C.; Vu, T.; Roberts, P.; Kaloshian, I.; Eulgem, T. The Synthetic Elicitor 2-(5-Bromo-2-Hydroxy-Phenyl)-Thiazolidine-4-Carboxylic Acid Links Plant Immunity to Hormesis. *Plant Physiol.* **2016**, *170*, 444–458. [CrossRef]
88. Speiser, A.; Silbermann, M.; Dong, Y.; Haberland, S.; Uslu, V.V.; Wang, S.; Bangash, S.A.K.; Reichelt, M.; Meyer, A.J.; Wirtz, M.; et al. Sulfur Partitioning between Glutathione and Protein Synthesis Determines Plant Growth. *Plant Physiol.* **2018**, *177*, 927–937. [CrossRef]
89. Bonet, B.; Corcoll, N.; Guasch, H. Antioxidant Enzyme Activities as Biomarkers of Zn Pollution in Fluvial Biofilms. *Ecotoxicol. Environ. Saf.* **2012**, *80*, 172–178. [CrossRef] [PubMed]
90. Gill, S.S.; Hasanuzzaman, M.; Nahar, K.; Macovei, A.; Tuteja, N. Importance of Nitric Oxide in Cadmium Stress Tolerance in Crop Plants. *Plant Physiol. Biochem.* **2013**, *63*, 254–261. [CrossRef] [PubMed]
91. Mittler, R.; Vanderauwera, S.; Gollery, M.; Van Breusegem, F. Reactive Oxygen Gene Network of Plants. *Trends Plant Sci.* **2004**, *9*, 490–498. [CrossRef] [PubMed]
92. Halliwell, B. Reactive Species and Antioxidants. Redox Biology Is a Fundamental Theme of Aerobic Life. *Plant Physiol.* **2006**, *141*, 312–322. [CrossRef]

International Journal of
Molecular Sciences

MDPI

Article

Nitrogen-Doped Carbon Dots Alleviate Pesticide Toxicity in Tomato by Regulating Antioxidant Systems

Xu Zhang [1,*], Yu Xin [2], Hao Wang [1], Yuting Dang [2], Wenhui Wang [1], Yi Gao [1], Yu Han [2], Rongrui Kang [1], Qinghua Shi [1] and Han Du [2,*]

[1] State Key Laboratory of Crop Biology, College of Horticulture Science and Engineering, Shandong Agricultural University, Taian 271018, China

[2] College of Food Science and Engineering, Shandong Agricultural University, Taian 271018, China

[*] Correspondence: zhangxu@sdau.edu.cn (X.Z.); duhan1211@sdau.edu.cn (H.D.)

Abstract: The overuse of pesticides has raised serious food-safety and environmental concerns. Carbon dots (CDs) can act as biostimulants by enhancing photosynthesis, thereby promoting plant growth and stress tolerance. However, their roles in plant pesticide detoxification remain unclear. This study synthesized nitrogen-doped carbon dots (N-CDs) with strong blue fluorescence, excellent biocompatibility, and no cytotoxicity observed in HEK 293T cells. The N-CDs were synthesized from 1.025 g citric acid and 0.379 g urea, producing particles with a size of around 2.42 nm and abundant hydrophilic groups. When applied to tomato plants, N-CDs (especially at 150 mg$\cdot$L^{-1}) significantly reduced chlorothalonil (CHT) residues affecting tomato, by up to 66%. Importantly, N-CDs also improved tomato plant growth, reversing the negative effects of CHT on key parameters such as height, leaf area, and biomass. Indeed, under CHT conditions, N-CDs significantly reduced the contents of malondialdehyde, superoxide, and hydrogen peroxide. In contrast, N-CDs significantly increased the activities of superoxide dismutase, peroxidases, catalase, and ascorbate peroxidase to 117.57%, 158.53%, 162.79%, and 152.23%, respectively. Notably, N-CDs dramatically changed the glutathione pool for tomato detoxification. Overall, this study synthesized the non-cytotoxic N-CDs that not only promote tomato growth but also alleviate CHT toxicity by strengthening the tomato's antioxidant defense system.

Keywords: tomato; pesticides; carbon dots; xenobiotic detoxification; antioxidants

1. Introduction

Pesticides are indispensable in modern agriculture for protecting crops against pests and diseases, thereby preventing substantial yield losses [1]. However, their frequent and excessive application has raised serious concerns regarding food safety, environmental contamination, and plant health [2]. A significant proportion of applied pesticides fails to reach the intended target pests and instead accumulates in ecosystems or enters the food chain. This issue is particularly acute in greenhouse cultivation, where limited exposure to UV light, rainfall, and airflow slows and impedes the natural degradation of pesticides, resulting in their persistence within edible plant tissues [3]. Conventional remediation methods such as washing and soaking are often ineffective for removing these internalized residues [3]. Therefore, minimizing pesticide residues has become an urgent priority that necessitates in-depth scientific study.

Pesticides, while essential for crop protection, can cause significant phytotoxicity in plants, including shoot tip atrophy, leaf necrosis, and growth inhibition. A key consequence of pesticide exposure is the excessive production of reactive oxygen species (ROS), such as superoxide anions ($O_2^{\bullet-}$) and hydrogen peroxide (H_2O_2) [4]. Excessive ROS disrupt photosynthesis and electron transport, leading to oxidative damage to lipids, proteins, and nucleic acids [5]. To mitigate these toxic effects, plants have evolved complex detoxification and antioxidant defense systems [6]. The detoxification of pesticides in plants generally proceeds through three coordinated phases. In Phase I, xenobiotics are modified by enzymes such as cytochrome P450s and peroxidases (POD) [7]. Phase II involves conjugation of the activated compounds with reduced glutathione (GSH) or sugars via enzymes like glutathione S-transferase (GST) and glycosyltransferase (GT), which reduces toxicity and enhances solubility [8]. In Phase III, the conjugated products are transported into vacuoles or the apoplast by ABC transporters for sequestration and degradation [9]. Parallel to this metabolic detoxification, ROS are scavenged by enzymatic antioxidants such as superoxide dismutase (SOD), catalase (CAT), POD, and ascorbate peroxidase (APX) [10]. Among these antioxidants, GSH plays a central role in both detoxification and maintaining redox homeostasis and cellular signaling [10]. It exists in both reduced GSH and oxidized GSH disulfide (GSSG) forms, with the GSH/GSSG ratio serving as a key indicator of oxidative stress [11]. GSH conjugates with electrophilic pesticide intermediates via its thiol group in a reaction catalyzed by GST [12]. The regeneration of GSH from GSSG is catalyzed by GSH reductase (GR), maintaining a functional detoxification cycle.

Although these intrinsic mechanisms provide plants with critical defense, they are often insufficient to fully counteract the damage caused by intensive pesticide use and ROS overaccumulation. This has driven the search for external strategies to support plant detoxification and stress management. Recent advancements in nanotechnology have revealed the potential of nanomaterials to promote plant growth and enhance stress tolerance [13]. Among them, carbon-based nanomaterials—particularly, carbon dots (CDs)—are emerging as promising candidates due to their high biocompatibility, water solubility, and environmental safety [14]. Although metal and metal oxide nanoparticles have been extensively studied, organic nanomaterials such as CDs remain relatively underexplored in agriculture applications. CDs, as zero-dimensional carbon nanomaterials, possess excellent photostability and a large surface area, making them suitable for diverse applications such as imaging, catalysis, and molecular delivery. Their role in agriculture, however, is only beginning to be explored. Emerging studies have shown that CDs can promote plant development by enhancing chlorophyll accumulation, improving root and leaf growth, and increasing the synthesis of proteins, sugars, and proline [15,16]. Notably, doping CDs with elements such as nitrogen and carbon further enhances their functional performance [17]. Nitrogen, a vital macronutrient involved in many physiological processes, not only improves the physicochemical properties of CDs but also contributes directly to plant growth and stress resilience [18]. These features make nitrogen-doped carbon dots (N-CDs) particularly attractive for mitigating environmental stress and improving crop resilience. However, until now, it remains unknown whether N-CDs are involved in tomato growth and pesticide detoxification processes.

Tomato (*Solanum lycopersicum* L.) is one of the most widely cultivated vegetable crops globally. To control fungal pathogens and maintain productivity, intensive fungicidal applications of chlorothalonil (CHT) are commonly employed. However, this fungicide often accumulates in leaves and fruits, inducing oxidative damage, disrupting photosynthesis, and impairing cellular metabolism in the plant. In this study, we synthesized and comprehensively characterized a series of N-CDs formulations (N-CDs1~N-CDs5) and evaluated their effects on CHT residue, photosynthetic efficiency, oxidative stress markers, antioxidant enzyme activities, and GSH metabolism in tomato. We aimed to elucidate the mechanisms by which N-CDs alleviate CHT-induced phytotoxicity and to assess their potential as effective, non-cytotoxic nanobiostimulants for CHT residue management in tomato.

2. Results

2.1. Synthesis of N-CDs with High Biological Safety

N-doped citric acid precursor N-CDs were synthesized from citric acid and urea, with urea serving as the nitrogen source. The morphology of the fabricated materials was characterized by TEM (Figure 1a), revealing a uniform size and high monodispersity. The size distribution of the N-CDs ranged from 1 to 4.5 nm, with an average particle size of (2.42 ± 0.24) nm (Figure 1b). The fluorescence 3D scanning spectrum showed that the N-CDs in aqueous solution emitted intense blue fluorescence under excitation light ranging from 315 to 500 nm (Figure 1c). Furthermore, the biosafety of N-CDs was evaluated using human embryonic kidney 293T (HEK 293T) cells. The histogram in Figure 1d shows the absorbance at 450 nm (indicative of cell viability) for cultures treated with N-CDs and the blank control over four days. The corresponding line graph illustrates the relative survival rate of the N-CD-treated cells compared to the control. Four days later, the cell survival rate in the N-CD-treated group decreased to 93.14%; however, no significant differences were observed between the treated and control groups. These results indicate that the as-prepared N-CDs in the experiment possessed high biocompatibility and are promising for practical applications.

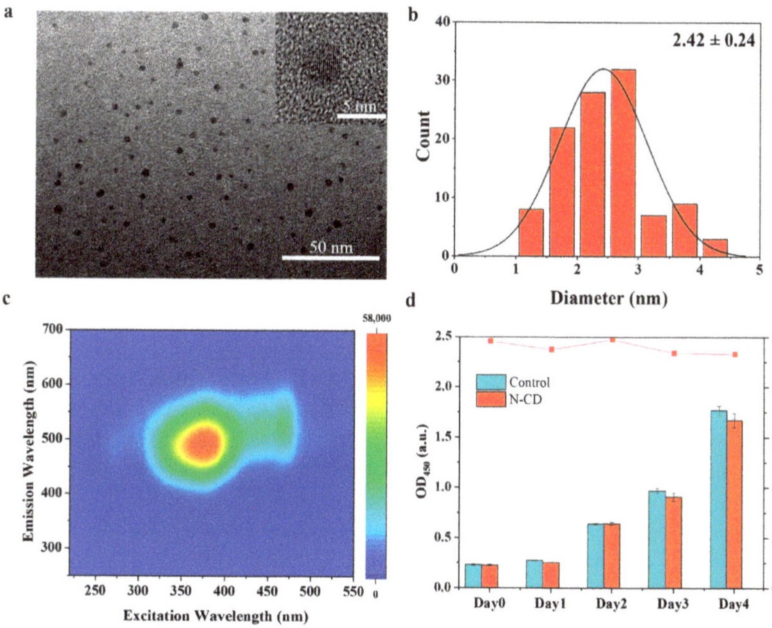

Figure 1. Verification of the morphological characteristics, fluorescence properties and safety of N-CDs. (**a**) Transmission electron microscopy (TEM) results. The scale bar represents 50 nm. (Inset: High-resolution TEM image with a scale bar representing 5 nm). (**b**) Size distribution of N-CDs. (**c**) Fluorescence 3D scanning spectra of N-CDs. (**d**) Cytotoxicity test.

2.2. Analysis of Elemental Groups of N-CDs

The ultraviolet-visible (UV/Vis) spectrum of the prepared N-CDs aqueous solution was investigated (Figure 2a). The spectrum (orange line) of N-CDs exhibited strong absorption bands at 235 nm and 340 nm. The first absorption peak at 235 nm was attributed to the

π-π^* transition of the sp^2 domain. The peak at 340 nm corresponded to n-π^* transitions of C=O or C=N bonds. The optimal excitation wavelength was 378 nm. Upon excitation at this wavelength, the N-CDs emitted the most intense blue fluorescence at 488 nm (Figure 2a, blue line).

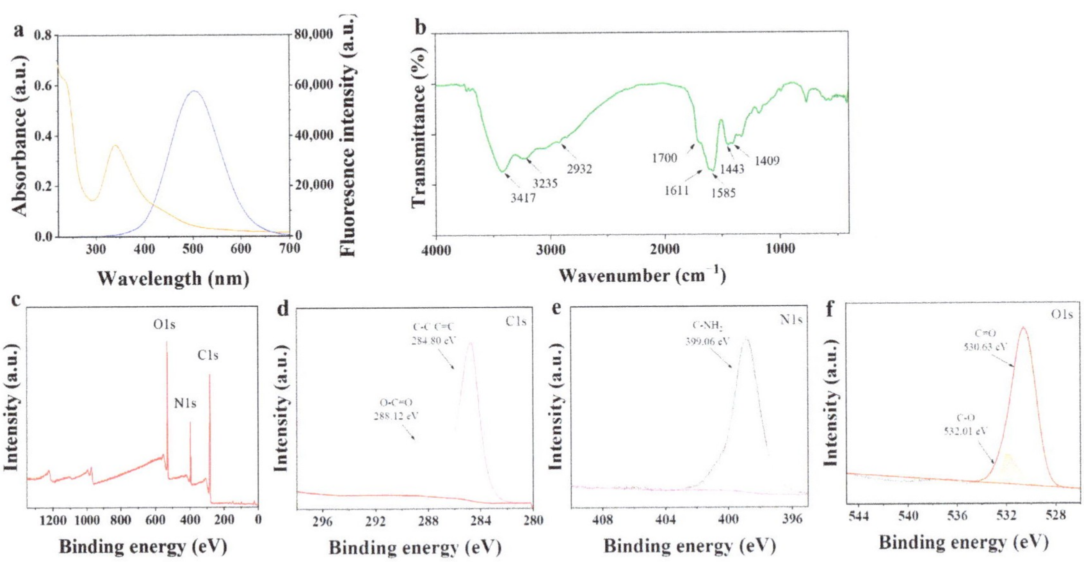

Figure 2. Characterization of elemental groups of N-CD. (**a**) UV/Vis absorption and fluorescence emission spectra of N-CDs (**b**) Fourier transform infrared spectroscopy (FT-IR) spectra. (**c**) X-ray photoelectron spectroscopy (XPS) spectrum of N-CDs. (**d**) The deconvolution graph of the C 1s spectrum of N-CDs. (**e**) The N 1s spectral deconvolution map of N-CDs. (**f**) The O 1S spectral deconvolution map of N-CDs.

The surface functional groups of the N-CDs were characterized by Fourier transform infrared (FT-IR) spectroscopy. As illustrated in Figure 2b, the broad peak near 3200 cm^{-1} was associated with f N–H stretching vibration. The strong peak at 3417 cm^{-1} was assigned to the stretching vibration of the intermolecular hydrogen bond O–H. The absorption peak at 2932 cm^{-1} corresponded to C–H stretching vibration. The peak at 1443 cm^{-1} was the fingerprint region of C–H. The characteristic absorption peak at 1700 cm^{-1} was associated with the stretching vibration of C=O and C=C. The absorption peak at 1611 cm^{-1} was related to the bending vibration of the primary amine (N–H), and the peak near 1585 cm^{-1} was also assigned to N–H bending. The peak at 1409 cm^{-1} was caused by C–N stretching vibration. These peaks demonstrate that the surface of the N-CDs was rich in carboxyl, ester, hydroxyl, and carbonyl.

Furthermore, the X-ray photoelectron spectroscopy (XPS) survey spectrum indicated that the N-CDs were composed primarily of carbon, nitrogen, and oxygen with an atomic ratio of 62:15:23 (Figure 2c). The chemical states of these elements were further analyzed by using fine XPS spectra. As shown in Figure 2d–f, the C 1s spectrum (Figure 2d) could be deconvoluted into two peaks at 284.8 eV (C–C/C=C) and 288.12 eV (C=O/C=N). The N 1s spectrum (Figure 2e) was fitted with a single peak at 399.06 eV (N–H). The O1s spectrum (Figure 2f) was fitted with two peaks of C=O at 530.63 eV and C–O at 532.01 eV. Combined with FT-IR results, these XPS data confirmed that the N-CDs possessed abundant hydrophilic functional groups.

2.3. N-CDs Treatment Decrease CHT Residue in Tomato

To analyze the effect of N-CDs on CHT residue in tomato leaves, different concentrations of N-CDs (N-CDs1~N-CDs5) were exogenously sprayed onto tomato leaves, followed by foliar application of CHT. Then, CHT residue in the tomato leaves was analyzed by high performance liquid chromatography (HPLC). Treatment with N-CDs1 to N-CDs5 significantly decreased CHT residue in tomato leaves compared to the control (Figure 3). Notably, the N-CDs3 treatment displayed the best effect, reducing CHT residues to 34% of the level in the control group (Figure 3). This result suggested that a concentration of 150 mg·L^{-1} N-CDs may be the most effective for decreasing CHT residue in tomato.

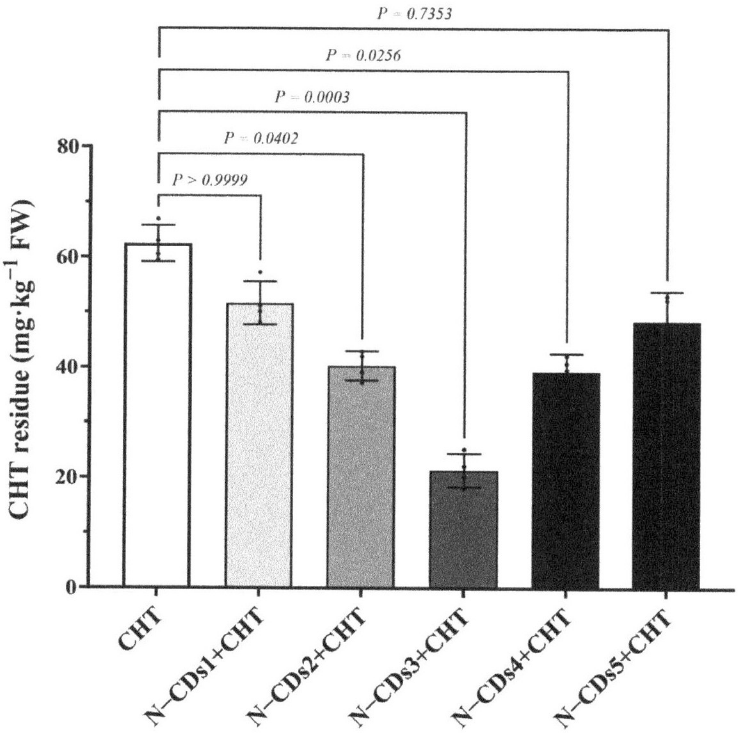

Figure 3. Effect of different concentrations of N-CD on chlorothalonil (CHT) residue in tomato. Two-week-old tomato plants were sprayed with N-CDs1 (50 mg·L^{-1}), N-CDs2 (100 mg·L^{-1}), N-CDs3 (150 mg·L^{-1}), N-CDs4 (200 mg·L^{-1}), and N-CDs5 (300 mg·L^{-1}). Two days later, the plants were sprayed with either water (control) or 37.6 mM CHT. The leaf samples were collected seven days after CHT treatment for analyzing CHT residues. Each dot represented a biological repetition. Statistical significance was determined using one-way ANOVA followed by *Kruskal-Wallis* testing. Data are the mean ± SD from at least four biological replicates. There was a significant difference when $p < 0.05$, but no significant difference when $p \geq 0.05$.

2.4. N-CDs Treatment Alleviated CHT-Induced Toxicity in Tomato

Thus, the 150 mg·L^{-1} N-CDs treatment was selected to analyze its effect on alleviating CHT-induced phytotoxicity in tomato plants. As shown in Figure 4a, N-CDs treatment enhanced tomato plant growth under both control and CHT conditions. Importantly, N-CDs-treated tomatoes under CHT stress exhibited growth similar to that of water-treated plants under control condition. In addition, plant height, leaf area, fresh weight, and dry weight were significantly increased by N-CDs treatment in both the control and CHT groups (Figure 4b–e). Interestingly, these growth indexes in CHT stressed plants treated with N-CDs recovered to levels comparable to those of the water treated control plants.

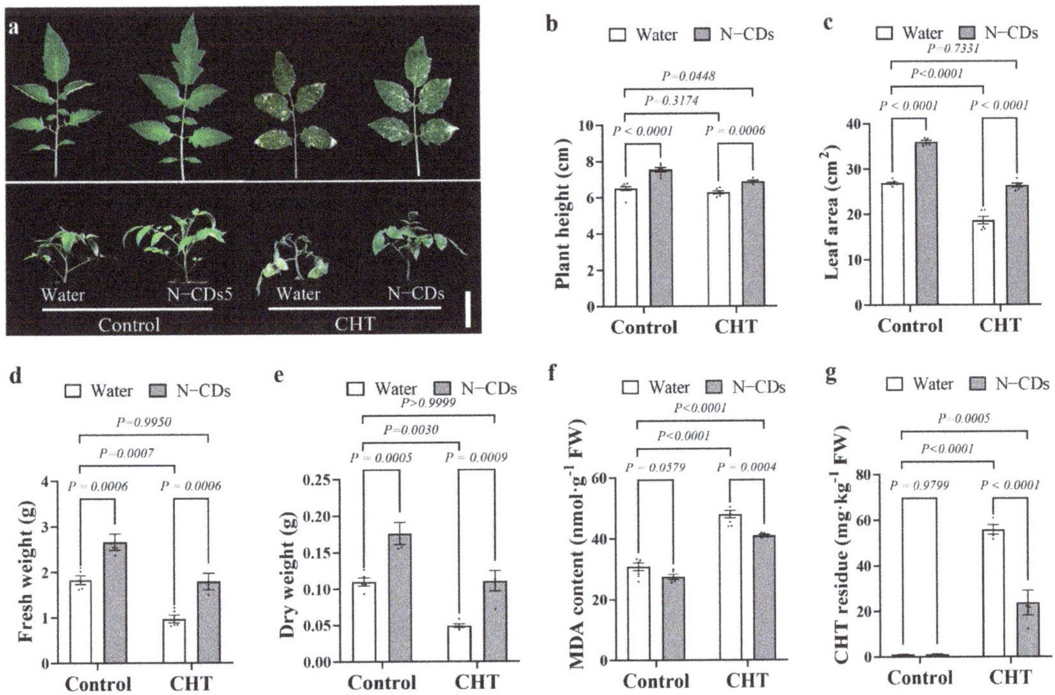

Figure 4. N-CDs treatment alleviated CHT-induced toxicity in tomato. (**a**) Phenotype of control and N-CDs sprayed tomato plants under control or CHT treatment. The scale bar represents 5 cm. Plant height (**b**), leaf area (**c**), dry weight (**d**), fresh weight (**e**), MDA concentration (**f**), and CHT residue (**g**) were measured in the indicated plants, respectively. Two-week-old tomato plants were sprayed with N-CDs (150 mg·L^{-1}). Two days later, the plants were sprayed with either water (control) or 37.6 mM CHT. The plant samples were collected seven days after CHT treatment, for analysis of the related indicators. Each dot represented a biological repetition. Statistical significance was determined using two-way ANOVA followed by *L.S.D.* testing. Data are the mean ± SD from at least four biological replicates. There was a significant difference when $p < 0.05$, but no significant difference when $p \geq 0.05$.

Then, malondialdehyde (MDA) content was measured to assess the extent of lipid peroxidation under CHT stress (Figure 4f). The accumulation of MDA was significantly reduced in N-CDs-treated tomatoes compared with water-treated tomatoes under CHT conditions. CHT residues were also significantly reduced in the N-CDs-treated tomatoes compared with the water-treated tomatoes (Figure 4f). These data indicated that N-CDs treatment could alleviate CHT-induced toxicity and decrease CHT residue accumulation in tomato plants.

2.5. N-CDs Treatment Alleviated CHT-Induced Depression in Photosynthesis in Tomato

Furthermore, photosynthetic capacity was also investigated (Figure 5). Under control conditions, total chlorophyll (chl) content, photosynthetic rate (Pn), the maximum quantum efficiency of PSII (Fv/Fm) level, and the efficiency of PSII photochemistry (Fq'/Fm') were significantly increased in N-CDs-treated tomatoes. Under CHT conditions, photosynthetic capacity in all the tomato plants was significantly inhibited. However, the total chl content and Pn of N-CDs-treated tomatoes under CHT conditions recovered to those in the water-treated control plants (Figure 5a,b). In addition, N-CDs treatment under CHT conditions significantly increased the *Fv/Fm* and *Fq'/Fm'* levels compared to the water-treated control (Figure 5c,d). These results meant that N-CDs treatment alleviated CHT-induced depression in photosynthesis in tomato, potentially by promoting chl synthesis.

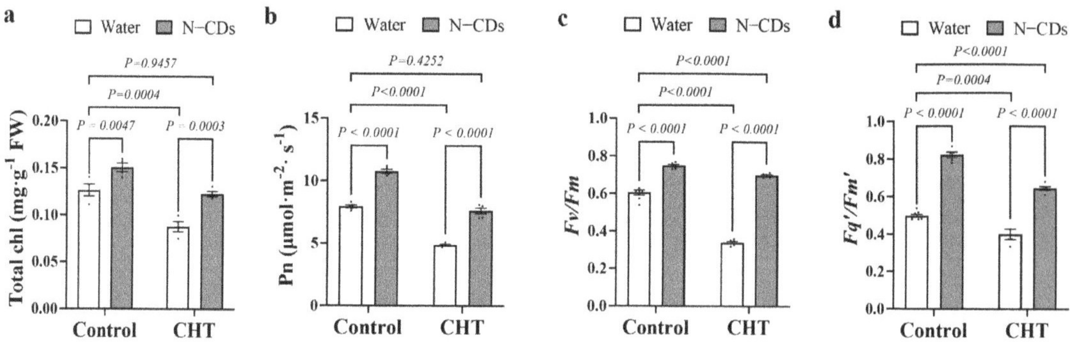

Figure 5. N-CDs treatment alleviated CHT-induced depression in photosynthesis in tomato. (**a**–**d**) represented total chl, Net photosynthetic rate (Pn), the maximum quantum yield of photosystem II (*Fv*/*Fm*), and the photochemical efficiency of photosystem II in the light (*Fv'*/*Fm'*), respectively. Two-week-old tomato plants were sprayed with N-CDs (150 mg·L^{-1}). Two days later, the plants were sprayed with either water (control) or 37.6 mM CHT. The leaves were used for analyzing the related indicators at seven days after CHT treatment. Each dot represented a biological repetition. Statistical significance was determined using two-way ANOVA followed by *L.S.D.* testing. Data are the mean ± SD from at least four biological replicates. There was a significant difference when $p < 0.05$, but no significant difference when $p \geq 0.05$.

2.6. N-CDs Treatment Repressed ROS Accumulation and Activated the Antioxidant Enzyme Activity in Tomato Under CHT Condition

To evaluate whether N-CDs mediates tomato detoxification by affecting ROS accumulation, the levels of $O_2^{\bullet-}$ and H_2O_2 were analyzed in tomato plants that were either water-treated or N-CDs-treated under both control and CHT conditions, respectively (Figure 6a,b). Under control conditions, the contents of $O_2^{\bullet-}$ and H_2O_2 showed no significant difference between water-treated and N-CDs-treated tomatoes. Under CHT conditions, the contents of $O_2^{\bullet-}$ and H_2O_2 were significantly enhanced in the leaves of all the tomato plants. However, the leaves of N-CDs-treated tomato plants accumulated significantly less $O_2^{\bullet-}$ and H_2O_2 than water-treated tomatoes.

Then, the activities of antioxidant enzymes, including SOD, POD, CAT, and APX, were measured to analyze the N-CDs-mediated ROS scavenging ability under CHT conditions. As shown in Figure 6c–f, under control conditions, the activities of SOD, CAT, and APX significantly increased in N-CDs-treated tomatoes compared with water-treated tomatoes. Importantly, under CHT conditions, the activities of SOD, POD, CAT, and APX were further activated in N-CDs-treated tomatoes compared with those in water-treated tomatoes (Figure 6c–f).

2.7. N-CDs Treatment Enhance the Detoxification Potential of Tomato

GSH and GST play crucial roles in the process of alleviating pesticide toxicity in plants. Therefore, the activity of GST, and the contents of GSH, GSSG, and GSH + GSSG, and the ratio of GSH/GSSG were analyzed, respectively. As shown in Figure 7c–e, CHT treatment significantly increased the content of GSSG and GSH + GSSG, but it drastically decreased the ratio of GSH/GSSG compared to the water-treated tomatoes. In contrast, under CHT conditions, N-CDs treatment significantly increased GST activity and GSH content by 50% and 75%, respectively, and reduced the content of GSSG by 14% compared to the water-treated tomatoes (Figure 7a–c). Notably, although N-CDs application did not affect the GSH content under control conditions, it increased the content of GST, leading to an increased ratio of GSH/GSSG (Figure 7a,b). Importantly, changes in GSH and GSSG resulted in a significant reduction in the ratio of GSH/GSSG under CHT conditions. However, N-CDs pretreatment significantly augmented the ratio of GSH/GSSG in N-CDs-treated tomatoes compared with the water-treated tomatoes under CHT conditions (Figure 7e).

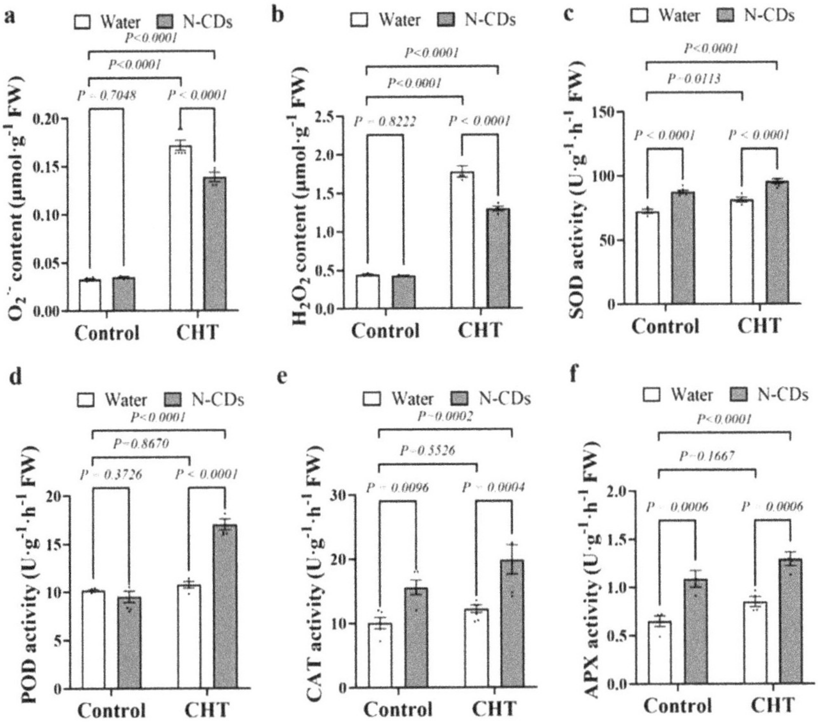

Figure 6. N-CDs treatment activated the antioxidant enzyme activity and repressed ROS accumulation in tomato. (**a**) H_2O_2 content, (**b**) $O_2^{\bullet-}$ content, (**c**) superoxide dismutase (SOD) activity, (**d**) peroxidase (POD) activity, (**e**) catalase (CAT) activity and (**f**) ascorbate peroxidase (APX) activity were measured in the indicated plants, respectively. Each dot represents a biological repetition. Statistical significance was determined using two-way ANOVA followed by *L.S.D.* testing. Data are the mean ± SD from at least four biological replicates. There was a significant difference when $p < 0.05$, but no significant difference when $p \geq 0.05$.

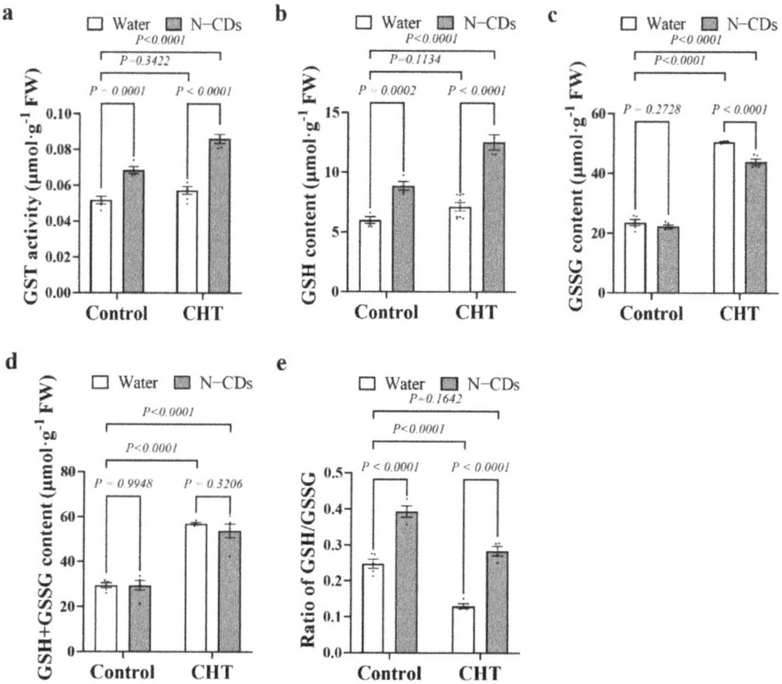

Figure 7. N-CDs treatments altered the synthesis and metabolism of glutathione in tomato. (**a**) Glutathione S-transferases (GST) activity, (**b**) glutathione (GSH) content, (**c**) oxidized glutathione (GSSG) content, (**d**) total glutathione content (GSH + GSSG), (**e**) ratio of GSH/GSSG was measured in the indicated plants, respectively. Each dot represented a biological repetition. Statistical significance was determined using two-way ANOVA followed by *L.S.D.* testing. Data are the mean ± SD from at least four biological replicates. There was a significant difference when $p < 0.05$, but no significant difference when $p \geq 0.05$.

3. Discussion

Pesticide residues accumulate in edible plant tissues and pose serious phytotoxic risks, including shoot apical atrophy, leaf necrosis, and impaired photosynthesis, ultimately reducing crop yield and threatening food safety [19]. Traditional approaches such as breeding for resistance or reducing application rates have only partially addressed this challenge. CDs offer high biocompatibility, water solubility, and environmental safety [20,21]. Previous studies have reported that the carbon dots could promote plant growth and improve plant stress tolerance by enhancing the photosynthetic capacity [17]. However, their ability to alleviate pesticide-induced phytotoxicity in plants had remained unexplored. Here, we synthesized N-CDs and found that the spraying N-CDs could dramatically reduce pesticide residues in tomato leaves (Figure 3). This result indicates that the alleviation of pesticide phytotoxicity by N-CDs may be attributed to their ability to enhance photosynthetic capacity. Generally, the smaller the particle size of carbon QDs, the more their fluorescence emission positions tend toward shorter wavelengths [22]. Our data confirm this trend (Figure 1). This ultra-small size was conducive to the efficient entry of N-CDs into plant cells to exert their potential biological functions. Previous studies have reported that citrazinic acid is a major contributor to the blue fluorescence of CDs derived from citric acid precursors [23]. In the presence of urea, green fluorescent molecular clusters such as 4-hydroxy-1*H*-pyrrole [3,4-*c*] pyridine-1,3,6(2*H*,5*H*)-trione (HPPT) can be generated during the preparation of CDs, with their abundance increasing as the proportion of urea rises [24]. In this work, the N-CDs synthesized from citric acid and urea emitted strong blue fluorescence (Figure 1), which may be attributed to differences in precursor ratios and synthesis conditions affecting the final product's properties. Furthermore, based on UV/Vis spectrum, FT-IR, and XPS analyses (Figure 2), it was determined that the fluorescence of N-CDs was primarily due to citrazinic acid. Some carbon-based (like graphene oxide) and metal-based (like zinc oxide and potassium permanganate) nanomaterials exhibit a certain degree of cytotoxicity, with cell survival rates lower than 80% [25,26]. Chemical modifications (such as fetal bovine serum protein and l-cysteine, etc.) can reduce cytotoxicity [27,28]. Importantly, N-CDs exhibited high biological safety in cytotoxicity assays (Figure 1d), which might be attributed to the reliable safety of the precursor substance itself [29], showing potential for application in practical agricultural production.

Plants mount defensive responses to various pesticides, such as CHT, which are known to induce the production of ROS [19]. Excessive ROS accumulation can lead to oxidative stress and damage plant cells, thereby limiting the growth and development of plants [30]. Consistent with this, the pesticide treatment led to excessive ROS accumulation in tomatoes, reduced photosynthetic capacity, and stunted plant growth (Figures 4–6). Importantly, the levels of ROS were significantly reduced in N-CDs-treated tomatoes (Figure 6a,b). Moreover, both photosynthetic performance and growth parameters recovered to normal levels in N-CDs-treated tomatoes (Figures 4 and 5). However, oxidative damage at the cellular level may reduce membrane integrity. We observed that under pesticide treatment, the MDA concentrations increased significantly in all plants but were markedly reduced in N-CDs-treated tomatoes (Figure 4f). This result suggests that N-CDs-treated tomatoes had a batter membrane stability. For sustaining redox homeostasis, plants can boost antioxidant enzyme activities to mitigate oxidative damage and enhance their stress adaptability [31]. As the initial defense against excessive ROS, SOD transforms $O_2^{\bullet-}$ into H_2O_2, which is then detoxified into water and oxygen by POD, CAT, and APX [32]. In this study, the activities of SOD, POD, CAT, and APX were more significantly activated in N-CDs-treated tomatoes than in water-treated tomatoes (Figure 6c–f). These results demonstrate that N-CDs treatment improved tomato tolerance to drought and salt stresses through decreasing ROS generation and maintaining a strong ROS-scavenging ability. Some studies have found that nanomaterials enhance photosynthesis [33] and improve salt tolerance [34] by activating the antioxidant enzyme system, which is consistent with our research results for plants under pesticide stress. These results indicate that the regulation of plant abiotic stress resistance by CDs may be inseparable from the antioxidant enzyme system. However, it remains unclear whether this impact was caused by the direct interaction of nanoparticles and antioxidant enzymes or was an indirect function of ROS-triggered signaling. In addition, previous studies have reported biostimulants capable of degrading CHT. For example, melatonin treatment reduced pesticide residues by 57% [6], while γ-aminobutyric acid (GABA) achieved a reduction of 45% [35]. Notably, in this study, N-CDs3 at a concentration of 150 $mg \cdot L^{-1}$ demonstrated the most significant efficacy, reducing pesticide residues by approximately 66% (Figure 3). This degradation performance was more effective compared to other biostimulants. The weakened decomposition effect at a concentration of 200–300 $mg \cdot L^{-1}$ may be due to the reduced spacing of N-CDs at high concentrations, resulting in a self-aggregation effect [36] and a decrease in the effective concentration of N-CDs entering the leaves. These results indicate that N-CDs treatment alleviated pesticide-induced phytotoxicity in tomato through activation of the antioxidant enzyme system, suppression of ROS generation, and stabilization of the plasma membrane at an appropriate level.

During pesticide detoxification in plants, GST plays a central role by catalyzing the conjugation of GSH to pesticide molecules [8]. We therefore profiled GST activity and the balance between reduced and oxidized GSH under CHT conditions. Under CHT conditions, the GSH/GSSG ratio decreased markedly, but N-CDs treatment restored this ratio to a healthier balance by enhancing antioxidant enzyme activity (Figures 6 and 7). These results imply that N-CDs facilitated the conjugation of toxic compounds with GSH, thereby accelerating detoxification in tomato plants. GST is known to catalyze the conjugation of GSH to xenobiotics. Previous studies have reported that H_2O_2 signaling promotes both GSH synthesis and GST induction via melatonin treatment [19]. Notably, our findings align with the concept that redox shifts can upregulate detoxification enzymes. However, unlike melatonin, which elevates GSH levels even under control conditions, N-CDs treatment did not alter GSH pools in control plants but still boosted GST activity (Figure 7). This distinction may stem from mild, localized ROS signaling generated by the CDs' surface, which selectively induces GST activity without broadly activating sulfur assimilation pathways. However, the roles played by Phase I (cytochrome P450s) and Phase III (ABC transporters) in the process of N-CDs alleviating pesticide toxicity still require further study.

Collectively, these results supported the conclusion that N-CDs mediated pesticide detoxification in tomato primarily through modulating redox homeostasis (Figure 8). In addition, N-CDs represent a novel agent of biostimulants that can fine-tune plant antioxidant defenses with minimal disruption under normal conditions. However, due to the potential species-specific toxicity of nanomaterials, future studies on plant seed and root will be needed to verify the plant-level safety of N-CDs. In this study, the growth-promoting and CHT-detoxifying effects of N-CDs were confirmed using the sequenced tomato cultivar 'Heinz 1706'. Whether these findings are applicable to other varieties and pesticides requires further validation.

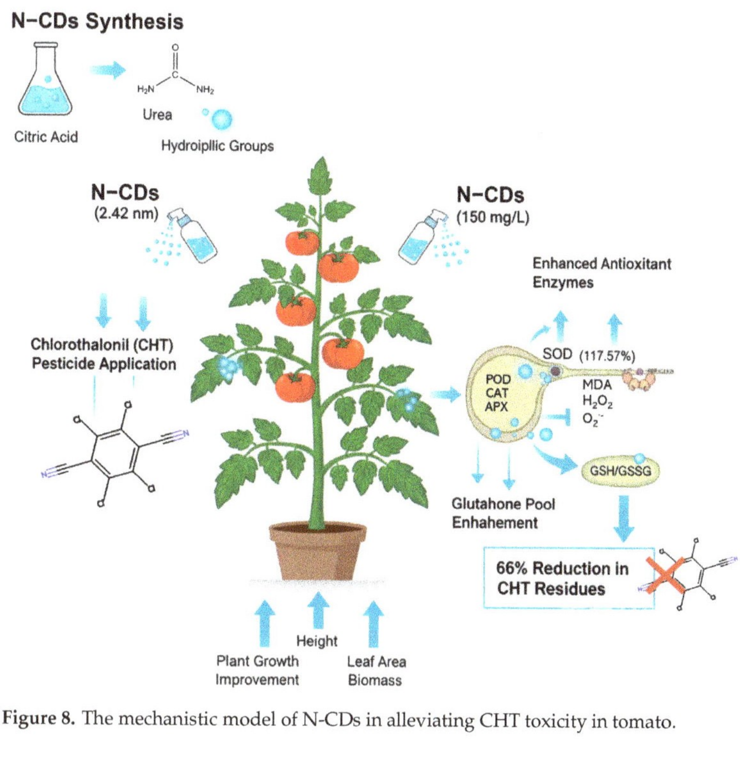

Figure 8. The mechanistic model of N-CDs in alleviating CHT toxicity in tomato.

4. Materials and Methods

4.1. The Synthesis, Characterization, and Cytotoxicity Test of N-CDs

The preparation of N-CDs followed our previous method with minor modifications [37]. Briefly, 1.025 g citric acid and 0.379 g urea underwent hydrothermal reaction at 160 °C for 6 h. The resulting product was dialyzed (3.5 kDa) for 48 h and stored at 4 °C in the dark. The N-CDs were characterization using the following techniques: TEM (JEM-F200, JEOL, Tokyo, Japan) for morphology and size; UV/Vis spectroscopy (Shimadzu, UV-2600i) and fluorescence spectroscopy (LF-1701009, Thermo Fisher Scientific, Waltham, MA, USA) for optical properties; FT-IR (Thermo Fisher Scientific, NICOLET iS20, Waltham, MA, USA) and XPS (K-Alpha, Thermo Scientific, Waltham, MA, USA) for identifying surface functional groups and elemental composition. Cytotoxicity was evaluated according to our previous method [38]. HEK 293T cells were treated with N-CDs at a concentration of 10 $mg \cdot mL^{-1}$, and cell viability was assessed using the CCK-8 assay at 0, 24, 48, 72, and 96 h. The measurements of cytotoxicity were performed in triplicate and expressed relative to untreated controls.

4.2. Plant Materials and Pesticide Treatment

Tomato cultivar 'Heinz 1706' seeds were germinated and grown in a growth chamber. The seeds were surface-sterilized by immersion in 75% ethanol for 1 min, followed by treatment with 8% NaClO solution for 15 min. Then, the seeds were rinsed five times with sterile water and sown on plates containing Morishige and Skoog (MS) medium supplemented with sucrose. Dry seeds were directly sown at the varying depth in a soil mixture containing 40% organic matter in pots of 8 × 8 × 8.5 cm. All seedlings were grown in a growth chamber at 28 °C with 16 h light and 22 °C with 8 h light, as previously described [39].

For pesticide treatment, two-week-old tomato plants were sprayed with different concentrations of N-CDs: N-CDs1 (50 $mg \cdot L^{-1}$), N-CDs2 (100 $mg \cdot L^{-1}$), N-CDs3 (150 $mg \cdot L^{-1}$), N-CDs4 (200 $mg \cdot L^{-1}$), and N-CDs5 (300 $mg \cdot L^{-1}$). Two days later, the plants were treated with either water (control) or 37.6 mM CHT. Leaf samples were collected 7 days after CHT treatment, flash-frozen in liquid nitrogen, ground to a fine powder, and stored at −80 °C until analysis. All measurements were performed in three independent experiments with at least four biological replicates.

4.3. Growth Related Parameters Analysis

Plants were photographed at the indicated time points. Plant height and the area of the fifth leaf were measured at the indicated time points using Image J software (ij154-win-java8). The fresh weight (FW) of shoot was recorded immediately. For dry weight (DW) determination, samples were first heated at 105 °C for 20 min and then dried at 75 °C until constant weight was achieved [30]. All measurements were performed in three independent experiments with at least four biological replicates.

4.4. Photosynthetic Related Parameters Analysis

Total chlorophyll content was determined according to the previously described method [40]. Briefly, total chlorophyll was extracted from 0.05 g of tomato leaf tissue using 80% acetone. The absorbance of the extract was measured using a multi-mode microplate reader (Synergy H4 Hybrid) at OD_{470}, OD_{647}, and OD_{663}. Total chlorophyll content was calculated using the following formula:

$$\text{chla}(\text{mg/L}) = 12.21 \times OD_{663} - 2.81 \times OD_{647} \tag{1}$$

$$\text{chlb}(\text{mg/L}) = 20.13 \times OD_{647} - 5.03 \times OD_{663} \tag{2}$$

$$\text{Total chl(mg/L)} = \frac{1000 \times OD_{647} - 3.27 \times \text{chla} - 104 \times \text{chlb}}{229} \quad (3)$$

Pn was measured with a Li-6400 portable gas-exchange system [41]. The Fv/Fm and the Fq'/Fm' in tomato leaves were assessed using an FMS-2 Pulse Modulated Fluorometer (Hansha Scientific Instruments, Beijing, China) [5]. All measurements were performed three independent experiments with at least four biological replicates.

4.5. ROS Contents and Antioxidant Enzyme Activities Analysis

The contents of $O_2^{\bullet-}$ and H_2O_2 and the activity of SOD, POD, CAT, and APX in tomato leaves were measured based on previously described methods [5]. Lipid peroxidation was evaluated by measuring MDA using the TBARS method [42]. In detail, the $O_2^{\bullet-}$ content was extracted from 0.05 g of tomato leaf tissue using a commercial assay kit (BC1295, Beijing Solarbio Science & Technology Co., Ltd., Beijing, China) by measuring absorbance at 530 nm. The concentration was calculated according to the following equation ($C_{standard}$ = 0.03125 μmol·mL^{-1}, V = 0.1 mL, and W = 0.05 g):

$$\text{Content} = 2 \times C_{\text{standard}} \frac{A_{\text{sample}} - A_{\text{blank}}}{A_{\text{standard}} - A_{\text{blank}}} \times \frac{V}{W} \quad (4)$$

The H_2O_2 content was extracted from 0.05 g of tomato leaf tissue using a commercial assay kit (BC3595, Beijing Solarbio Science & Technology Co., Ltd.) by detecting absorbance at 415 nm. The content was also calculated by Equation (4) ($C_{standard}$ = 2 μmol·mL^{-1}, V = 0.1 mL, and W = 0.05 g):

In addition, 0.1 g of tomato leaf samples were homogenized in 0.05 mol·L^{-1} phosphate buffer (pH 7.8) and centrifuged to obtain the crude enzyme extract. The supernatant was subsequently used for the analysis of enzyme activities and MDA content.

SOD activity was assayed by adding 25 μL of crude enzyme extract to 18 μL of 0.6 mg/mL nitrobluetetrazolium (NBT), followed by absorbance measurement at 560 nm. SOD activity was calculated as follows:

$$\text{SOD content} = \left(A_{\text{light}} - A + A_{\text{dark}}\right) \times \frac{V}{0.5 \times A_{\text{light}} \times W \times V_t}, \quad (5)$$

where A_{light} = absorbance of the light control, A_{dark} = absorbance of the dark control, A = absorbance of the sample, W = 0.1 g, V = 0.6 mL, and V_t = 0.025 mL.

POD activity was measured by mixing 20 μL of crude enzyme extract with 28 μL of 100% guaiacol. Absorbance at 470 nm was recorded at 1-min intervals for a total of 16 readings. POD activity was calculated using Formula (6). CAT activity was determined by adding 5 μL of crude enzyme extract to 50 μL of 0.1 mol/L hydrogen peroxide. The decrease in absorbance was monitored at 240 nm. CAT activity was also calculated according to Formula (6). APX activity was assessed by incubating 5 μL of crude enzyme extract with 100 μL of 0.88 mg/mL ascorbate hydrogen peroxide. Absorbance was measured at 560 nm. APX activity was calculated as follows:

$$\text{Total POD activity} = \frac{\Delta A \times V}{a \times W \times t}, \quad (6)$$

where $\Delta A = [|A_{t1} - A_{t0}| + |A_{t2} - A_{t1}| + \ldots + |A_{t15} - A_{t14}|]/15$, V = 0.6 mL, a = 0.01 mL, W = 0.1 g, and t = 1 min.

The MDA content was determined by mixing 300 μL of crude enzyme extract with 0.67% thiobarbituric acid (TBA). Absorbance was measured at 450, 532, and 600 nm. The MDA content was calculated using the following equation:

$$\text{MDA content} = C \times \frac{V}{W}, \quad (7)$$

where C = 6.45 × (A_{532} − A_{600}) − 0.56 × A_{450}, V = 3 mL, and W = 0.1 g.

All spectrophotometric measurements were performed using a Synergy H4 Hybrid Reader (BioTek Instruments, Winooski, VT, USA). All measurements were performed in three independent experiments with at least four biological replicates.

4.6. Analysis of CHT Content by HPLC

The CHT content in tomato leaves was quantified by HPLC according to a previously reported method with minor modifications [43]. In detail, 0.2 g of sample was extracted with 2.0 mL of acetonitrile and 0.2 g of NaCl, vortexed thoroughly, and incubated at room temperature for 30 min. The supernatant was collected and evaporated to near dryness under a gentle stream of nitrogen at 60 °C. The residue was reconstituted in 2.0 mL of n-hexane. The extract was purified using an alumina-N solid-phase extraction column cartridge preconditioned sequentially with n−hexane/acetone and n−hexane. After loading the sample, the cartridge was rinsed, and the eluate was collected. The eluate was evaporated at 50 °C and reconstituted in 2.0 mL mobile phase. The final solution was filtered through a 0.22 μm membrane filter prior to HPLC injection. A stock solution of CHT (1.0 mg·mL^{-1}) was prepared in acetonitrile and stored at −20 °C. Working standards of 0.05, 0.1, 0.5, 1, 2, 5, 10, 50, 100, and 200 mg·L^{-1} CHT was prepared by gradient dilution of the stock solution with the mobile phase before analysis. HPLC analysis was performed under the following conditions: column: ZORBAX Eclipse XDB-C18 (250 mm × 4.6 mm, 5 μm); mobile phase: acetonitrile/water (55:45, *v*/*v*); flow rate: 0.6 mL·min^{-1}; detection: 232 nm; injection volume: 10 μL; column temperature: 30 °C. All measurements were performed in three independent experiments with at least four biological replicates.

4.7. Data Analysis

Data were compiled in Microsoft Excel. Means and standard deviations were calculated from biolobical replicates using Graphpad Prism software. Statistical significance was determined using one-way ANOVA followed by Kruskal-Wallis testing or two-way ANOVA followed by *L.S.D.* testing, also conducted in Graphpad Prism software (version 10.6.1). All data are presented as mean ± SD from at least four biological replicates. *, **, ***, and **** indicate significant differences at $p < 0.05$, $p < 0.01$, $p < 0.001$, and $p < 0.0001$, respectively.

5. Conclusions

In conclusion, this study demonstrates that N-CDs are non-cytotoxic and effective nanobiostimulants for enhancing plant growth and mitigating pesticide-induced stress in tomato. The hydrothermally synthesized N-CDs (~2.4 nm), rich in hydrophilic groups and exhibiting strong blue fluorescence, dramatically reduced CHT residues by up to 66% at 150 mg·L^{-1}. Under CHT conditions, N-CD treatment restored plant growth parameters including height, leaf area, and biomass to levels comparable with the water-treated tomato. These improvements coincided with reduced levels of oxidative stress markers (MDA, $O_2^{\bullet-}$, H_2O_2) and enhanced the activities of antioxidant enzymes (SOD, POD, CAT, APX), alongside an expansion of the GSH pool. Collectively, our findings indicated that N-CDs strengthen the antioxidant defense system to promote detoxification in tomato.

Author Contributions: H.D., Q.S. and X.Z. conceived and designed the experiments. X.Z., Y.X., H.W., Y.D., W.W., Y.G., Y.H. and R.K. performed the experiments. X.Z. and H.D. analyzed the data and wrote the manuscript. All authors have read and agreed to the published version of the manuscript.

Funding: This work was supported by the Shandong Provincial Natural Science Foundation (ZR2024QC072) and the Key R&D Program of Shandong Province, China (2024TZXD055 and 2024LZGC013).

Institutional Review Board Statement: Not applicable.

Informed Consent Statement: Not applicable.

Data Availability Statement: Data is contained within the article.

Conflicts of Interest: The authors declare no conflicts of interest.

References

1. Deutsch, C.A.; Tewksbury, J.J.; Tigchelaar, M.; Battisti, D.S.; Merrill, S.C.; Huey, R.B.; Naylor, R.L. Increase in crop losses to insect pests in a warming climate. *Science* **2018**, *361*, 916–919. [CrossRef] [PubMed]
2. Gonçalves, A.M.M.; Rocha, C.P.; Marques, J.C.; Gonçalves, F.J.M. Fatty acids as suitable biomarkers to assess pesticide impacts in freshwater biological scales—A review. *Ecol. Indic.* **2021**, *122*, 107299. [CrossRef]
3. Chung, S.W.C. How effective are common household preparations on removing pesticide residues from fruit and vegetables? A review. *J. Sci. Food Agric.* **2018**, *98*, 2857–2870. [CrossRef]
4. Ahammed, G.J.; Wang, Y.Q.; Mao, Q.; Wu, M.J.; Yan, Y.R.; Ren, J.J.; Wang, X.J.; Liu, A.R.; Chen, S.C. Dopamine alleviates bisphenol A-induced phytotoxicity by enhancing antioxidant and detoxification potential in cucumber. *Environ. Pollut.* **2020**, *259*, 113957. [CrossRef]
5. Zhang, X.; Bao, Z.L.; Gong, B.; Shi, Q.H. *S-adenosylmethionine synthetase 1* confers drought and salt tolerance in transgenic tomato. *Environ. Exp. Bot.* **2020**, *179*, 104226. [CrossRef]
6. Kanwar, M.K.; Xie, D.L.; Yang, C.; Ahammed, G.J.; Qi, Z.Y.; Hasan, M.K.; Reiter, R.J.; Yu, J.Q.; Zhou, J. Melatonin promotes metabolism of bisphenol A by enhancing glutathione-dependent detoxification in *Solanum lycopersicum* L. *J. Hazard. Mater.* **2020**, *388*, 121727. [CrossRef]
7. Lu, K.; Song, Y.Y.; Zeng, R.S. The role of cytochrome P450-mediated detoxification in insect adaptation to xenobiotics. *Curr. Opin. Insect Sci.* **2021**, *43*, 103–107. [CrossRef]
8. Vaish, S.; Gupta, D.; Mehrotra, R.; Mehrotra, S.; Basantani, M.K. Glutathione S-transferase: A versatile protein family. *3 Biotech* **2020**, *10*, 321. [CrossRef]
9. Chen, S.C.; Yan, Y.R.; Wang, Y.Q.; Wu, M.J.; Mao, Q.; Chen, Y.F.; Ren, J.J.; Liu, A.R.; Lin, X.M.; Ahammed, G.J. *Trichoderma asperellum* reduces phoxim residue in roots by promoting plant detoxification potential in *Solanum lycopersicum* L. *Environ. Pollut.* **2020**, *259*, 113893. [CrossRef] [PubMed]
10. Yu, G.B.; Zhang, Y.; Ahammed, G.J.; Xia, X.J.; Mao, W.H.; Shi, K.; Zhou, Y.H.; Yu, J.Q. Glutathione biosynthesis and regeneration play an important role in the metabolism of chlorothalonil in tomato. *Chemosphere* **2013**, *90*, 2563–2570. [CrossRef] [PubMed]
11. Foyer, C.H.; Kunert, K. The ascorbate-glutathione cycle coming of age. *J. Exp. Bot.* **2024**, *75*, 2682–2699. [CrossRef] [PubMed]
12. Flohe, L.; Toppo, S.; Orian, L. The glutathione peroxidase family: Discoveries and mechanism. *Free Radical Bio. Med.* **2022**, *187*, 113–122. [CrossRef]
13. Lin, Z.H.; Xu, D.H.; Zhao, Y.M.; Sheng, B.; Wu, Z.J.; Wen, X.B.; Zhou, J.; Chen, G.; Lv, J.; Wang, J.; et al. Micro/Nanoplastics in plantation agricultural products: Behavior process, phytotoxicity under biotic and abiotic stresses, and controlling strategies. *J. Nanobiotechnol.* **2025**, *23*, 231. [CrossRef]
14. Li, Y.D.; Xu, X.K.; Wu, Y.; Zhuang, J.L.; Zhang, X.J.; Zhang, H.R.; Lei, B.F.; Hu, C.F.; Liu, Y.L. A review on the effects of carbon dots in plant systems. *Mater. Chem. Front.* **2020**, *4*, 437–448. [CrossRef]
15. Kou, E.F.; Yao, Y.Y.; Yang, X.; Song, S.W.; Li, W.; Kang, Y.Y.; Qu, S.N.; Dong, R.Y.; Pan, X.Q.; Li, D.N.; et al. Regulation Mechanisms of Carbon Dots in the Development of Lettuce and Tomato. *Acs Sustain. Chem. Eng.* **2021**, *9*, 944–953. [CrossRef]
16. Swift, T.A.; Fagan, D.; Benito-Alifonso, D.; Hill, S.A.; Yallop, M.L.; Oliver, T.A.A.; Lawson, T.; Galan, M.C.; Whitney, H.M. Photosynthesis and crop productivity are enhanced by glucose-functionalised carbon dots. *New Phytol.* **2021**, *229*, 783–790. [CrossRef] [PubMed]

17. Liu, H.X.; Zhong, X.; Pan, Q.; Zhang, Y.; Deng, W.T.; Zou, G.Q.; Hou, H.S.; Ji, X.B. A review of carbon dots in synthesis strategy. *Coordin. Chem. Rev.* **2024**, *498*, 215468. [CrossRef]
18. Li, Y.J.; Tang, Z.H.; Pan, Z.Y.; Wang, R.G.; Wang, X.; Zhao, P.; Liu, M.; Zhu, Y.X.; Liu, C.; Wang, W.C.; et al. Calcium-Mobilizing Properties of-Derived Carbon Dots Confer Enhanced Environmental Adaptability in Plants. *Acs Nano* **2022**, *16*, 4357–4370. [CrossRef]
19. Peng, X.H.; Wang, N.N.; Sun, S.S.; Geng, L.J.H.; Guo, N.; Liu, A.R.; Chen, S.C.; Ahammed, G.J. Reactive oxygen species signaling is involved in melatonin-induced reduction of chlorothalonil residue in tomato leaves. *J. Hazard. Mater.* **2023**, *443*, 130212. [CrossRef]
20. Ghirardello, M.; Ramos-Soriano, J.; Galan, M.C. Carbon Dots as an Emergent Class of Sustainable Antifungal Agents. *Acs Nano* **2025**, *19*, 24377–24403. [CrossRef]
21. Delgado-Martín, J.; Delgado-Olidén, A.; Velasco, L. Carbon Dots Boost dsRNA Delivery in Plants and Increase Local and Systemic siRNA Production. *Int. J. Mol. Sci.* **2022**, *23*, 5338. [CrossRef] [PubMed]
22. Zhu, L.L.; Shen, D.K.; Liu, Q.; Luo, K.H.; Li, C. Mild Acidolysis-Assisted Hydrothermal Carbonization of Lignin for Simultaneous Preparation of Green and Blue Fluorescent Carbon Quantum Dots. *Acs Sustain. Chem. Eng.* **2022**, *10*, 9888–9898. [CrossRef]
23. Sharma, A.; Gadly, T.; Neogy, S.; Ghosh, S.K.; Kumbhakar, M. Molecular Origin and Self-Assembly of Fluorescent Carbon Nanodots in Polar Solvents. *J. Phys. Chem. Lett.* **2017**, *8*, 5861–5864; Erratum in *J. Phys. Chem. Lett.* **2017**, *8*, 1044–1052. [CrossRef]
24. Kasprzyk, W.; Swiergosz, T.; Bednarz, S.; Walas, K.; Bashmakova, N.V.; Bogdal, D. Luminescence phenomena of carbon dots derived from citric acid and urea—A molecular insight. *Nanoscale* **2018**, *10*, 13889–13894. [CrossRef] [PubMed]
25. Xie, J.N.; Li, H.L.; Zhang, T.R.; Song, B.K.; Wang, X.H.; Gu, Z.J. Recent Advances in ZnO Nanomaterial-Mediated Biological Applications and Action Mechanisms. *Nanomaterials* **2023**, *13*, 1500. [CrossRef]
26. Leskinen, J.; Tissari, J.; Uski, O.; Viren, A.; Torvela, T.; Kaivosoja, T.; Lamberg, H.; Nuutinen, I.; Kettunen, T.; Joutsensaari, J.; et al. Fine Particle Emissions in three Different Combustion Conditions of a Wood Chip-fired Appliance-Particulate Physico-chemical Properties and Induced Cell Death. *Atmos. Environ.* **2014**, *86*, 129–139. [CrossRef]
27. Mu, L.; Gao, Y.; Hu, X.G. L-Cysteine: A Biocompatible, Breathable and Beneficial Voating for Graphene Oxide. *Biomaterials* **2015**, *52*, 301–311. [CrossRef]
28. Hussain, Z.; Al-Shadidi, J.R.M.H.; Jagal, J.; Rawas-Qalaji, M. Overcoming stealthing effect of PEGylation on chitosan nanoparticles: Optimization, characterization, and assessment of cytotoxicity, apoptosis, cellular internalization, and antimetastatic efficacy. *Int. J. Biol. Macromol.* **2025**, *321*, 146514. [CrossRef]
29. Hu, X.G.; Li, D.D.; Gao, Y.; Mu, L.; Zhou, Q.X. Knowledge gaps between nanotoxicological research and nanomaterial safety. *Environ. Int.* **2016**, *94*, 8–23. [CrossRef]
30. Zhang, X.; Chen, L.C.; Shi, Q.H.; Ren, Z.H. *SlMYB102*, an R2R3-type MYB gene, confers salt tolerance in transgenic tomato. *Plant Sci.* **2020**, *291*, 110356. [CrossRef]
31. Zhu, J.K. Abiotic Stress Signaling and Responses in Plants. *Cell* **2016**, *167*, 313–324. [CrossRef]
32. Willems, P.; Mhamdi, A.; Stael, S.; Storme, V.; Kerchev, P.; Noctor, G.; Gevaert, K.; Van Breusegem, F. The ROS Wheel: Refining ROS Transcriptional Footprints. *Plant Physiol.* **2016**, *171*, 1720–1733. [CrossRef]
33. Guirguis, A.; Yang, W.; Conlan, X.A.; Cahill, D.M.; Wang, Y. Boosting Plant Photosynthesis with Carbon Dots: A Critical Review of Performance and Prospects. *Small* **2023**, *19*, 202300671. [CrossRef]
34. Chen, Q.; Cao, X.F.; Li, Y.Y.; Sun, Q.; Dai, L.N.; Li, J.W.; Guo, Z.J.; Zhang, L.; Ci, L.J. Functional carbon nanodots improve soil quality and tomato tolerance in saline-alkali soils. *Sci. Total Environ.* **2022**, *830*, 154817. [CrossRef] [PubMed]
35. Shan, Q.; Liu, M.H.; Li, R.; Shi, Q.H.; Li, Y.; Gong, B. γ-Aminobutyric acid (GABA) improves pesticide detoxification in plants. *Sci. Total Environ.* **2022**, *835*, 155404. [CrossRef]
36. Ru, Y.; Waterhouse, G.I.N.; Lu, S.Y. Aggregation in carbon dots. *Aggregate* **2022**, *3*, e296. [CrossRef]
37. Du, H.; Zhang, X.; Yao, M.R.; Yang, Q.L.; Wu, W. Aptamer-guided luminous microsphere for anchoring and lightening. *Sens. Actuators B Chem.* **2022**, *366*, 131938. [CrossRef]
38. Du, H.; Ping, T.; Wu, W.; Yang, Q.L. Sandwich Fluorescence Detection of Foodborne Pathogen with CD Fluorescence Signal Amplification in Food Samples. *Foods* **2022**, *11*, 945. [CrossRef]
39. Zhang, X.; Xu, Z.X.; Chen, L.C.; Ren, Z.H. Comprehensive Analysis of Multiprotein Bridging Factor 1 Family Genes and Negatively Regulate the Resistance to in Tomato. *BMC Plant Biol.* **2019**, *19*, 437. [CrossRef] [PubMed]
40. Zhang, S.; Wang, S.J.; Lv, J.L.; Liu, Z.B.; Wang, Y.; Ma, N.N.; Meng, Q.W. SUMO E3 Ligase SlSIZ1 Facilitates Heat Tolerance in Tomato. *Plant Cell Physiol.* **2018**, *59*, 58–71. [CrossRef] [PubMed]
41. Kong, F.Y.; Deng, Y.S.; Zhou, B.; Wang, G.D.; Wang, Y.; Meng, Q.W. A chloroplast-targeted DnaJ protein contributes to maintenance of photosystem II under chilling stress. *J. Exp. Bot.* **2014**, *65*, 143–158. [CrossRef] [PubMed]

42. Liu, N.; Gong, B.; Jin, Z.Y.; Wang, X.F.; Wei, M.; Yang, F.J.; Li, Y.; Shi, Q.H. Sodic alkaline stress mitigation by exogenous melatonin in tomato needs nitric oxide as a downstream signal. *J. Plant Physiol.* **2015**, *186*, 68–77. [CrossRef] [PubMed]
43. Mujahid, M.; Latif, S.; Ahmed, M.; Shehzadi, W.; Imran, M.; Ahmad, M.; Asari, A.; Jehangir, M.; Mahmud, Z. Modified matrix solid phase dispersion–HPLC method for determination of pesticide residue in vegetables and their impact on human health: A risk assessment. *Front. Chem.* **2022**, *10*, 1084350. [CrossRef] [PubMed]

International Journal of *Molecular Sciences*

MDPI

Article

The Metabolic Screening in Wounded Scales of *Hippeastrum* × *hybridum* Hort. Bulbs

Wiesław Wiczkowski [1], Lesław B. Lahuta [2], Dorota Szawara-Nowak [1], Karolina Stałanowska [2], Marian Saniewski [3], Agnieszka Marasek-Ciołakowska [3], Justyna Góraj-Koniarska [3] and Marcin Horbowicz [2,*]

1 Institute of Animal Reproduction and Food Research of the Polish Academy of Sciences, Biotransformation and Bioavailability of Phytochemicals Team, Trylińskiego 18, 10-683 Olsztyn, Poland; w.wiczkowski@pan.olsztyn.pl (W.W.); d.szawara-nowak@pan.olsztyn.pl (D.S.-N.)
2 Department of Plant Physiology, Genetics and Biotechnology, University of Warmia and Mazury, Oczapowskiego 1A, 10-719 Olsztyn, Poland; lahuta@uwm.edu.pl (L.B.L.); karolina.stalanowska@uwm.edu.pl (K.S.)
3 The National Institute of Horticultural Research, Konstytucji 3 Maja 1/3, 96-100 Skierniewice, Poland; marian.saniewski@inhort.pl (M.S.); agnieszka.marasek@inhort.pl (A.M.-C.); justyna.goraj@inhort.pl (J.G.-K.)
* Correspondence: marcin.horbowicz@uwm.edu.pl

Abstract: The aim of this study was to determine changes in primary and secondary metabolites after four days of storage of mechanically wounded *Hippeastrum* × *hybridum* Hort. (amaryllis) bulbs. Mechanically wounded scales of amaryllis bulbs stored for four days change color from white to orange-red, which is the plant's protective response to stress caused by damage. During this process, changes in the content of primary metabolites (carbohydrates, amino acids, organic acids) and secondary metabolites (flavonoids, phenolic acids, anthocyanins) were analyzed. The resulting color of *Hippeastrum* scales is due to the presence of several pigmented phenolic compounds: flavonoids, phenolic acids and anthocyanins. In particular, the increase in anthocyanin and luteolin content in stored scales probably affected the intensity of their color. The decrease in L-phenylalanine content in the *Hippeastrum* scales to trace levels indicates high intensity of phenolic compound biosynthesis. The increased content of 4-coumaric, ferulic and sinapic acids suggests that the lignification process also occurs. Moreover, the observed significant decrease in glucose, galactose, and sucrose levels indicates intense respiration processes and biosynthesis of various metabolites, which may contribute to counteracting wound-related stress. Mechanical wounding and storage of wounded *Hippeastrum* bulbs for several days may be a simple way to improve the quality of pharmaceutical products manufactured from these bulbs, but this requires further investigations.

Keywords: *Hippeastrum*; amaryllis; bulb scales; wounding stress; primary metabolites; secondary metabolites

1. Introduction

Wounding or mechanical damage is the loss of physical integrity of a plant's cells, tissues or organs. Wounding poses a serious threat to plant tissues which can open the way for pathogen attack or the result of pest feeding, while weather conditions are responsible for mechanical damage [1]. As a result, wounding can lead to water loss and the occurrence of drought stress. Plant defense against wounding stress can be divided into synthesis of secondary metabolites and specific proteins [2]. Proteins encoded by wound-induced genes may be involved (1) in the production of toxic compounds or reduce the digestibility of plant tissue for herbivores; (2) in the activation of wound defence pathways; or (3) in the repair of damaged tissue by adapting the plant's metabolism to the increased demand for energy and repair components [1,3]. In response to mechanical injury, plants can lead to changes in metabolite content by synthesising and accumulating defense compounds [4–6]. Such compounds are various polyphenols and alkaloids, which act as antimicrobial and antifungals agents [7,8].

The tissues of Amaryllidaceae, to which *Hippeastrum* × *hybridum* Hort. (amaryllis) belongs, contain numerous bioactive compounds, including specific alkaloids found exclusively in this plant subfamily [8]. Therefore amaryllis species are a valuable source of bioactive compounds with therapeutic applications [9–11]. A recently published paper showed that 48 alkaloids with potential pharmacological significance were identified in *Hippeastrum* species native to Bolivia [9]. Lycorine and homolycorine alkaloids were dominant in *H. chionedyanthum* and *H. haywardii* species, and these species have particularly high levels of lycorine, which is a promising anti-cancer compound [9]. In turn, the species *H. evansiarum* and *H. mollevillquense* contained significant amounts of galantamine-type alkaloids, which are important in the treatment of Alzheimer's disease [9].

Besides a large group of alkaloids, *Hippeastrum* species also contain many phenolic compounds that are also medically important [10,11]. The phenolic compounds present in *Hippeastrum* also have anti-fungal properties in plant diseases [12,13]. One of the most common diseases that amaryllis plants suffer from is red spotting caused by infection with the fungus *Phoma narcissi* Aderh. (syn. *Stagonospora curtisii* (Berk.) Sacc.) [12]. *Phoma narcissi* is a worldwide-known pathogen of *Hippeastrum*, *Narcissus*, *Hymenocallis* and various species of Amaryllidaceae causing red or reddish-brown spots on various organs [13]. The plant's defense against the threat of *Phoma narcissi* involves producing a mixture of orange-colored chalcones and flavanes, which can then be oxidised to red dimers or polymers [13,14]. These compounds prevent penetration of injured tissues by *Phoma narcissi*, *Botrytis cinarea*, *Fusarium oxysporum* and *Phoma poolensis* [13,15–18]. It has been shown that compounds with antifungal properties are present in the red-dyed tissues. As the content of these compounds increases, the infection caused by *Phoma narcissi* has been inhibited [15].

Detailed analysis showed that this pigment, which is absent in unwounded tissue, is a mixture of one orange chalcone and three colorless flavans [14]. Colorless flavans can be oxidised to red dimers or polymers. The light absorption maximum of the compounds from the reddish colored organs extracted with methanol/HCl (99:1) was 476 nm indicating the absence or low content of anthocyanins among them [12,14]. In contrast, the maximum absorption of the red compounds in the amaryllis cv. Red Lion flower was 516 nm, indicating the presence of anthocyanins.

The red color formation reaction observed in wounded *Hippeastrum* scales was accompanied by an increase in methyl jasmonate (JA-Me) content, while reducing the JA-Me content lowered the plant's ability to produce red pigment [19]. Similarly, mechanical wounding or JA-Me application induced activity of the chalcone synthase in white spruces needles, catalyzing the first step in flavonoid biosynthesis [4].

Phenolic compounds ubiquitous in plant tissues are generally divided into two main groups: flavonoids and non-flavonoids [20]. Phenolic compounds are secondary metabolites that do not directly affect plant growth and development, but they regulate metabolic pathways through signal transduction [21]. These compounds protect plants from disease/damage and also affect the aroma and flavour and especially the colour of the plant [22,23]. The biosynthesis of phenolic compounds is dependent on the availability of L-phenylalanine (L-phe), the main amino acid used as a precursor for the biosynthesis of phenylpropanoids, and L-phe is biosynthesized by the shikimic acid pathway [24]. Abiotic stress in plants induces the biosynthesis and accumulation of polyphenols, which help them adapt to stress conditions [25]. Polymers formed from phenylpropanoids, such as lignin, suberin, and tannins, contribute to the stability and resistance of plants to mechanical or environmental damage, such as drought and wounding [26–30].

Flavonoids are closely associated with plant defense against various stresses [31]. It is known that certain flavonoids induced in plants during pathogen attacks or after mechanical injury play an important role in plant resistance against microbial infections [32]. Flavonoid accumulation is one of the main responses of *Dracaena cochinchinensis* [33] and *Marchantia polymorpha* L. [34] to wound stress. Also, in mechanically crushed red onion peel, the total flavonoid content was twice as high as compared to the control. Among these flavonoids, the highest increase in content in crushed onion peel was for quercetin and quercetin 4′-*O*-glucopyranoside [35]. In addition, the antioxidant activity and total phenolic content were increased by very fine milling of the red rice grains [36].

The aim of this study was to determine changes in primary and secondary metabolites after four-day storage of mechanically wounded scales of *Hippeastrum* bulbs. The main purpose was to determine the content of compounds that may affect the color appearing after storage of wounded amaryllis bulbs. This study is a continuation of our earlier research [12,13,15–19]. Amaryllidaceae tissues are potentially important as a source of compounds with pharmacological significance. Therefore, the study also aimed to determine how mechanical wounding of *Hippeastrum* bulbs affects the ingredients they contain.

2. Results

Immediately after mechanical wounding, the scales were white with a slight grayish tint (Figure 1A), and after 4 days of storage in daylight (Figure 1B), they were intensely red-orange. However, scales that were cut and stored in the dark had a lighter orange-red color than those stored in daylight (Figure 1C).

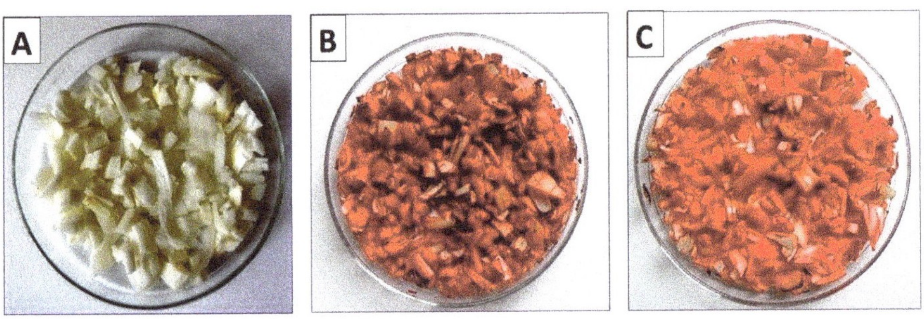

Figure 1. Appearance of *Hippeastrum* (amaryllis) bulb scales immediately after mechanical wounding (**A**) and after wounding and 4 days of storage: in daylight (**B**) and in darkness (**C**).

In mechanically cut scales of amaryllis bulb, which were then stored in daylight, clear changes were observed in most of the secondary metabolites. In these tissues, a significant increase in the content of total phenolic compounds, i.e., phenolic acids, anthocyanins and flavonoids, was observed after four days of storage (Figure 2). A particularly large increase in content occurred in the case of anthocyanins. At the same time, a decrease in the total content of carbohydrates and amino acids was noted, while the content of organic acids remained unchanged.

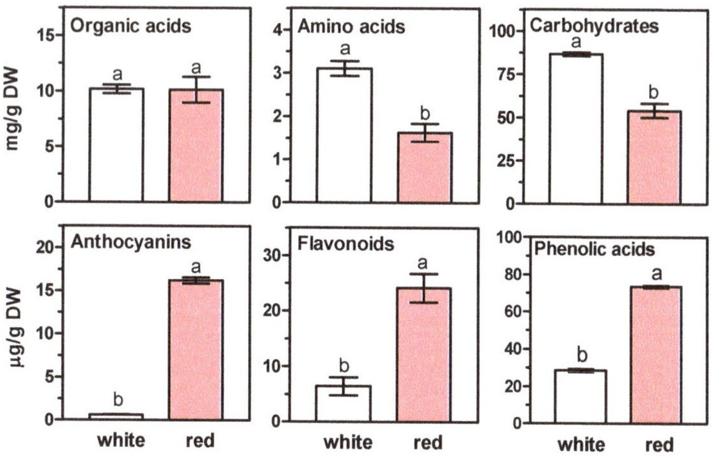

Figure 2. Total contents of organic acids, amino acids, carbohydrates, anthocyanins, flavonoids, and phenolic acids in scales of *Hippeastrum* (amaryllis) bulbs immediately after mechanical wounding (white) and after wounding and 4 days of storage under daylight conditions (red). Bar results (means ± SD) marked with the same letter do not differ at the significance level of $p < 0.05$ according to Duncan's test.

The following phenolic acids were identified in the scales of the *Hippeastrum* bulbs: ferulic, 4-coumaric, protocatechuic, caffeic, caftaric, ferulic, ellagic and synapic (Table 1).

Table 1. Total phenolic acid content (µg/g DW ± SD), as well as their aglycones (A), esters (E), and glycosides (G) in scales of *Hippeastrum* (amaryllis) bulb after mechanical wounding (white) and after 4 days of storage under ambient light conditions (red). Tr, traces—below 0.01 µg/g DW. Means in the rows marked with the same letter do not differ at the significance level of $p < 0.05$ according to Duncan's test.

	White			Red		
			Ferulic acid			
Total		22.70 ± 0.55 [b]			50.73 ± 1.16 [a]	
A/E/G	0.58 ± 0.09 [c]	22.03 ± 0.43 [b]	0.09 ± 0.03 [d]	1.27 ± 0.48 [c]	49.08 ± 0.61 [a]	0.38 ± 0.08 [c]
			4-Coumaric acid			
Total		0.70 ± 0.03 [b]			3.77 ± 0.07 [a]	
A/E/G	0.03 ± 0.01 [d]	0.66 ± 0.03 [c]	Tr	1.10 ± 0.01 [b]	2.63 ± 0.05 [a]	0.04 ± 0.01 [d]
			Protocatechuic acid			
Total		0.10 ± 0.02 [b]			0.37 ± 0.05 [a]	
A/E/G	Tr	Tr	0.07 ± 0.01 [c]	0.06 ± 0.02 [c]	0.15 ± 0.01 [a]	0.16 ± 0.02 [a]
			Caffeic acid			
Total		2.31 ± 0.10 [b]			11.88 ± 0.36 [a]	
A/E/G	0.04 ± 0.01 [e]	2.14 ± 0.08 [b]	0.12 ± 0.02 [d]	0.63 ± 0.01 [c]	11.06 ± 0.35 [a]	0.19 ± 0.01 [d]
			Caftaric acid			
Total		0.03 ± 0.01 [a]			0.09 ± 0.03 [a]	
A/E/G	Tr	0.02 ± 0.01 [a]	Tr	0.07 ± 0.02 [a]	0.02 ± 0.02 [a]	Tr
			Ellagic acid			
Total		1.77 ± 0.19 [b]			4.36 ± 0.21 [a]	
A/E/G	0.77 ± 0.01 [b]	0.33 ± 0.11 [cd]	0.67 ± 0.07 [bc]	3.46 ± 0.20 [a]	0.29 ± 0.01 [d]	0.61 ± 0.01 [c]
			Synapic acid			
Total		1.06 ± 0.24 [b]			2.25 ± 0.02 [a]	
A/E/G	Tr	1.06 ± 0.24 [b]	Tr	Tr	2.25 ± 0.02 [a]	Tr

The phenolic acid that was quantitatively predominant was ferulic acid, whose total content was approximately 80% in freshly cut scales of amaryllis bulbs. After storage, the total content of this acid was about 70% of all acids. Most of the phenolic acids (ferulic, 4-coumaric, caffeic, and synapic) occurred in ester form. Ellagic acid, on the other hand, occurred mainly in the form of aglycone. Four-day storage of the cutting scales resulted in a significant increase in the content of the particular forms of phenolic acids and their total contents (Table 1). Among the acids and their forms present, the highest 5-fold increase after 4 days of storage occurred for caffeic acid esters and a 4-fold increase for 4-coumaric acid esters as well as the ellagic acid. Cutting of amaryllis bulbs and storing them for 4 days had no significant effect on the content of phenolic acid glycosides.

The flavonoid content in the scales of amaryllis bulb was significantly lower than the phenolic acid content (Table 2).

Table 2. Total contents of flavonoids (μg/g DW ± SD), as well as their aglycones (A), esters (E), and glycosides (G) in scales of *Hippeastrum* (amaryllis) bulb after mechanical wounding (white) and after 4 days of storage in ambient light conditions (red). Tr, traces—below 0.01 μg/g DW. Means in the rows marked with the same letter do not differ at the significance level of $p < 0.05$ according to Duncan's test.

	White			Red		
			Quercetin			
Total		0.02 ± 0.01 ^{b}			0.36 ± 0.02 ^{a}	
A/E/G	Tr	Tr	0.02 ± 0.01 ^{b}	0.33 ± 0.01 ^{a}	Tr	0.03 ± 0.01 ^{b}
			Apigenin			
Total		Tr			0.05 ± 0.01	
A/E/G	Tr	Tr	Tr	0.05 ± 0.01	Tr	Tr
			Kaempferol			
Total		0.02 ± 0.01 ^{b}			0.78 ± 0.09 ^{a}	
A/E/G	Tr	Tr	0.02 ± 0.01 ^{b}	0.70 ± 0.07 ^{a}	0.07 ± 0.02 ^{b}	Tr
			Luteolin			
Total		0.07 ± 0.03 ^{b}			13.05 ± 1.64 ^{a}	
A/E/G	0.03 ± 0.01 ^{c}	Tr	0.04 ± 0.01 ^{c}	12.43 ±1.62 ^{a}	0.56 ± 0.03 ^{b}	0.06 ± 0.01 ^{c}
			Catechin			
Total		6.33 ± 1.54 ^{a}			4.12 ± 0.55 ^{a}	
A/E/G	0.23 ± 0.21 ^{bc}	0.08 ± 0.02 ^{c}	6.01 ± 1.31 ^{a}	0.71 ± 0.39 ^{bc}	0.27 ± 0.07 ^{c}	3.14 ± 0.09 ^{a}
			Naringenin			
Total		0.02 ± 0.01 ^{b}			5.83 ± 0.28 ^{a}	
A/E/G	0.02 ± 0.01 ^{c}	Tr	Tr	5.70 ± 0.27 ^{a}	0.12 ± 0.01 ^{b}	Tr

Flavonoids were present in these tissues mainly in the form of aglycones, with the exception of catechin, which was present as glycosides. The content of particular flavonoids increased after 4 days of storage of the wounded scales compared to the content immediately after wounding. The exception was catechin, whose glycoside content decreased, but not significantly. In wounded and stored amaryllis scales, the main flavonoid in terms of quantity was luteolin aglycone. The content of this flavonoid before storage was less than 0.1 μg/g DW, while after storage it increased to over 12 μg/g DW (Table 2). A similar great increase in content occurred in the case of naringenin aglycone, from 0.02 μg/g DW after cutting to 5.70 μg/g DW after 4 days of storage.

Eight anthocyanins were detected in the scales of amaryllis bulbs: cyanidin and pelargonidin aglycones, cyanidin, delphinidin, and pelargonidin monoglucosides, and cyanidin, peonidin, and delphinidin diglucosides (Table 3).

Table 3. Anthocyanin content (μg/g DW ± SD) in scales of *Hippeastrum* (amaryllis) bulb after mechanical wounding (white) and after 4 days of storage under ambient light conditions (red). Tr—traces, below 0.01 μg/g DW. Means in the rows marked with the same letter do not differ at the significance level of $p < 0.05$ according to Duncan's test.

Determined Anthocyanin	White	Red
Cyanidin	Tr	5.75 ± 0.11
Pelargonidin	0.02 ± 0.01 ^{b}	2.66 ± 0.07 ^{a}
Cyanidin 3-glucoside	0.31 ± 0.04 ^{b}	1.14 ± 0.03 ^{a}
Delphinidin 3-glucoside	Tr	0.42 ± 0.05
Pelargonidin 3-glucoside	Tr	1.20 ± 0.01
Cyanidin diglucoside	0.02 ± 0.01 ^{b}	2.29 ± 0.01 ^{a}
Peonidin diglucoside	Tr	1.54 ± 0.07
Delphinidin diglucoside	0.29 ± 0.01 ^{b}	1.20 ± 0.05 ^{a}

Four days of storage of the wounded scales resulted in a rapid increase in anthocyanin content, especially cyanidin and pelargonidin aglycones. After storage, the main quantitatively anthocyanins were cyanidin and pelargonidin aglycones, which accounted for more than half of their total content.

In fresh scales of amaryllis bulb, malic acid and citric acid were the quantitatively dominant organic acids (Table 4). After wounding of the scales and their subsequent storage, the organic acid content did not change significantly, with the exception of acetic acid, whose level decreased slightly. After 4 days storage of scales, the content of eight of the twelve amino acids (valine, serine, leucine, isoleucine, proline, L-phenylalanine, aspartic acid, and γ-aminobutyric acid) decreased (Table 4). Particularly large decreases in content occurred for serine, aspartic acid and L-phenylalanine.

Table 4. Contents (mg/g DW ± SD) of organic acids, amino acids, carbohydrates, and other polar compounds in scales of *Hippeastrum* (amaryllis) bulb after mechanical wounding (white) and after 4 days of storage in ambient light conditions (red). Means in the rows marked with the same letter do not differ at the significance level of $p < 0.05$ according to Duncan's test.

Determined Metabolite	White	Red
	Organic acids	
Acetic acid	0.17 ± 0.02 ^{a}	0.07 ± 0.01 ^{b}
Succinic acid	0.08 ± 0.01 ^{a}	0.09 ± 0.01 ^{a}
Malic acid	3.27 ± 0.09 ^{a}	4.57 ± 0.53 ^{a}
Citric acid	6.68 ± 0.32 ^{a}	5.42 ± 0.61 ^{a}
	Amino acids	
Valine	0.18 ± 0.02 ^{a}	0.10 ± 0.01 ^{b}
Alanine	0.08 ± 0.01 ^{a}	0.08 ± 0.01 ^{a}
Serine	0.45 ± 0.01 ^{a}	0.19 ± 0.03 ^{b}
Leucine	0.14 ± 0.01 ^{a}	0.06 ± 0.01 ^{b}
Isoleucine	0.23 ± 0.01 ^{a}	0.09 ± 0.02 ^{b}
Proline	0.12 ± 0.01 ^{a}	0.06 ± 0.01 ^{b}
Hydroxyproline	0.49 ± 0.12 ^{a}	0.33 ± 0.07 ^{a}
L-Phenylalanine	0.13 ± 0.03 ^{a}	0.01 ± 0.01 ^{b}
Aspargic acid	0.70 ± 0.03 ^{a}	0.42 ± 0.06 ^{b}
Glutamic acid	0.14 ± 0.09 ^{a}	0.04 ± 0.02 ^{a}
Asparagine	0.06 ± 0.01 ^{a}	0.12 ± 0.02 ^{a}
γ-Amino butyric acid (GABA)	0.38 ± 0.07 ^{a}	0.13 ± 0.04 ^{b}

Table 4. *Cont.*

Determined Metabolite	White	Red
	Carbohydrates	
Fructose	17.48 ± 0.15 [a]	16.28 ± 1.87 [a]
Galactose	3.26 ± 0.29 [a]	1.81 ± 0.22 [b]
Glucose	17.69 ± 0.66 [a]	7.81 ± 0.71 [b]
myo-Inositol	0.20 ± 0.02 [a]	0.22 ± 0.04 [a]
Sucrose	27.66 ± 1.35 [a]	13.69 ± 1.62 [b]
1-Kestose	20.40 ± 0.35 [a]	14.38 ± 0.36 [b]
	Others	
Glycerol	0.22 ± 0.04 [a]	0.26 ± 0.06 [a]
Phosphoric acid	3.42 ± 0.09 [a]	5.06 ± 0.67 [a]
Total polar compounds	103.63 ± 1.62 [a]	71.29 ± 6.09 [b]

The content of most carbohydrates present in the wounded scales of amaryllis bulb decreased after 4 days of storage. This was the case for galactose, glucose, sucrose, and kestose (Table 4). The sucrose and galactose contents were twice as low, while the glucose content was two and a half times lower after storing cut scales compared to those before storage. In contrast, fructose and *myo*-inositol contents did not change significantly during the time of storage.

3. Discussion

Plant wounding can be caused by pest feeding, as well as by weather conditions or human activity, leading to the loss of physical integrity of plant cells, tissues, or organs [1]. As a result, this process can lead to water loss and the occurrence of drought stress [2,3]. Plants can respond to wounding and infection by synthesizing and accumulating defense compounds [4–6].

The production of colorful compounds in mechanically wounded *Hippeastrum* bulbs is a response to the threat caused by tissue injury [13–19]. The compounds produced at that time help prevent *Phoma narcissi* fungal disease [17]. Detailed analysis showed that this pigment, which is absent in unwounded tissue, contains a mixture of flavonoids. According to Wink and Lehmann [14] this pigment is a mixture of an orange chalcone: 3,2′4′-trihydroxy-4-methoxychalcone and three colourless flavans: 7,4′-dihydroxy-8-methylflavan, 7,3′-dihydroxy-4′-methoxyflavan and 7-hydroxy-3′,4′-methylene dioxyflavan. Our current investigation demonstrated that the content of almost all phenolic compounds present in the examined tissues significantly increased (Figure 2; Tables 1–3). At the same time, the content of L-phenylalanine in wounded amaryllis bulbs decreased to trace levels, which may indicate its use in the intensive biosynthesis of these phenolic compounds (Table 4).

Plant wounding induced the transcription of major enzymes of the phenylpropanoid pathway, which are critical for the synthesis of phenolic compounds and the biosynthesis of suberin, lignin, and flavonoids [4,29]. Among the secondary metabolites, the 3-*O*-rhamnosides of quercetin and myricetin content were higher in wounded *Catharanthus roseus* roots than in the control [37]. In addition, in the wounded leaves the levels of gallic acid, salicylic acid and daidzein were increased [37]. However, in *Arabidopsis* leaves, the salicylic acid content increased rapidly after wounding, and 24 h after wounding, there was an increase in the content of salicylic acid glucoside and its ester derivative [36].

The results of our study indicate that biosynthesis and significant accumulation of flavonoids and phenolic acids are important components of the response to wound stress, confirming earlier reports [4,29,36,37]. Also, in wounding carrot roots and cassava roots there was an increased accumulation of flavonoids caused by the up-regulation of genes involved in their biosynthesis [38,39]. Moreover, wounding induced the phenylpropanoid pathway and biosynthesis of luteolin, apigenin and isoricardin C in *Marchantia polymorpha* L., while α-aminooxy-β-phenylpropionic acid, an inhibitor of phenylalanine ammonialyase activity, inhibited production of the mentioned phenolic compounds [33].

The more than 200-fold increase in luteolin content demonstrated in our study appears to be a specific indicator of the response of the *Hippeastrum* scales to wound stress (Table 2). The increase in luteolin content relates almost exclusively to its aglycone, which may indicate its involvement in the coloring of wounded scales. Luteolin has an intense yellow color, so a significant increase in its content in wounded amaryllis scales may affect their color. There is little data on the occurrence and content of flavonoids in *Hipeastrum* tissues. The derivatives of 7-hydroxy flavanone and the flavonol rutin were identified in *Hippeastrum vittatun* [10] and in *Hippeastrum stapfianum* [40], and recently, the flavonol quercetin 3-O-rhamnose was identified in *Hippeastrum stapfianum* [41]. However, there is no data available on the flavonoid content in *Hippeastrum x hybridum* bulb tissues.

In addition, increased content of phenolic acids observed in the scales of amaryllis can also affect the intensity and tone of their color because these acids are usually yellow, creamy or pale yellow. However, wounded scales contained mainly phenolic acid esters rather than their aglycones. The available literature has no data on the presence and content of phenolic acids *Hippeastrum* × *hybr*. Hort. bulb tissues.

It is widely known that light is the main factor influencing anthocyanin biosynthesis in plants [42,43]. Light affects this accumulation by regulating the expression of genes in the anthocyanin synthesis pathway [44,45]. Recently published data show that storing cut lily bulbs in ambient light caused them to turn purple-red, a color resulting from the accumulation of anthocyanins [46]. The authors suggest that this phenomenon is the result of increased activity of key enzymes and gene expression in the anthocyanin synthesis pathway. The color and quality of hyacinth bulbs were also affected by the light conditions during their storage [47]. Under the influence of blue light, the bulbs were smaller, but they were colored by the presence of anthocyanins and chlorophyll. Red light, on the other hand, did not cause any coloring of the hyacinth bulbs [47].

The results of our study show that anthocyanin content in freshly cut scales of the *Hippeastrum* bulb was very low (Table 3). Analyses of anthocyanin content in the underground parts of bulbous plants are not performed, which does not mean that they are unnecessary. This is demonstrated by the results of our research, which clearly show that after four days of storage, the wounded scales had a significant increase in the content of all detected anthocyanins. It seems that storage process in the presence of daylight is responsible for the increase in the content of these pigments. The color of the wounded scales stored in darkness was lighter than that of the bulbs stored in the daylight conditions (Figure 1). Cyanidin has a characteristic red color, and its glucoside is also red, while pelargonidin is orange. Therefore, it is likely that the greatly increased anthocyanin content led to a darker or more intense color of the wounded and stored scales of amaryllis bulbs. Unfortunately, no analysis was carried out on the presence and content of anthocyanins in the tissues of the scales stored in darkness.

In soybean leaves, rapid accumulation of GABA was found in response to mechanical stimulation or damage [48]. However, the results of our study showed that the GABA content in wounded and stored scales of *Hippeastrum* bulb decreased. The marked decrease in glucose, galactose and sucrose levels in wounded scales of the amaryllis bulbs observed in our study indicates a high intensity of respiratory processes. These processes are a source of energy and many important metabolites that may be involved in counteracting wound stress [1–3]. These metabolites are essential to counteracting the risk of disease and water loss associated with stress caused by wounding. The demonstrated lack of reduction in fructose content may indicate the intensity of the release of this carbohydrate as a result of hydrolysis of 1-kestose (a di-fructose derivative of glucose), or other fructans that could not be identified and determined by the method applied. It has been shown that the use of starch and sucrose may be the material basis for the response to mechanical damage in *Aquilaria sinensis* [49]. However, in wounded leaves and roots of *Catharanthus roseus*, the relative sugar content appeared to be higher compared to the unwounded control [37].

Stress induced by wounding triggers the repair of plant tissue damage, in which lignin plays a role. The primary substrates for lignin biosynthesis are monosaccharides or closely related compounds [50,51]. The lignification process involves the polymerization of monolignols, i.e., 4-coumaryl, coniferyl, and sinapyl alcohols [49]. The monolignol biosynthesis is common to the general phenylpropanoid pathway. Lignin biosynthesis uses L-phenylalanine for the biosynthesis of 4-coumaric acid and CoA esters of 4-coumaric, ferulic, and sinapic acids [50]. These acids are then converted to the corresponding alcohols and finally polymerize to form lignin [51]. The increase in phenolic acid contents and significant decrease in most carbohydrate contents in wounded scales demonstrated in our study indicate that the lignification process has been probably initiated. However, this assumption requires further examination.

A recently published article showed that the tissues of *Hippeastrum* (amaryllis) bulbs contain forty-eight alkaloids that may be useful in pharmacology [9]. Beyond the many alkaloids that are typical for the Amaryllidaceae family, other research shows that *Hippeastrum* has the ability to make a variety of flavonoids [10,11]. The results of our study indicate that stress caused by wounding and subsequent storage of wounded scales leads to a significant increase in the content of phenolic compounds in them. Since these compounds may also have medical significance, wounding and storing wounded amaryllis bulbs for several days may be one way to improve the quality of the pharmaceutical products made from them.

4. Materials and Methods

4.1. Plant Material

For the experiment, bulbs of *Hippeastrum* × *hybr.* Hort. (amaryllis) were used. The plant specimens were sourced from our internal cultivation program. The white bulb scales were cut into small pieces, about 2–4 mm square. One portion was frozen at −20 °C, while other samples of the cut scales were stored at ambient conditions, at temperature 18–22 °C in closed Petri dishes with the addition of small amounts of water to prevent them from drying. Then, after four days of storage, the red wounded scales were frozen. Subsequently, these scales and the white scales frozen prior to storage were freeze-dried and powdered, and the samples were used for the analysis of primary and secondary metabolites. Pictures were taken immediately after cutting the scales and four days of storage in daylight and/or darkness. The plant samples were freeze-dried for 48 h, during which the temperature of the condenser was 218 K (−55 °C) and the final pressure was 600 pA (0.06 mBar). This process was carried out in a laboratory freeze-dryer (Labconco, Kansas City, MO, USA).

4.2. Determination of Organic Acids and Amino Acids

Primary polar metabolites were extracted with a mixture of methanol/water (1:1, *v*/*v*), containing ribitol as an internal standard. The obtained supernatant was extracted with chloroform to remove lipid compounds, and the upper layer (methanol/water fraction) was dried in a speed vacuum rotary evaporator (JWElectronic, Warsaw, Poland). The compounds in dry residues were derivatized in two steps using *O*-methoxyamine hydrochloride in pyridine and a mixture of *N*-methyl-*N*-(trimethyl-silyl)-trifluoroacetamide in pyridine (both chemicals were from Sigma-Aldrich, Merck, Burlington, MA, USA). All further details of the method of the analysis have been described previously [52,53]. The identification and quantitative analyses of metabolites was confirmed by gas chromatography coupled with mass spectrometry (GCMS-QP2010 Plus, Shimadzu, Japan). Polar metabolites were identified by comparing the retention indices and mass spectra collected in the NIST 05 library (Shimadzu, Kyoto, Japan) and the internal collection of mass spectra obtained for the original standards purchased from SIGMA [54].

4.3. Determination of Soluble Carbohydrates

Soluble carbohydrates were extracted with a mixture of ethanol/water (1:1, *v*/*v*), containing xylitol as an internal standard, according to the method earlier described [55]. The carbohydrates were derivatized in a mixture of 1-(trimethylsilyl)imidazole in pyridine and then were separated on the GC capillary column ZEBRON ZB-1 (length, 15 m; diameter 0.25 mm; film thickness 0.10 μm; 100% dimethylpolysiloxane (Phenomenex, Torrance, CA, USA)).

4.4. Determination of Flavonoids and Phenolic Acids

The content of phenolic acids and flavonoids was determined according to the method described in detail by Dębski et al. [56]. Briefly, crude flavonoid extracts were obtained by stirring samples overnight with a mixture of methanol, water, and formic acid. Then, the free forms of phenolic acids and flavonoids were isolated from crude extracts with diethyl ether. Next, after the free forms were isolated, esters present in the extracts were hydrolyzed with 4 M NaOH and extracted with diethyl ether. Next, the glycosides were hydrolyzed with 6 M HCl and obtained free forms were extracted with diethyl ether. The free compounds and compounds released from bound forms were dissolved in methanol, centrifuged and subjected to analysis on an HPLC–MS/MS system (QTRAP 5500 ion trap mass spectrometer, AB SCIEX, Vaughan, ON, Canada) equipped with a HALO C18 column (2.7 μm particles, 0.5 × 50 mm, Eksigent, Vaughan, ON, Canada). The contents of phenolic compounds obtained by alkaline and acid hydrolysis were presented as their free, ester, and glycosidic forms or as a total of the all forms.

4.5. Determination of Anthocyanins

The extraction of anthocyanins and determination of their content was carried out using a method described in detail by Wiczkowski et al. [57]. Briefly, anthocyanins were extracted with 0.4% trifluoroacetic acid in methanol by vortexing and sonication, and the obtained extracts were centrifuged, and the supernatants were combined. The analyses were carried out using an LC-200 Eksigent HPLC system coupled with a Triple TOF 5600+ mass spectrometer (AB SCIEX, Vaughan, ON, Canada). Chromatographic separation of anthocyanins was carried out on the HALO C18 column (2.7 μm, 100 × 0.5 mm, Eksigent, Vaughan, ON, Canada) with a solvent gradient system consisting of solvent A (0.95% formic acid aqueous solution) and solvent B (0.95% formic acid in acetonitrile). Identification of the anthocyanins was based on a comparison of their retention time and MS/MS fragmentation spectrum (m/z values) with data of standards analysis and the published, or/and on the interpretation of the fragmentation spectrum obtained. Further details on the method of anthocyanin analysis were described in the cited reference [57].

4.6. Statistical Evaluation of Results

Results of measurements were subjected to analysis of variance (ANOVA), followed by Duncan's multiple range test. The results of three replicates of primary and secondary metabolites were used for statistical analysis. p values of <0.05 were considered to be statistically significant for means.

5. Conclusions

Mechanical wounding of the tissues of *Hippeastrum* × *hybridum* Hort. (amaryllis) bulbs and their subsequent storage has a significant impact on the levels of the compounds contained therein. The color of mechanically wounded and stored amaryllis scales is the result of a mixture of many colorful phenolic compounds: flavonoids, phenolic acids and anthocyanins. Likely, the significantly increased content of several anthocyanins and luteolin in the wounded and stored scales resulted in a more intense color. A large decrease in content of L-phenylalanine to trace amounts indicates the intensity of biosynthesis of these compounds. The increased content of 4-coumaric, ferulic and sinapic acids, which are substrates for the biosynthesis of the appropriate alcohols used for production of lignin, suggests that this process also occurs. In addition, the significant decline in glucose, galactose and sucrose levels observed in our study indicates vigorous respiratory processes. These processes are both a source of energy and various metabolites that may contribute to counteracting wound-related stress. Mechanical wounding of the bulbs and their subsequent storage for several days may be a simple way to improve the quality of natural medicines made from amaryllis bulbs. However, this requires further detailed studies, especially with regard to compounds with pharmacological significance, such as alkaloids.

Author Contributions: Conceptualization, M.S. and J.G.-K.; methodology, W.W. and L.B.L.; software, W.W. and L.B.L.; investigation, K.S. and D.S.-N.; writing—original draft preparation, M.H. and M.S.; writing—review and editing, M.H.; visualization, L.B.L., A.M.-C. and M.H.; supervision, M.S.; project administration, A.M.-C. All authors have read and agreed to the published version of the manuscript.

Funding: This research received no external funding. This study was carried out as part of the statutory activities of the National Institute of Horticultural Research in Skierniewice, Poland (ZBS/7/2021).

Institutional Review Board Statement: Not applicable.

Informed Consent Statement: Not applicable.

Data Availability Statement: All data are presented in the manuscript.

Conflicts of Interest: The authors declare no conflicts of interest.

Abbreviations

The following abbreviations are used in this manuscript:

JA-Me	Jasmonic acid methyl ester; methyl jasmonate
L-phe	L-phenylalanine
SD	Standard deviation
DW	Dry weight
Tr	Traces
GABA	γ-Amino butyric acid

References

1. León, J.; Rojo, E.; Sánchez-Serrano, J.J. Wound signalling in plants. *J. Exp. Bot.* **2001**, *52*, 1–9. [CrossRef]
2. Hoermayer, L.; Friml, J. Feeling the danger: Local wound signaling in plants. *Cell Res.* **2024**, *34*, 761–762. [CrossRef]
3. Ryan, C. The systemin signaling pathway: Differential activation of plant defensive genes. *Biochim. Biophys. Acta* **2000**, *1477*, 112–121. [CrossRef] [PubMed]
4. Richard, S.; Lapointe, G.; Rutledge, R.G.; Seguin, A. Induction of chalcone synthase expression in white spruce by wounding and jasmonate. *Plant Cell Physiol.* **2000**, *41*, 982–987. [CrossRef]
5. Grayer, R.J.; Kokubun, T. Plant-fungal interactions: The search for phytoalexins and other antifungal compounds from higher plants. *Phytochemistry* **2001**, *56*, 253–263. [CrossRef] [PubMed]

6. Hasegawa, M.; Mitsuhara, I.; Seo, S.; Imai, T.; Koga, J.; Okada, K.; Yamane, H.; Ohashi, Y. Phytoalexin accumulation in the interaction between rice and the blast fungus. *Mol. Plant Microbe Interact.* **2010**, *23*, 1000–1011. [CrossRef] [PubMed]
7. Suzuki, N.; Ron Mittler, R. Reactive oxygen species-dependent wound responses in animals and plants. *Free Radic. Biol. Med.* **2012**, *53*, 2269–2276. [CrossRef]
8. Ka, S.; Koirala, M.; Merindol, N.; Desgagne-Penix, I. Biosynthesis and Biological Activities of Newly Discovered Amaryllidaceae Alkaloids. *Molecules* **2020**, *25*, 4901. [CrossRef]
9. Rodríguez-Escobar, M.L.; Lara, R.F.; Atahuachi, M.; Fuentes, A.F.; Maldonado, C.; Bastida, J.; Tallini, R.R.; Torras-Claveria, L. Alkaloid Profile Characterisation and Bioactivity Evaluation of Bolivian *Hippeastrum* Species (Amaryllidaceae) as Cholinesterase Inhibitors. *Life* **2025**, *15*, 719. [CrossRef]
10. Khalifa, M.F.; Fahim, J.R.; Allam, A.E.; Shoman, M.E.; El Zawily, A.; Kamel, M.S.; Shimizu, K.; Attia, E.Z. Studies on the nonalkaloidal secondary metabolites of *Hippeastrum vittatum* (L'Her.) Herb. Bulbs. *ACS Omega* **2023**, *8*, 26749–26761. [CrossRef]
11. Ibrakaw, A.S.; Akinfenwa, A.O.; Hussein, A.A. A comprehensive review of non-alkaloidal metabolites from the subfamily Amaryllidoideae (Amaryllidaceae). *Open Chem.* **2023**, *21*, 20220252. [CrossRef]
12. Saniewska, A.; Prus-Głowacki, W. Mycelial growth, pathogenicity and electrophoretic characteristics of some enzymes among isolates of *Phoma narcissi* (Aderh.) Boerema, de Gruyter et Noordel., comb. nov. from *Hippeastrum*, *Narcissus* and *Hymenocallis*. *Phytopathol. Pol.* **1998**, *15*, 5–13.
13. Saniewska, A.; Budzianowski, J. The nature of red pigment formed in wounded and infected *Hippeastrum* bulb scale response to wounding tissues by *Stagonospora curtissi* (Berk.) Sacc. (*Phoma narcissi*). *Acta Hortic.* **1997**, *430*, 843–848. [CrossRef]
14. Wink, M.; Lehmann, P. Wounding-and Elicitor-induced Formation of Coloured Chalcones and Flavans (as Phytoalexins) in *Hippeastrum* × *hortorum*. *Bot. Acta* **1996**, *109*, 412–421. [CrossRef]
15. Saniewska, A.; Saniewski, M. Studies in the factor responsible for reddish coloration of mechanically injured tissue and tissue infected by fungus, *Stagonospora curtisii* (Berk.) Sacc. in *Hippeastrum* × *hybr.* hort. *Folia Hortic.* **1992**, *4*, 3–9.
16. Saniewski, M.; Urbanek, H.; Puchalski, J. Wound-induced phenolic metabolism in scales of *Hippeastrum* × *hybr.* hort. *Acta Hortic.* **1992**, *325*, 303–306. [CrossRef]
17. Saniewska, A.; Horbowicz, M.; Saniewski, M. The effect of salicylic acid and acetylsalicylic acid on red pigment formation in mechanically wounded scales of *Hippeastrum* × *hybr.* hort. and on the growth and development of *Phoma narcissi*. *Acta Agrobot.* **2005**, *58*, 81–90. [CrossRef]
18. Saniewski, M.; Horbowicz, M.; Saniewska, A. The effect of ruthenium red, a Ca++ channel blocker, on a red pigment formation in mechanically wounded scales of *Hippeastrum* × *hybr.* hort., and on the growth and development of *Phoma narcissi*. *J. Fruit Ornam. Plant Res.* **2006**, *14*, 211–222.
19. Wilmowicz, E.; Frankowski, K.; Grzegorzewska, W.; Kęsy, J.; Kućko, A.; Banach, M.; Szmidt-Jaworska, A.; Saniewski, M. The role of jasmonates in the formation of a compound of chalcones and flavans with phytoalexin-like properties in mechanically wounded scales of *Hippeastrum* × *hybr.* bulbs. *Acta Biol. Cracov. Bot.* **2014**, *56*, 54–58. [CrossRef]
20. Agati, G.; Azzarello, E.; Pollastri, S.; Tattini, M. Flavonoids as antioxidants in plants: Location and functional significance. *Plant Sci.* **2012**, *196*, 67–76. [CrossRef]
21. Mandal, S.M.; Chakraborty, D.; Dey, S. Phenolic acids act as signaling molecules in plant-microbe symbioses. *Plant Signal. Behav.* **2010**, *5*, 359–368. [CrossRef]
22. Lea, A.G.H. Flavor, Color, and Stability in Fruit Products: The Effect of Polyphenols. In *Plant Polyphenols: Synthesis, Properties, Significance*; Hemingway, R.W., Laks, P.E., Eds.; Basic Life Sciences; Springer: Boston, MA, USA, 1992; Volume 59, pp. 827–847. [CrossRef]
23. Rahim, M.A.; Zhang, X.; Busatto, N. Phenylpropanoid biosynthesis in plants. *Front. Plant Sci.* **2023**, *14*, 1230664. [CrossRef]
24. Hahlbrock, K.; Scheel, D. Physiology and molecular biology of phenylpropanoid metabolism. *Annu. Rev. Plant Physiol. Plant Mol. Biol.* **1989**, *40*, 347–369. [CrossRef]
25. Ray, A.; Kundu, S.; Mohapatra, S.S.; Sinha, S.; Khoshru, B.; Keswani, C.; Mitra, D. An Insight into the Role of Phenolics in Abiotic Stress Tolerance in Plants: Current Perspective for Sustainable Environment. *J. Pure Appl. Microbiol.* **2024**, *18*, 64–79. [CrossRef]
26. Vogt, T. Phenylpropanoid biosynthesis. *Mol. Plant* **2010**, *3*, 2–20. [CrossRef] [PubMed]
27. Yang, W.L.; Bernards, M.A. Metabolite profiling of potato (*Solanum tuberosum* L.) tubers during wound-induced suberization. *Metabolomics* **2007**, *3*, 147–159. [CrossRef]
28. Boerjan, W.; Ralph, J.; Baucher, M. Lignin biosynthesis. *Annu. Rev. Plant Biol.* **2003**, *54*, 519–546. [CrossRef]
29. Zhang, X.; Liu, C.-J. Multifaceted regulations of gateway enzyme phenylalanine ammonia-lyase in the biosynthesis of phenylpropanoids. *Mol. Plant* **2015**, *8*, 17–27. [CrossRef]
30. Kumar, S.; Pandey, A.K. Chemistry and Biological Activities of Flavonoids: An Overview. *Sci. World J.* **2013**, *2013*, 162750. [CrossRef]
31. Mattern, L.; Kneusel, R.F. Phenolic compounds in plant disease resistance. *Phytoparasitica* **1988**, *16*, 9–17. [CrossRef]

32. Liu, Y.; Gao, S.; Zhang, Y.; Zhang, Z.; Wang, Q.; Xu, Y.; Wei, J. Transcriptomics and Metabolomics Analyses Reveal Defensive Responses and Flavonoid Biosynthesis of *Dracaena cochinchinensis* (Lour.) Under Wound Stress in Natural Conditions. *Molecules* **2022**, *27*, 4514. [CrossRef]
33. Yoshikawa, M.; Luo, W.; Tanaka, G.; Konishi, Y.; Matsuura, H.; Takahashi, K. Wounding stress induces phenylalanine ammonia lyases, leading to the accumulation of phenylpropanoids in the model liverwort *Marchantia polymorpha*. *Phytochemistry* **2018**, *155*, 30–36. [CrossRef] [PubMed]
34. AbouZid, S.F.; Elsherbeiny, G.M. Increase in flavonoids content in red onion peel by mechanical shredding. *J. Med. Plants Res.* **2008**, *2*, 258–260.
35. Chen, Q.M.; Fu, M.R.; Yue, F.L.; Cheng, Y.Y. Effect of superfine grinding on physicochemical properties, antioxidant activity and phenolic content of red rice (*Oryza sativa* L.). *Food Nutr. Sci.* **2015**, *6*, 1277–1284. [CrossRef]
36. Ogawa, T.; Ara, T.; Aoki, K.; Suzuki, H.; Shibata, D. Transient increase in salicylic acid and its glucose conjugates after wounding in Arabidopsis leaves. *Plant Biotechnol.* **2010**, *27*, 205–209. [CrossRef]
37. Chen, Q.; Lu, X.; Guo, X.; Liu, J.; Liu, Y.; Guo, Q.; Tang, Z. The specific responses to mechanical wound in leaves and roots of *Catharanthus roseus* seedlings by metabolomics. *J. Plant Interact.* **2018**, *13*, 450–460. [CrossRef]
38. An, F.; Cui, M.; Chen, T.; Cheng, C.; Liu, Z.; Luo, X.; Xue, J.; Tang, Y.; Cai, J.; Chen, S. Flavonoid accumulation modulates the responses of cassava tuberous roots to postharvest physiological deterioration. *Postharvest Biol. Technol.* **2023**, *198*, 112254. [CrossRef]
39. Heredia, J.B.; Cisneros-Zevallos, L. The effect of exogenous ethylene and methyl jasmonate on pal activity, phenolic profiles and antioxidant capacity of carrots (*Daucus carota*) under different wounding intensities. *Postharvest Biol. Technol.* **2009**, *51*, 242–249. [CrossRef]
40. Gomes-Copeland, K.K.P.; Meireles, C.G.; Gomes, J.V.D.; Torres, A.G.; Sinoti, S.B.P.; Fonseca-Bazzo, Y.M.; Magalhães, P.d.O.; Fagg, C.W.; Simeoni, L.A.; Silveira, D. *Hippeastrum stapfianum* (kraenzl.) R.S.Oliveira & Dutilh (Amaryllidaceae) ethanol extract activity on acetylcholinesterase and PPAR-alpha/gamma receptors. *Plants* **2022**, *11*, 3179. [CrossRef]
41. Estrada, O.; Alves, R.; Togawa, G.; Fagg, C.W.; Almeida, G.; Fonseca-Bazzo, Y.; Magalhães, P.d.O.; Silveira, D.A. A new chemical study of *Hippeastrum stapfianum* (kraenzl.) RS Oliveira & dutilh (amaryllidaceae). *Biochem. Syst. Ecol.* **2025**, *121*, 105015. [CrossRef]
42. Mancinelli, A.L. Interaction between light quality and light quantity in the photoregulation of anthocyanin production. *Plant Physiol.* **1990**, *92*, 1191–1195. [CrossRef]
43. Lu, Y.; Yu, Y.; Xuan, Y.; Kari, A.; Yang, C.; Wang, C.; Zhang, C.; Gu, W.; Wang, H.; Hu, Y.; et al. Integrative transcriptome and metabolome analysis reveals the mechanisms of light-induced pigmentation in purple waxy maize. *Front. Plant Sci.* **2023**, *14*, 1203284. [CrossRef]
44. Matsushita, K.; Sakayori, A.; Ikeda, T. The effect of high air temperature on anthocyanin concentration and the expressions of its biosynthetic genes in strawberry Sachinoka. *Environ. Control Biol.* **2016**, *54*, 101–107. [CrossRef]
45. Yang, J.F.; Li, B.; Shi, W.; Gong, Z.; Chen, L.; Hou, Z. Transcriptional activation of anthocyanin biosynthesis in developing fruit of blueberries (*Vaccinium corymbosum* L.) by preharvest and postharvest UV irradiation. *J. Agric. Food Chem.* **2018**, *66*, 10931–10942. [CrossRef]
46. Fan, W.; Bai, P.; Chen, R.; Guo, T.; Tian, Y.; Tian, H.; Ren, H. Postharvest Light Irradiation Induces Anthocyanin Accumulation in Fresh-Cut Lily Bulb (*Lilium davidii* var. *unicolor*) Scales. *J. Food Process. Pres.* **2024**, *2024*, 7984106. [CrossRef]
47. Gude, H.; Dijkema, M.H.G.E. The effects of light quality and cold treatment on the propagation of hyacinth bulbs. *Acta Hortic.* **1992**, *325*, 157–164. [CrossRef]
48. Wallace, W.; Secor, J.; Schrader, L.E. Rapid accumulation of 4-aminobutyric acid and alanine in soybean leaves in response to an abrupt transfer to lower temperature, darkness, or mechanical manipulation. *Plant Physiol.* **1984**, *75*, 170–175. [CrossRef]
49. Liu, P.; Zhang, X.; Yang, Y.; Sui, C.; Xu, Y.; Wei, J. Interxylary phloem and xylem rays are the structural foundation of agarwood resin formation in the stems of *Aquilaria sinensis*. *Trees* **2019**, *33*, 533–542. [CrossRef]
50. Whetten, R.W.; MacKay, J.J.; Sederoff, R.R. Recent advances in understanding lignin biosynthesis. *Annu. Rev. Plant Phys.* **1998**, *49*, 585–609. [CrossRef]
51. Zhong, R.; Ye, Z.-H. Transcriptional regulation of lignin biosynthesis. *Plant Signal Behav.* **2009**, *4*, 1028–1034. [CrossRef]
52. Lisec, J.; Schauer, N.; Kopka, J.; Willmitzer, L.; Fernie, A.R. Gas chromatography mass spectrometry-based metabolite profiling in plants. *Nat. Protoc.* **2006**, *1*, 387–396. [CrossRef]
53. Szablińska-Piernik, J.; Lahuta, L.B. Changes in Polar Metabolites during Seed Germination and Early Seedling Development of Pea, Cucumber, and Wheat. *Agriculture* **2023**, *13*, 2278. [CrossRef]
54. Szablińska-Piernik, J.; Lahuta, L.B. Metabolite profiling of semi-leafless pea (*Pisum sativum* L.) under progressive soil drought and subsequent re-watering. *J. Plant Physiol.* **2021**, *256*, 153314. [CrossRef]
55. Lahuta, L.B. Biosynthesis of raffinose family oligosaccharides and galactosyl pinitols in developing and maturing seeds of winter vetch (*Vicia villosa* Roth.). *Acta Soc. Bot. Pol.* **2006**, *75*, 219–227. [CrossRef]

56. Dębski, H.; Wiczkowski, W.; Horbowicz, M. Effect of elicitation with iron chelate and sodium metasilicate on phenolic compounds in legume sprouts. *Molecules* **2021**, *26*, 1345. [CrossRef]
57. Wiczkowski, W.; Szawara-Nowak, D.; Romaszko, J. The impact of red cabbage fermentation on bioavailability of anthocyanins and antioxidant capacity of human plasma. *Food Chem.* **2016**, *190*, 730–740. [CrossRef] [PubMed]

International Journal of
Molecular Sciences

MDPI

Article

Metabolomics Combined with Photosynthetic Analysis Reveals Potential Mechanisms of Phenolic Compound Accumulation in *Lonicera japonica* Induced by Nitrate Nitrogen Supply

Yiwen Cao [1,2,3], Yating Yang [1], Zhengwei Tan [2,3], Xihan Feng [1], Zhiyao Tian [1], Tianheng Liu [1], Yonghui Pan [1], Min Wang [1], Xiaoyu Su [2,3], Huizhen Liang [2,3,*] and Shiwei Guo [1,*]

1 Jiangsu Provincial Key Laboratory for Solid Organic Waste Utilization, Key Laboratory of Organic-Based Fertilizers of China, Jiangsu Collaborative Innovation Center for Solid Organic Wastes, Nanjing Agricultural University, Weigang 1, Nanjing 210095, China; 2020203071@stu.njau.edu.cn (Y.C.)

2 Institute of Chinese Herbal Medicines, Henan Academy of Agricultural Sciences, Zhengzhou 450002, China; suxiaoyu_2014@163.com (X.S.)

3 Provincial Key Laboratory of Conservation and Utilization of Traditional Chinese Medicine Resources, Zhengzhou 450002, China

* Correspondence: lianghuizhen@hnagri.org.cn (H.L.); sguo@njau.edu.cn (S.G.)

Abstract: Mineral nutrition is of vital importance in plant growth and secondary metabolites accumulation, and thereby in the nutritional value of plants. In *Lonicera japonica*, a preference to nitrate (NO_3^-—N) in comparison to ammonium (NH_4^+—N) was found in our previous study, which can be revealed from the rapid growth rate of *L. japonica* under NO_3^-—N. This study assessed whether a preference for nitrogen sources could invoke metabolic reprogramming and interrelationships between factors. NO_3^-—fed plants exhibited substantial enhancement of carbon stimulation, which was strongly and positively correlated with mesophyll conductance. As a result, the elevated carbon flux by NO_3^- supplement was shuttled to phenolic metabolites synthesis, including flavones and caffeoylquinic acids compounds. Notably, the stimulation was triggered by changes in the NO_3^- and C/N ratio and was mediated by the induction of several enzymes in the phenylpropanoid pathway. On the contrary, NH_4^+ plants showed an increment in the content of nitrogen, carbohydrates, and amino acids (mainly a strong increase in citrulline and theanine). Within secondary metabolism, NH_4^+ may involve active lignin metabolism, showing a dramatic increment in hydroxy—ferulic acid and lignin content. This work provides significant insights regarding the mechanisms of *L. japonica* in response to diverse nitrogen regimes and effective strategies of nitrogen fertilizer input for *L. japonica*.

Keywords: nitrate; caffeoylquinic acids; phenolic compounds; mesophyll conductance; photosynthesis; *Lonicera japonica*

1. Introduction

Caprifoliaceae, a plant family of dicotyledons, comprises ~800 species with worldwide distribution [1]. *Lonicera japonica*, commonly known as honeysuckle, is perhaps the most popular plant in the Caprifoliaceae family. The pharmacopeia of countries, such as China and Japan, has recorded the medicinal applications of this species [2], which include treating different types of viral infections, such as SARS coronavirus, H1N1 flu virus, and the influenza A virus [3]. Additionally, the active compounds extracted from *L. japonica* have extensive worldwide applications in industries such as cosmetics, pharmaceuticals, and food [4]. Improving the growth of *L. japonica*, especially the accumulation of bioactive substances, would be a key approach in securing commercial *L. japonica* productivity.

Phytochemical studies have shown that *L. japonica* has a complex composition of secondary substances, including flavonoids and (poly)phenolic acids, which are responsible for its pharmacological activity and nutritional values [2,4]. In particular, the abundant chlorogenic acid (CGA) and its derivatives are important phenolic components in *L. japonica*, playing a vital role in pharmacological activities [5]. Phenolic metabolites are the most prevalent secondary substances in plants, and researchers have recently focused significant attention on studying them [6–8]. Previous studies have extensively examined the effects of the variability of environmental factors, e.g., salt [9], drought [10], nutrition [11], CO_2 concentration [12], and light sources [13], on the accumulation of phenolic compounds.

During these episodes, nitrogen (N) serves as an indispensable nutrient and is available to plants mainly in the forms of nitrate (NO_3^-) and ammonium (NH_4^+) in agricultural soils [14]. NO_3^- is the primary form of N utilized by plants. Although ammonium can also be absorbed, sole NH_4^+ application is usually detrimental to plant growth and disrupts metabolism [15]. The N assimilation characteristics of plants exhibit notable differences when N is provided in the form of NO_3^- or NH_4^+. NO_3^- is first reduced to NH_4^+ and then incorporated into amino acids through catalysis by glutamine synthetase/glutamate synthase (GS/GOGAT). In this context, utilization of NO_3^- or NH_4^+ is represented by quite different demands for energy and carbon (C) skeletons [16]. Moreover, transformations between N− and C−based metabolites are commonly observed. The C−based substances primarily consist of phenolics, e.g., phenolic acids, flavonoids, and anthocyanins [17].

Agricultural practices through the choice of N source have been found to have considerable diverse effects on the phenolic substance. Generally, woody plants showed a better growth performance in a condition of nitrite, i.e., a study on *Cyclocarya paliurus* showed that sole NO_3^- supplementation significantly improved the accumulation of polyphenolic compounds [18]. With regard to herbaceous plants, Lang et al. [19] observed that phenolic acids in grape leaves were significantly altered by fertilization with different N forms, and nitrate application largely improved phenolic acid content. Similarly, Zhu et al. [20] found that a NH_4^+/NO_3^- ratio of 1:3 resulted in the highest accumulation of rosmarinic acid and flavonoids. On the contrary, Prinsi et al. [21] showed an increased content of phenolic acid in the condition of NH_4^+. Moreover, the phenolic compound biosynthesis pathways were regulated by the N form at the gene level. Based on transcriptome profiles of *Camellia sinensis*, it was shown that chalcone isomerase and chalcone synthase were induced in plants supplied with NO_3^- in comparison with NH_4^+ [22].

According to Domínguez et al. [23], the impact of N source on the phenolics accumulation depended largely on plant preference for N forms. For instance, NH_4^+-sensitive spinach exhibited increased levels of total soluble phenol in shoots, while no significant elevation was observed in ammonium-tolerant peas when comparing plants grown with NO_3^- and those grown with NH_4^+. N form can also regulate phenolic metabolic processes by influencing the synthesis of precursor amino acids and carbon–nitrogen metabolism, or as a signal transmitter. Aromatic amino acids are especially closely related to the synthesis of phenolic metabolites [24]. Li et al. [25] found that the increase in CO_2 led to a decrease in the contents of phenylalanine and tyrosine in strawberries, thereby reducing the accumulation of metabolites such as anthocyanins, eugenol, and lignin. Moreover, the carbon–nutrition balance (CNB) hypothesis is often applied to explain the influence of N availability on plant phenolic compounds accumulation [26]. Based on the CNB hypothesis, the biosynthesis of phenolics is constrained by N availability and net photosynthetic rate (P_n) in plants, e.g., the reduced N content of leaf tissues; additionally, a considerable P_n content would increase phenolics content. In agreement with this, a recent study on *Andrographis paniculata* suggested that the accumulation of carbohydrates, P_n, and photosynthetic N use efficiency largely stimulates the biosynthesis of andrographolide (C−based secondary metabolites) [27]. In this context, a comprehensive investigation of metabolic profiles when exposed to different N forms could provide valuable insights.

A previous study of *L. japonica* showed a preference for NO_3^- over NH_4^+, indicated by the rapid growth rate and higher photosynthetic efficiency under NO_3^- nutrition [28]. However, it is not sufficiently known how the increased photosynthesis under NO_3^- conditions affects leaf metabolism or, especially, the biosynthesis of phenolic metabolites (CGA, 3,5−di−caffeoylquinic acid (3,5−di−CQA) and 4,5−di−caffeoylquinic acid (4,5−di−CQA)) in *L. japonica*. Therefore, a hypothesis was proposed that the increased C/N ratio under NO_3^- conditions might induce the accumulation of specialized phenolic components in *L. japonica* compared with NH_4^+. The availability of omics in conjunction with the basic phenotype, such as photosynthetic parameters, helps to facilitate a more nuanced analysis [29–31]. The research findings can provide a scientific basis for enhancing *L. japonica* growth so that it can be produced in agricultural settings.

2. Results

2.1. Plant Growth and Physiology

In the present study, *L. japonica* grown in NH_4^+ nutrition showed significant growth depression. Sole NO_3^- and mixed N treatments significantly increased leaf biomass and C accumulation in comparison with sole NH_4^+ (Table 1), but had little effect on the area of the newly expanded leaf. On the contrary, the leaf N content was greater in the presence of NH_4^+, with an approximately 14% increase compared with sole NO_3^- treatment. Further analysis revealed no significant difference in NH_4^+ content among the treatments, while NO_3^- content was higher in the sole NO_3^- treatment compared with the mixed N and sole NH_4^+ treatments (Table 1), suggesting a higher N turnover in the presence of NH_4^+ nutrition. As a result, the C/N ratio was statistically different among the treatments.

Table 1. Effects of nitrogen (N) source on physiological traits of *Lonicera japonica* leaves.

Treatment	Leaf Biomass (g Leaf^{-1} FW)	Leaf Area (cm^2)	C Content (g m^{-2} DW)	N Content (g m^{-2} DW)	NH_4^+ (μmol g^{-1} FW)	NO_3^- (μmol g^{-1} FW)	C/N Ratio
A	0.16 ± 0.02 b	16.6 ± 1.9 a	12.5 ± 1.0 b	1.7 ± 0.1 a	4.9 ± 0.4 a	17.0 ± 1.4 b	7.2 ± 0.2 c
AN	0.22 ± 0.05 a	17.8 ± 2.5 a	15.6 ± 1.0 a	1.8 ± 0.1 a	4.6 ± 0.5 a	18.2 ± 1.5 b	8.7 ± 0.7 b
N	0.22 ± 0.03 a	16.3 ± 1.5 a	16.3 ± 1.5 a	1.5 ± 0.1 b	5.0 ± 0.3 a	25.1 ± 1.1 a	10.7 ± 0.5 a

Note: C, carbon. A, AN, and N represent treatments with NH_4^+ alone, a mixed N supply, and NO_3^- alone, respectively. Data are means ± SD. Different letters indicate statistically significant differences ($p < 0.05$).

In comparison with NH_4^+ treatment, P_n was relatively higher in NO_3^- and mixed N treatments, with approximately 1.3−fold greater P_n (Table 2). In agreement with this, the NH_4^+−grown plants exhibited an overall depression of stomatal conductance (g_s), electron transfer rate (ETR), and g_m (Table 2). Correlation analysis indicated a significant and positive correlation between leaf C accumulation and P_n (R^2 = 0.49, $p < 0.01$), ETR (R^2 = 0.48, $p < 0.01$), and g_m (R^2 = 0.73, P < 0.01) (Figure 1). Moreover, photorespiration occurs in conjunction with photosynthesis [16]. Here, we used the value of Γ^* to assess the alteration in leaf photorespiration, and the significantly upregulated Γ^* indicated a higher photorespiration rate under the NO_3^- condition (Table 2). Notably, the benign electron transfer rate and CO_2 assimilation rate suggested a different function of photorespiration in plants grown in nitrate from that of plants exposed to stress conditions (Table 2).

Table 2. Effects of N source on photosynthetic traits of *L. japonica* leaves.

Treatment	P_n (μmol m^{-2} s^{-1})	g_s (mol m^{-2} s^{-1})	g_m (mol m^{-2} s^{-1})	ETR (μmol m^{-2} s^{-1})	Γ* (μmol mol^{-1})
A	14.0 ± 0.5 b	0.23 ± 0.04 b	0.12 ± 0.01 c	129 ± 6 b	43.8 ± 0.7 b
AN	19.3 ± 1.5 a	0.35 ± 0.12 a	0.21 ± 0.01 a	168 ± 19 a	43.1 ± 1.0 b
N	18.2 ± 1.3 a	0.29 ± 0.06 ab	0.19 ± 0.01 b	157 ± 13 a	46.2 ± 0.8 a

P_n, saturated photosynthetic rate; g_s, stomatal conductance; g_m, mesophyll conductance; ETR, transpiration rate; Γ*, CO_2 compensation points in the absence of daytime respiration. A, AN, and N represent treatments with NH_4^+ alone, a mixed N supply, and NO_3^- alone, respectively. Data are means ± SD. Different letters indicate statistically significant differences ($p < 0.05$).

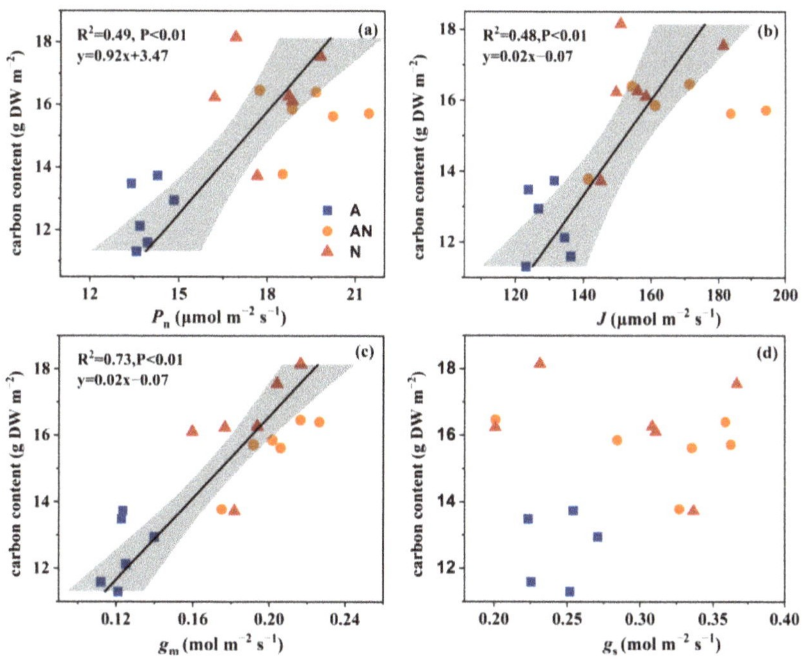

Figure 1. Correlation analyses of the leaf carbon content and net photosynthetic rate (P_n) (**a**), electron transfer rate (ETR) (**b**), mesophyll conductance (g_m) (**c**), and stomatal conductance (g_s) (**d**). A (blue squares), AN (orange circles), and N (red triangles) in the plot represent the sole NH_4^+ supply, mixed N supply and sole NO_3^- supply, respectively. The data were fitted by linear regressions.

2.2. Metabolomic Profile Analyses

The impact of the form of N on leaf metabolic profiles was investigated using non−targeted LC−MS analysis, and 509 metabolites were detected and mapped. An unsupervised PCA of overall metabolomic profiles revealed clear divisions among the three treatments (Figure 2a). Moreover, plants grown with NH_4^+ exhibited a higher similarity to plants grown under mixed N conditions compared with those grown with NO_3^-. A Venn diagram revealed that a total of 76 DEMs were identified under NO_3^- treatment compared with NH_4^+ treatment, while 23 and 62 DEMs were detected in the comparison of mixed N and NH_4^+ and NO_3^-, respectively (Figure 2b).

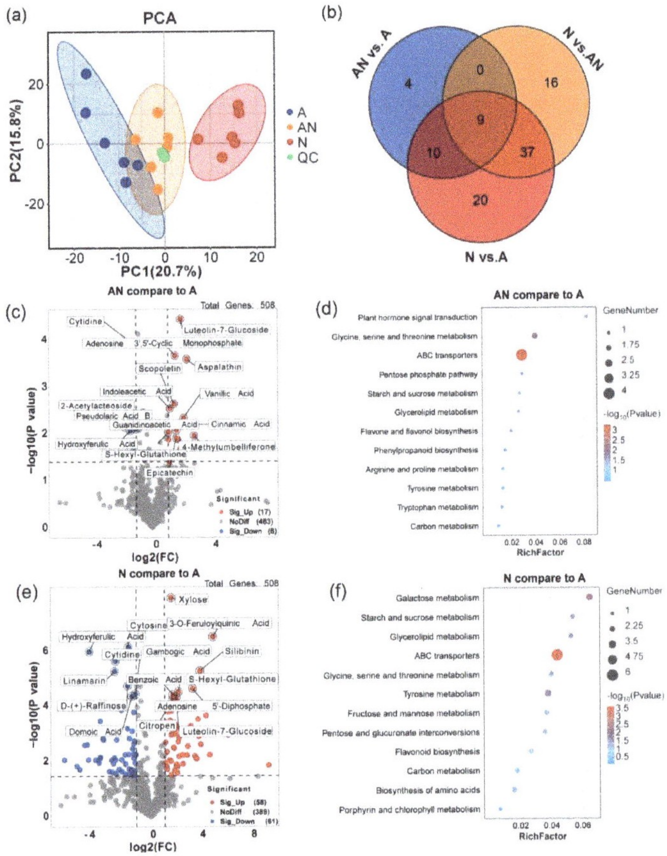

Figure 2. Characterization and analysis of metabolomic profiles of *L. japonica* under different nitrogen sources. Principal component analyses (PCAs) of all detected peaks under different N forms conducted by Liquid Chromatograph Mass Spectrometry/Mass Spectrometry (LC−MS/MS) (**a**). Venn diagram showed the overlap of the differential expressed metabolites (DEMs) in treatment AN versus A, treatment N versus A, and treatment N versus AN (**b**). Volcano plot of metabolomic profiles in the comparison of treatment AN and A (**c**) and treatment N and A (**e**). The red dots denote upregulated metabolites, the blue dots represent downregulated metabolites, while the grey dots correspond to metabolites detected but not significantly changed. Bubble plot of KEGG enrichment analysis of the DEMs in treatment AN versus A (**d**) and treatment N versus A (**f**). The size of bubble represents the number of DEMs enriched in the corresponding pathway and the color of the bubble denotes the significance level of the enrichment analysis, the red color indicates the pathway was significantly enriched, while the blue color revealed the pathways that were not that significant between the treatments. A, AN, and N represent the sole NH_4^+ supply, mixed N supply, and sole NO_3^- supply, respectively.

2.3. Key Differential Metabolites and Pathways

Taking sole NH_4^+ treatment as the control treatment, mixed N and NO_3^- treatments significantly induced the synthesis of phenolic compounds of *L. japonica* (Figure 2c,e). Specifically, replacing ammonium with nitrate led to a gradual increment in 4−methylumbelliferone, cinnamic acid, vanillic acid, aspalathin, and peonidin−3−O−glucoside (Tables S3 and S4). Moreover, dramatic and exclusive increments in the relative contents of N−formylmethionine, caffeoylquinic acid, and 3−O−feruloylquinic acid were observed in NO_3^- treatment compared to NH_4^+, by factors of 308, 120, and 21, respectively

(Table S3). In contrast, the contents of hydroxyferulic acid, linamarin, and cytidine were significantly decreased under mixed N and NO_3^- treatments.

KEGG enrichment analysis revealed that, among the 23 DEMs identified from mixed N and NH_4^+ treatments, 4 were mapped to ABC transporters, 6 were involved in the C and amino acid metabolism, such as glycine, serine and threonine metabolism, starch and sucrose metabolism, and pentose phosphate pathway, and 10 were involved in the biosynthesis of secondary metabolites, phenylpropanoid pathway, and flavone and flavanol biosynthesis (Figure 2d). Similarly, DEMs between NO_3^- and NH_4^+ treatment were mainly involved in ABC transporters, galactose metabolism, starch and sucrose metabolism, flavonoid biosynthesis, and glycine, serine, and threonine metabolism (Figure 2f). Taken together, these results revealed a notable variation and interaction in amino acid metabolism, C metabolism, and phenolic metabolism among different N forms.

2.4. Amino Acid Metabolism

Regarding the detected individual amino acids, most of them showed an increased relative content in plants fed with NH_4^+ (Figure 3). Compared with NO_3^- treatment, the content of theanine, serine, glutamine, citrulline, and alanine was substantially higher in NH_4^+ conditions, by a relative increment for more than 2−fold; meanwhile, mixed N supply decreased the content of serine and alanine to a lesser extent when compared to NH_4^+ treatment (Figure 3b). On the contrary, sole NO_3^- supply induced the increases in valine and isoleucine accumulation by factors of 1.3 and 1.5, respectively. Furthermore, the soluble protein content in plant grown with NO_3^- exhibited a notable elevation in comparison with mixed N and NH_4^+ treatment (Figure 4a).

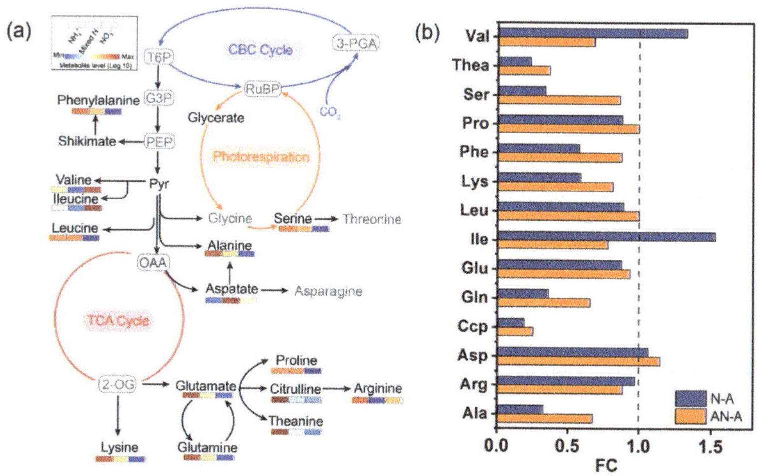

Figure 3. Expression profiles of the amino acid involved in the primary metabolism pathway. The scholar diagram displays the expression pattern of amino acids (**a**). Metabolites in black were present in the pathway, and those in grey were absent from our metabolic profile; the grey letter in the grey box represents the undetected metabolites expected for amino acid. The heatmap corresponding to each amino acid indicates relative expression levels in sole NH_4^+ supply, mixed N supply, and sole NO_3^- supply. Relative values of each amino acid content in treatment AN and N compared to treatment A (**b**). T6P, triose−6−phosphate; G3P, glucose−3−phosphate; PEP, phosphoenolpyruvate; 3−PGA, 3−phosphoglycerate; RuBP, ribulose 1,5−bisphosphate; OAA, oxaloacetate; 2−OG, 2−oxoglutarate; FC, fold change; Val, valine; Thea, theanine; Ser, serine; Pro, proline; Phe, phenylalanine; Lys, lysine; Leu, leucine; Ile, isoleucine; Glu, glutamate; Gln, glutamine; Ccp, citrulline; Asp, aspartate; Arg, arginine; Ala, alanine.

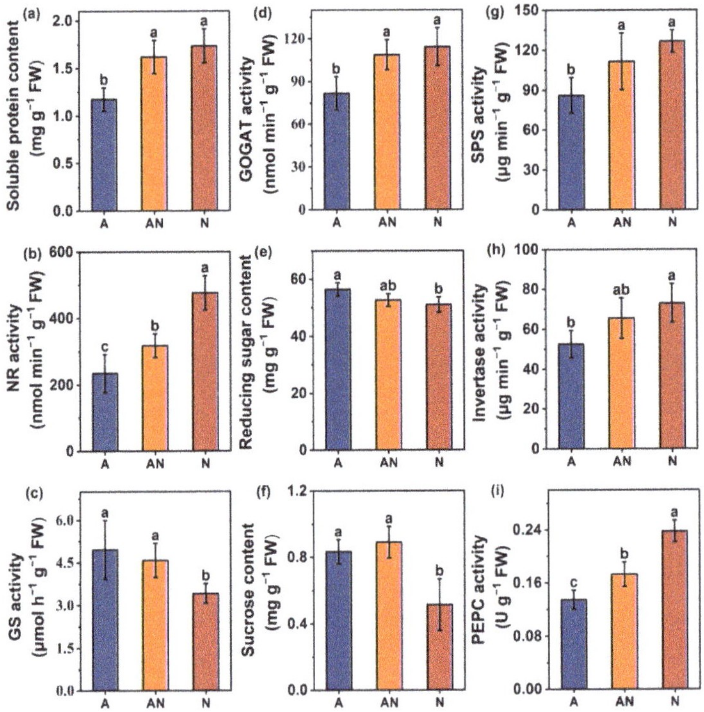

Figure 4. Effects of different N forms on soluble protein content (**a**), nitrate reductase (NR) activity (**b**), glutamine synthetase (GS) activity (**c**), glutamate synthetase (GOGAT) activity (**d**), reducing sugar content (**e**), sucrose content (**f**), sucrose phosphate synthase (SPS) activity (**g**), invertase activity (**h**), and phosphoenolpyruvate carboxylase (PEPC) activity (**i**). A, AN, and N represent treatments with sole NH_4^+, mixed N, and sole NO_3^- supply, respectively. Data are means ± SD. Different letters indicate statistically significant differences ($p < 0.05$).

The activities of N assimilation enzymes, including NR, NiR, GS, and GOGAT, were measured (Figure 4). In general, the N form had a significant effect on the activity of the measured enzymes. NR activity progressively increased in plants grown with mixed N and NO_3^- treatments (Figure 4b). GOGAT had a similar pattern, with increased activity under NO_3^- supply (Figure 4c). However, GS activity was, on average, 24% and 31% lower in plants grown with NO_3^- than in plants grown with mixed N and NH_4^+, respectively (Figure 4d). NiR activity was not significantly affected by the N source (Figure S2).

2.5. C Metabolism

NO_3^- and NH_4^+ treatments increased the contents of reducing sugars, especially sucrose, by 10.6% and 62.7%, respectively (Figure 4e,f). SPS, which is a key enzyme in sucrose synthesis, was inhibited by 31.7% under NH_4^+ treatment compared with NO_3^- treatment, and by 22.7% compared with mixed N treatment (Figure 4g). The hydrolysis of sucrose into reducing sugar is catalyzed by invertase. Invertase activity was similar in NO_3^- and mixed N treatments, and was, on average, 40.3% and 25% higher than in NH_4^+ treatment (Figure 4h). The PEPC enzyme plays a key role in maintaining C assimilation and utilization. In the present study, PEPC activity showed a clear pattern, decreasing with the gradual replacement of NO_3^- with NH_4^+ (Figure 4i).

2.6. Biosynthesis and Accumulation of Phenolic Metabolites

Altering the N source significantly affected the pattern of phenolic metabolism in *L. japonica* (Figure 5). Importantly, the relative contents of cinnamic and caffeoylquinic acids were largely increased by factors of 15 and 120, respectively, by NO_3^- supply (Figure 5, Tables S3 and S4). According to the Chinese pharmacopoeia, caffeoylquinic acid derivatives, including CGA, 3,5−di−CQA, and 4,5−di−CQA, are the index compositions representing the medicinal quality of *L. japonica*. In the present study, similar increases in the contents of CGA, 3,5−di−CQA, and 4,5−di−CQA were observed in plants grown with NO_3^- (Figure S2). Luteolin−7−glucoside is a flavonoid that exhibits important pharmacological activity in *L. japonica*. Compared with NH_4^+ treatment, mixed N and NO_3^- treatments induced significant accumulation of luteolin−7−glucoside by 3.54−fold and 3.46−fold, respectively, on average (Tables S3 and S4). Importantly, the relative contents of epicatechin, catechin, and (especially) hydro−ferulic acid were increased by sole NH_4^+ supply compared with sole NO_3^- supply (Figure 5).

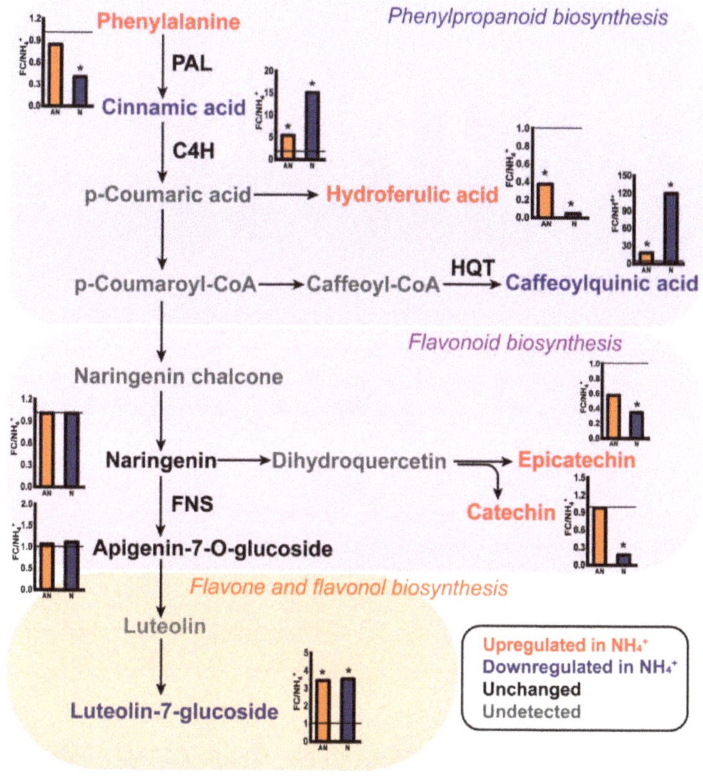

Figure 5. Expression pattern of metabolic profiles and genes involved in phenolic metabolism. Red coloring reflects significantly ($p < 0.05$) increased metabolites under sole NH_4^+ condition compared to mixed N and sole NO_3^- condition, while blue indicates the significantly decreased ones. Metabolites in black were present in the pathways but were not significantly changed, and those in grey were absent from our metabolic profile. The green letter denotes the relative key enzymes involved in the pathway. The column plot shows the relative value of each metabolite in treatments AN and N compared to treatment A. Asterisks indicate values determined to be significantly different ($p < 0.05$) between treatment AN, N and A. PAL, phenylalanine ammonia−lyase; C4H, cinnamate 4−hydroxylase; HQT, hydroxycinnamoyl–CoA quinate transferase; FNS, flavone synthase.

To determine whether the variation in these metabolites was regulated at the transcription level, we measured the relative expression of key enzymes involved in phenolic metabolism, including PAL, C4H, HQT, and FNS (Figure 6). PAL is the first enzyme in the phenylpropanoid pathway, converting phenylalanine to cinnamic acid. Compared with the sole NH_4^+ treatment, the relative expression of PAL was upregulated by 3.6−fold and 2.3−fold in mixed N and sole NO_3^- treatments, respectively (Figure 6a). The C4H enzyme catalyzes the conversion of cinnamic acid to p−coumaric acid. No significant changes were observed in the relative expression of the C4H enzyme (Figure 6b). As the rate−limiting enzyme catalyzing caffeoyl−CoA conversion into caffeoylquinic acid, the relative expression of HQT enzyme increased gradually with the application of NO_3^- (Figure 6c). FNS is the rate−limiting enzyme involved in flavone and flavanol biosynthesis. At the transcription level, FNS expression increased by 1.80−fold under mixed N treatment, and by 1.24−fold under NO_3^- treatment (Figure 6d). Taken together, these results indicate that the relative expression of enzymes involved in phenolic metabolism might be inhibited by sole NH_4^+ application.

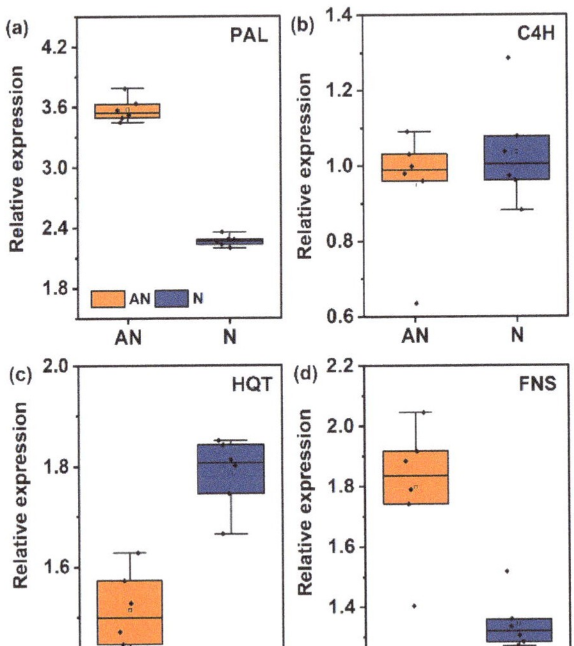

Figure 6. Effects of N form on the relative expression of genes involved in phenolic metabolism pathway. PAL, phenylalanine ammonia−lyase (**a**); C4H, the cinnamate 4−hydroxylase (**b**); HQT, hydroxycinnamoyl–CoA quinate transferase (**c**); FNS, flavone synthase (**d**). The sole NH_4^+ supply was set as the control treatment to calculate the relative gene expression levels in treatments AN (mixed N) and N (sole NO_3^- supply).

3. Discussion

3.1. Photosynthesis and C/N Metabolism in L. japonica in Response to N Source

The applied N sources—NH_4^+, mixed N, and NO_3^-—exerted a considerable influence on biomass, photosynthesis, and metabolism at a single−leaf level in *L. japonica* (Tables 1 and 2, Figure 2). NO_3^- supplementation significantly promoted photosynthetic

CO_2 assimilation by *L. japonica*, accompanied by improved C content and C/N ratio (Table 1). Generally, changes in the photosynthetic carbon assimilation rate can regulate metabolic processes by affecting leaf carbohydrate accumulation [14]. Research by Lu et al. [29] showed that carbon starvation mediated by decreased CO_2 conductance can reduce the biosynthesis of carbohydrates and organic acids. In the present study, correlation analyses showed a strong and positive relationship between leaf C content and g_m ($R^2 = 0.73$, $p < 0.01$, Figure 1), rather than g_s. Moreover, Xu et al. [32] found that the enhanced g_m largely elevated the C utilization efficiency of *Euonymus japonicus* seedlings. In that condition, plants were forced to adjust their allocation of C and nutrient investments due to the varied g_m [29]. Along this line, SPS and NI enzymes, which exert in key enzymes coordinating C metabolism in *L. japonica* among different N sources, were evaluated here [33]. The activation of SPS and invertase activities in plants grown with NO_3^- indicated a higher turnover of sucrose and thus more utilization of C (Figure 4) [30]. Furthermore, we observed an increase in the accumulation of sucrose and reducing sugars in plants grown with NH_4^+, which could be explained by impaired C utilization by leaves [34].

Metabolic adaptation to N sources has been studied in multiple species and involves N−assimilating enzymes, including mainly NR, NiR, GS, and GOGAT [35–37] (Figures 4 and S1). In the present study, NR activity was positively correlated with the proportion of NO_3^- application (Figure 4), similar to a previous study [38]. The GS/GOGAT enzyme system catalyzes the conversion of N from its inorganic form to organic forms, in turn producing a wide variety of amino acids [35]. However, when plants are faced with NH_4^+ stress, PEPC enzymes are required to ensure GS/GOGAT carbon demand via regulation of TCA cycle−associated anaplerotic routes [37,39]. In this context, the increase in GOGAT activity under mixed N and NO_3^- treatments are associated with sufficient provision of C skeleton via active C metabolism (Figure 4). Masumoto and Gantt [40] observed the inhibition of GOGAT activity in the chloroplast PEPC gene silencing line *Osppc4*, which further affected C metabolism.

Along this line, *L. japonica* showed a general increment in amino acids in NH_4^+ nutrition (Figure 3; Table S3), in accordance with the higher N content compared to mixed N and NO_3^- treatment. At the level of individual amino acids, mainly citrulline and theanine (which are N−rich amino acid) revealed a great accumulation in NH_4^+ condition, by relative increments of 5.28−fold and 4.31−fold in relation to NO_3^-. Selective preference for NH_4^+ in citrulline and theanine production has been confirmed in study by Ruan et al. [41]. Similarly, the relative contents of alanine and glutamine were increased by 3.04−fold and 2.76−fold in NH_4^+ treatment compared to NO_3^- (Figure 3). According to Krapp, [42], N−rich amino acids are commonly synthesized to quickly metabolize inorganic N, especially in sole NH_4^+−grown plants. Taken together, the results indicate that *L. japonica* can assimilate NH_4^+ by enhancing GS activity and accumulating the specialized amino acids as endogenous N storage stores to avoid the toxic effects of NH_4^+.

Moreover, contrary to the decrease in amino acid content, soluble protein content in leaves significantly increased under N treatment. Liu et al. [43] observed a similar result in their analysis of apples. Proteins are the main constituents of many enzymes, and their synthesis is initiated from N−formylmethionine [44]. Notably, a significant 308−fold increase in the relative content of N−formylmethionine was observed with N treatment compared with A treatment (Table S3). Although more research is required to determine the mechanism of action, it is most likely associated with the N allocation pattern in plants grown with NO_3^-, under which conditions the protein was preferentially synthesized to sustain metabolism.

3.2. Phenolic Metabolism Patterns in *L. japonica* in Response to the N Source

Phenolic compounds have received considerable attention over the past few decades, since phenolic metabolites have a wide variety of structures and abundance and they have been associated with various beneficial effects on health [45]. The most visible impact of nitrate application on metabolism in the present study was a general increase in the levels of C−rich phenolic metabolites. This pattern of accumulation was most noticeable in phenylpropanoid derivatives (cinnamic acid, caffeoylquinic acid, 3−O−feruloylquinic acid), including coumarin pathway intermediates such as vanillic acid and 4−methylumbelliferone (Tables S3 and S4). The response was especially evident in plants with reduced N uptake. A correlation analysis indicated the potential stimulation of these substances by photosynthesis improvement (Figure 7), confirming earlier observations [30,31]. Kong et al. [46] have also observed the accumulation of phenolic acids, such as ferulic acid, vanillic acid, and caffeic acid, with increased irradiance levels.

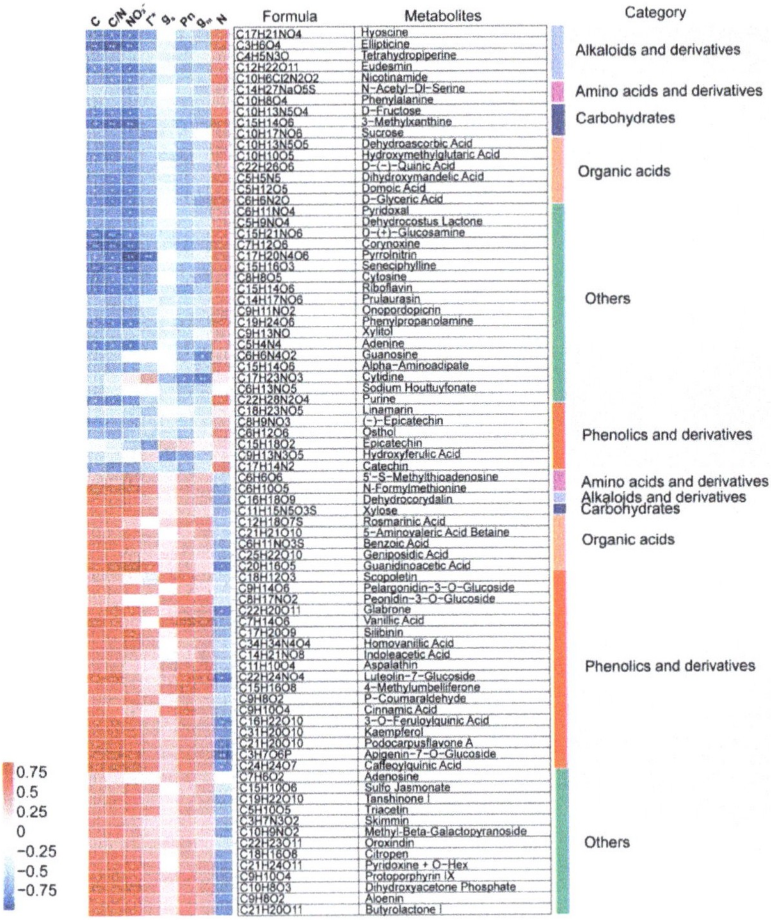

Figure 7. Correlation of DEMs (Tables S3 and S4) with physiological (C, N, C/N, and NO_3^-) photosynthetic traits (P_n, g_s, g_m, Γ^*). The correlations were estimated by the linear model within 95% confidence level. Colors denote Pearson's rank correlation between metabolites and traits. Red coloring reflects positive correlation, while blue indicates the negative correlation. The significant correlations are marked with asterisk (*, $p < 0.05$ and $R^2 > 0.70$; **, $p < 0.05$ and $R^2 > 0.80$).

Caffeoylquinic acid is a bioactive phenolic substance in *L. japonica* [1,3]. The plasticity of caffeoylquinic acid in plants in response to various environmental situations has been widely reported in recent years [17,47,48]; however, the effect of nitrate on these compounds is still controversial. In the present study, nitrate enhanced the accumulation of caffeoylquinic acid in comparison with ammonium (Figure 5), and a notable and positive relationship between caffeoylquinic acid and C/N ratio and endogenous NO_3^- status was observed (Figure 7). According to Fritz et al. [49], NO_3^- can stimulate phenylpropanoid metabolism by inducing several enzymes involved in the initial steps of the phenylpropanoid pathway. Here, transcriptome profiles showed increased expression of PAL and HQT enzymes in N and AN treatments (Figure 6). These treatments resulted in the accumulation of downstream molecules (cinnamic acid and caffeoylquinic acid), revealing changes in relative gene expression in relation to NO_3^- application. Hydroferulic acid was the most abundant phenolic acid in sole NH_4^+-grown plants (Tables S3 and S4). As the precursor of the lignin biosynthesis pathway, its abundance may suggest the promotion of higher lignification by NH_4^+. The results were further confirmed by the significantly increased amounts of lignin in plants grown exclusively on NH_4^+ (Figure S3). Lignin, the aromatic heteropolymer in plants, is usually synthesized during plant development or stress response [50]. For instance, lignin biosynthesis pathway genes were upregulated under Cu stress, by inducing lignin polymerization enzymes (e.g., peroxidase) [51]. In such cases, NH_4^+ may trigger mild oxidative stress, leading to the accumulation of reactive oxygen signaling molecules (such as H_2O_2) and activating the lignin biosynthesis. Our findings on NH_4^+ are in accordance with those of Kováčik and Klejdus, [17], who suggested this enhanced lignification could serve as a protective mechanism in the presence of exclusive NH_4^+ supplement. Taken together, the enhanced lignin biosynthesis could be responsible for the retardation of growth mediated by decreased photosynthesis (Table 1).

Flavonoids, a class of low−molecular−weight phenolic chemicals, are frequently studied due to their variety of biological activities [52]. Many previous studies have observed the effect of NO_3^- on the induction of flavonoid accumulation [53,54]. This response was particularly distinct when nitrate was applied under low levels of N [55]. According to Duan et al. [53], the increased accumulation of flavonoids under NO_3^- supplementation was due to the enhanced C/N ratio and C metabolism. In our study, specific flavonoids in *L. japonica* such as apigenin−7−O−glucoside and especially luteoline−7−glucoside increased in response to C/N ratio under NO_3^- supplement (Figure 7). In accordance with this, the FNS enzyme, which regulates the biosynthesis of luteoline−7−glucoside, was observed to have a higher expression in treatment AN and N in comparison to treatment A (Figure 6). However, a close look may reveal an inconsistent variation because the contents of catechin and epicatechin were considerably decreased in the N and AN treatments compared to treatment A (Figure 7, Tables S3 and S4). Similarly, theanine, the synthetic precursor of catechin biosynthesis, showed a significant decrement in the N and AN treatments (Figure 3). Catechins are constituents of tea−flavor−related compounds; a lot of research on catechins has been reported in tea (*Camellia sinensis* L.) plants, revealing decreased biosynthesis of catechins in a condition of NH_4^+ [41,56]. Current evidence has suggested that the inconsistent results on *L. japonica* and tea may be ascribed to the N preference of plants, since tea has a NH_4^+ preference. Such variability could also be reflective of the different patterns of C allocation in NH_4^+ conditions, which have a low leaf C accumulation [57]. Compared with luteoline−7−glucoside, catechin and epicatechin, when synthesized, are subject to the cost of a lower C skeleton. NH_4^+ assimilation requires C skeletons, and sufficient C skeletons could be related to NH_4^+ tolerance in vascular plants.

4. Materials and Methods

4.1. Plant Material and Growth Condition

The experiment was performed at the greenhouse of Nanjing Agricultural University in Nanjing, China. The environmental temperature was set to 26 ± 3 °C in the day and 18 ± 3 °C at night; humidity was set to 50 ± 10%. Photosynthetic photon flux density (PPFD) was maintained at 1000 μmol (photons) m^{-2} s^{-1} at the leaf level with SON−T AGRO 400 W bulbs (Philips, Hamburg, Germany), and the photoperiod was set to 14 h light/10 h dark. One−year−old *Lonicera japonica* plant seedlings with 3–4 leaves were used in this study. For hydroponic culture experiments, seedlings were cultured with a 50% nutrient solution (for composition, see Table S1) for 2 weeks, followed by an adaptation period of the next 2 weeks with a 100% solution. Then, the homogeneous seedlings were transplanted to 12 L plastic containers. Based on our previous experiment, *L. japonica* was treated with different sources of 2.8 mM N: $(NH_4)_2SO_4$ alone, $Ca(NO_3)_2$ alone, and a combination of 50% $(NH_4)_2SO_4$ and 50% $Ca(NO_3)_2$. $CaCl_2$ was added to the $(NH_4)_2SO_4$ and mixed N treatments to adjust Ca concentration and dicyandiamide (0.1%) was used as a nitrification inhibitor. The nutrition solution was aerated through the entire culture period and was renewed every 4 days, whereas the pH was adjusted daily to 6.0 ± 0.1 with 1 M HCl and 1 M NaOH.

In the experimental design, there were three experimental units (the plastic box) per treatment and eight subunits (plants) within each experimental unit. Four weeks after the N form treatment, six samples from each treatment were selected for the photosynthetic parameters, gene expression, and metabolomic measurement. The other plants were selected and sampled for the growth and physiological biochemical index.

4.2. Gas Exchange and Fluorescence Measurements

A portable LI−6800 photosynthesis system (LI−COR) was used for the measurements of gas exchange and chlorophyll fluorescence parameters on new fully expanded leaves. During determination, photosynthesis was initiated under controlled conditions within the leaf chamber: PPFD, 1000 μmol m^{-2} s^{-1} (10%: 90%, blue light: red light); leaf temperature, 25 °C; CO_2 concentration, 400 μmol mol^{-1}. For fluorescence measurement, an LI−6800 system equipped with a multiphase flash fluorometer chamber was used. The multiphase flash methodology was based on references published by Han et al. [58]. g_m was calculated as previously described [28]. Six seedlings with similar growth from each treatment were selected for analysis.

4.3. Measurement of the CO_2 Compensation Point in the Absence of Mitochondrial Respiration (Γ*)

The Γ* was measured after dark adaptation (at least 4 h) and six seedlings from each treatment were used for analysis. Prior to the initial measurements, the conditions within the leaf chamber were set at 600 μmol m^{-2} s^{-1} PPFD and 100 μmol mol^{-1} CO_2 concentration. Leaves were acclimated for 30 min inside the leaf chamber. P_n/C_i response curves were conducted at a series of CO_2 concentrations (50, 80, 100 μmol mol^{-1}) and PPFDs (150, 300, 600 μmol m^{-2} s^{-1}). Their linear regressions were then predicted to converge at one point where the x−axis and y−axis of the point were defined as the photo−compensation point (C_i*) and the day respiration rate (R_d). The Γ* was then calculated according to Sun et al. [59]:

$$\Gamma^* = C_i{}^* + \frac{R_d}{g_m} \quad (1)$$

4.4. Measurement of Leaf Area and Biomass

Leaf area was captured using a camera, and the measurement for the leaf area was conducted using IMAGE−PRO plus 6.0 software. The fresh new fully expanded leaf was separated from the seedlings and weighed as leaf fresh biomass (FW). Then, the leaf was dried at 105 °C for 15 min followed by further drying at 65 °C in an oven until the weight was unchanged. The weight was recorded as leaf dry biomass (DW). Leaf water content was calculated as leaf water content (%) = (FW − DW)/FW × 100%. The dry samples were preserved for the determination of later physiological parameters. Six seedlings with similar growth from each treatment were selected for analysis.

4.5. Physiology Measurement

Milled dry leaves samples were sieved (<180 μm) and weighed to approximately 10–15 mg to measure leaf C content and N content using an N/C analyzer (multi−EA 5000; Analytikjena, Jena, Germany), and the C/N ratio was calculated accordingly. Sucrose was determined by colorimetric resorcinol method and the reducing sugar was assayed using the dinitro−salicylic acid method. Six replicates from each treatment were selected for analysis.

For NH_4^+ determination, samples were homogenized with pre−cooled dilute H_2SO_4 (0.3 mM, *w:v*, 1:10). The mixture was separated at 15,000× *g* for 20 min, and NH_4^+ content was determined according to Sun et al. [59] at a wavelength of 625 nm, with modified phenol−hypochlorite method. For the measurement of NO_3^-, 0.50 g samples were homogenized with 5 mL deionized water. The mixture was boiled in water for 30 min, then centrifuged at 12,000× *g* for 20 min to collect the supernatant. NO_3^- content was determined at a wavelength of 410 nm according to the sulfuric acid–salicylic acid method. Enzyme activity, including nitrate reductase (NR), nitrite reductase (NiR), glutamine synthase (GS), glutamate synthase (GOGAT), sucrose phosphate synthase (SPS), phosphoenolpyruvate carboxylase (PEPC), and invertase, was assayed using the kit provided by Comin Biotechnology Co., Ltd. (Huizhou, China). Six replicates from each treatment were selected for analysis.

Metabolites assay was conducted by Agilent 1260 high−performance liquid chromatography (Santa Clara, CA, USA). Leaf samples were soaked in 75% methanol (*w:v*, 1:100), and the mixture was sonicated for 30 min and followed by filtration using a syringe (0.45 μm) into a vial. For the determination of chlorogenic acid (CGA), 3,5−di−caffeoylquinic acid (3,5−di−CQA), and 4,5−di−caffeoylquinic acid (4,5−di−CQA), the mobile phases consisted of phase A (0.1% phosphoric acid) and phase B (acetonitrile). The mass spectrometry parameters were set according to Cao et al. [28]. Six replicates from each treatment were selected for analysis.

4.6. Metabolomic Analyses

Metabolomic analyses were performed by Bio−Profile Biotechnology Co., Ltd. (Shanghai, China). Firstly, fresh samples (100 mg) were pulverized with liquid N, and a pre−cooled mixture of 200 μL deionized water and 800 μL methanol/acetonitrile (*v:v*, 1:1) was added. The mixture was then vortexed and incubated for 1 h. Next, tubes containing the mixture were centrifuged at 6000× *g* and 4 °C for 30 min. The supernatant was carefully transferred into a new centrifuge tube and dried using a vacuum centrifuge. Then, the samples were dissolved in a 100 μL solution consisting of an equal volume of acetonitrile and water followed by another round of centrifugation at 16,000× *g* and 4 °C for 30 min. Finally, the supernatant was collected for ultra−high−performance liquid chromatography using Agilent 1290 Infinity LC system, coupled with AB Sciex triple time−of−flight mass spectrometry 5600 (Framingham, MA, USA). Six replicates from each treatment were selected for analysis.

For metabolite separation, samples were then analyzed using an ACQUIY UPLC BEH column (2.1 mm × 100 mm, 1.7 μm). Gradient elution analysis was performed using a mobile phase consisting of two phases: phase A, 25 mM ammonium acetate, and 25 mM ammonium hydroxide in water; phase B, acetonitrile. The flow rate was set as 0.5 mL min^{-1}. After equilibration, each sample was injected with a volume of 5 μL solutions. A gradient profile was employed as follows: from 0 to 0.5 min, the mobile phase was consistent of 95% B; from 0.5 to 7.0 min, the proportion changed gradually from 95% B to 65% B; from 7.0 to 9.0 min, the proportion decreased gradually from 65% B to 40%; from 9.0 to 10.0 min, it remained constant at 40% B; from 10.0 to 12.0 min, the ratio increased from 40% B to 95% B; from 12.0 to 16.0 min, the mobile phase was consistent of 95% B. Acquisition of mass spectrum could be found in the Supplementary Materials (Method S1).

At the end of the assay, the determination of the row data with a unique signal intensity was performed with format conversion, peak alignment retention time correction, and peak area extraction using MS−DIAL software (version 3.98). Accurate mass number matching (molecular weight error < 25 PPM) and secondary spectrogram matching formula were conducted, searching public databases such as HMDB (http://www.hmdb.ca) and Mass Bank (http://www.massbank.jp) on 1 July 2022. Default values (>50% within the treatments) in the extracted peaks were deleted. Then, the positive and negative ion peaks were integrated and applied for pattern recognition with SIMCA−P 14.1 software. Data processing was conducted by Pareto scaling (Par) and the evaluation of the data can be found in the Supplementary Materials (Method S2). A multivariate statistical analysis was performed, which involved unsupervised principal component analysis (PCA) and supervised orthogonal partial least square discriminant analysis (OPLS−DA). For the identification of the most relevant metabolites to the treatments, a threshold of variable importance in the project (VIP) > 1 with a p value < 0.05 was set for the selection of differentially expressed metabolites (DEMs).

4.7. Total RNA Extraction and qPCR Analysis

Leaf samples were pulverized with liquid N, and the total RNA from each analyzed sample was extracted using Trizol regent (ThermoFisher Scientific, Waltham, MA, USA). The synthesis of cDNA and the analysis of qPCR were performed according to the kit's manufacturer protocols (Vazyme, Nanjing, China). The qPCR was carried out using Applied Biosystems 7500. Six replicates from each treatment were selected for analysis.

All the procedures were conducted following user protocols. Primer sequences for four phenolic metabolism pathway genes (PAL, phenylalanine ammonia−lyase; FNS, flavone synthase; HQT, hydroxycinnamoyl–CoA quinate transferase; C4H, cinnamate 4−hydroxylase; listed in Table S2) were synthesized by GenScript Technology (Nanjing, China).

4.8. Statistical Analyses

All the statistical analyses were performed using SPSS 25.0 software. One−way ANOVA was used to determine the difference between mean values from the treatments, and the significant differences were assessed using the least significant difference test (LSD, $p < 0.05$). Bioinformatic analyses were performed using the OmicStudio tools (https://www.omicstudio.cn/tool, accessed on 1 February 2024). The scatter plot, bar plot, and regression analyses were generated with Origin Pro 2021 software. Image layout was performed using Adobe Illustrator 2021.

5. Conclusions

Elevated CO_2 conductance, especially g_m, was found to facilitate P_n and C accumulation in NO_3^-−grown plants. Increases in SPS, PEPC, and invertase activities indicated good C utilization in the NO_3^- treatment. On the other hand, NH_4^+−grown plants showed increases in carbohydrate and amino acid (with strong increases in citrulline and theanine) synthesis, which may mitigate the ammonium effect. Variations in C accumulation had a significant impact on the levels of C−rich phenolic metabolites. For example, caffeoylquinic acid biosynthesis was mainly activated in plants grown with NO_3^-, and was triggered by changes in NO_3^- content and the C/N ratio and mediated by the induction of PAL and HQT enzymes in the biosynthetic pathway, while plants grown with NH_4^+−N tended to synthesize fewer C−containing compounds, e.g., catechins and epicatechins. Notably, NH_4+ may activate the lignin biosynthesis pathway, leading to a significant increase in hydroxyferulic acid and lignin accumulation. In conclusion, nitrate addition results in substantial improvements in the growth and biosynthesis of bioactive components of *L. japonica*. Responses under field conditions need to be further quantified to identify N manipulations that can be targeted for the industrial cultivation of *L. japonica*.

Supplementary Materials: The following supporting information can be downloaded at: https://www.mdpi.com/article/10.3390/ijms26094464/s1.

Author Contributions: S.G. and Y.C. conceived the idea and designed the experiment; Y.C., Y.Y., Y.P., X.F., Z.T. (Zhiyao Tian) and T.L. completed the experiments; Y.C. and Y.Y. analyzed the data and wrote the manuscript; S.G., Z.T. (Zhengwei Tan), X.S. and Y.P. helped in revising the manuscript; S.G., H.L. and M.W. provided funding support. All the authors contributed critically to the drafts and gave final approval for publication. All authors have read and agreed to the published version of the manuscript.

Funding: This work was financially supported by the following organizations: Science and Technology Development Programs of Henan [252102111079]; National Natural Science Foundation of China [32072673]; Key Research and Development Program of Henan [241111310200, 231111110800, 251111310600, 251111111800]; Henan Center for Overseas Scientists [GZS2024025]; Central government funds to guide local science and technology development [Z20241471131]; Independent Innovation Projects of Henan Academy of Agricultural Sciences [2025ZC45]; the National College Students Innovation and Entrepreneurship Training Program [202310307028Z].

Institutional Review Board Statement: Not applicable.

Informed Consent Statement: Not applicable.

Data Availability Statement: The data underlying this article are available in the article and in its online Supplementary Materials.

Conflicts of Interest: The authors declare no conflicts of interest.

Abbreviations

The following abbreviations are used in this manuscript:

C	carbon
CNB	carbon–nutrition balance
C4H	cinnamate 4−hydroxylase
CGA	chlorogenic acids
DEMs	differential expressed metabolites
ETR	electron transfer rate
FC	fold change
FNS	flavone synthase
g_m	mesophyll conductance

GS	glutamine synthetase
GOGAT	glutamate synthetase
HQT	hydroxycinnamoyl–CoA quinate transferase
N	nitrogen
NR	nitrate reductase
NiR	nitrite reductase
OPLS−DA	orthogonal partial least squares discriminant analyses
PAL	phenylalanine ammonia−lyase
PCA	principal component analyses
PEPC	phosphoenolpyruvate carboxylase
P_n	net photosynthetic rates
PPFD	photosynthetic photon flux density
SPS	sucrose phosphate synthase
VIP	variable importance in the project
Γ^*	CO_2 compensation point in the absence of mitochondrial respiration

References

1. Pu, X.; Li, Z.; Tian, Y.; Gao, R.; Hao, L.; Hu, Y.; Song, J. The honeysuckle genome provides insight into the molecular mechanism of carotenoid metabolism underlying dynamic flower coloration. *New Phytol.* **2020**, *227*, 930–943. [CrossRef]
2. Schierenbeck, K. Japanese honeysuckle (*Lonicera japonica*) as an invasive species; history, ecology, and context. *Crit. Rev. Plant Sci.* **2004**, *23*, 391–400. [CrossRef]
3. Lee, Y.; Chang, M.; Yeh, Y.; Huang, C.; Lin, F.; Huang, J. Honeysuckle Aqueous Extracts Induced Let-7a Suppress EV71 Replication and Pathogenesis In Vitro and In Vivo and Is Predicted to Inhibit SARS-CoV-2. *Viruses* **2021**, *13*, 308. [CrossRef] [PubMed]
4. Li, R.; Kuang, X.; Wang, W.; Wan, C.; Li, W. Comparison of chemical constitution and bioactivity among different parts of *Lonicera japonica* Thunb. *J. Sci. Food Agric.* **2020**, *100*, 614–622. [CrossRef]
5. Feng, Y.; Zhang, G.; Zhu, P.; Zhu, W.; Li, Y.; Fan, X. Metabolite profiles and antibacterial and antioxidant activities of leaf extracts of five Lonicera species: A comparative study. Chem. *Biol. Tech. Agri.* **2023**, *10*, 91. [CrossRef]
6. Albuquerque, B.; Heleno, S.; Oliveira, M.; Barros, L.; Ferreira, I. Phenolic compounds: Current industrial applications, limitations and future challenges. *Food Func.* **2021**, *12*, 14–29. [CrossRef]
7. Vogt, T. Phenylpropanoid biosynthesis. *Mol. Plant.* **2010**, *3*, 2–20. [CrossRef]
8. Cheynier, V. Phenolic compounds: From plants to foods. *Phytochem. Rev.* **2012**, *11*, 153–177. [CrossRef]
9. Bistgani, Z.; Hashemi, M.; DaCosta, M.; Craker, L.; Maggi, F.; Morshedloo, M. Effect of salinity stress on the physiological characteristics, phenolic compounds and antioxidant activity of *Thymus vulgaris* L. and *Thymus daenensis* Celak. *Ind. Crops Prod.* **2019**, *135*, 311–320. [CrossRef]
10. Nichols, S.; Hofmann, R.; Williams, W. Physiological drought resistance and accumulation of leaf phenolics in white clover interspecific hybrids. *Environ. Exp. Bot.* **2015**, *119*, 40–47. [CrossRef]
11. Chandler, S.; Dodds, J. The effect of phosphate, nitrogen and sucrose on the production of phenolics and solasodine in callus cultures of *Solanum laciniatum. Plant Cell Rep.* **1983**, *2*, 205–208. [CrossRef] [PubMed]
12. Sobuj, N.; Virjamo, V.; Zhang, Y.; Nybakken, L.; Julkunen, R. Impacts of elevated temperature and CO_2 concentration on growth and phenolics in the sexually dimorphic *Populus tremula* (L.). *Environ. Exp. Bot.* **2018**, *146*, 34–44. [CrossRef]
13. Ghasemzadeh, A.; Jaafar, H.; Rahmat, A.; Wahab, P.; Halim, M. Effect of different light intensities on total phenolics and flavonoids synthesis and anti-oxidant activities in young ginger varieties (*Zingiber officinale* Roscoe). *Int. J. Mol. Sci.* **2010**, *11*, 3885–3897. [CrossRef]
14. Tcherkez, G.; Carroll, A.; Abadie, C.; Mainguet, S.; Davanture, M.; Zivy, M. Protein synthesis increases with photosynthesis via the stimulation of translation initiation. *Plant Sci.* **2020**, *291*, 110352. [CrossRef]
15. Guo, S.; Zhou, Y.; Gao, Y.; Li, Y.; Shen, Q. New insights into the nitrogen form effect on photosynthesis and photorespiration. *Pedosphere* **2007**, *17*, 601–610. [CrossRef]
16. Wang, D.; Liu, H.; Li, S.; Zhai, G.; Shao, J.; Tao, Y. Characterization and molecular cloning of a serine hydroxymethyltransferase 1 (OsSHM1) in rice. *J. Integr. Plant Biol.* **2015**, *57*, 745–756. [CrossRef]
17. Kováčik, J.; Klejdus, B. Induction of phenolic metabolites and physiological changes in chamomile plants in relation to nitrogen nutrition. *Food Chem.* **2014**, *142*, 334–341. [CrossRef]
18. Qin, J.; Yue, X.; Ling, Y.; Zhou, Y.; Li, N.; Shang, X.; Fang, S. Nitrogen form and ratio impact phenolic accumulation and relative gene expression in *Cyclocarya paliurus. Trees* **2021**, *35*, 685–696. [CrossRef]

19. Lang, C.; Merkt, N.; Klaiber, I.; Pfannstiel, J.; Zörb, C. Different forms of nitrogen application affect metabolite patterns in grapevine leaves and the sensory of wine. *Plant Physiol. Biochem.* **2019**, *143*, 308–319. [CrossRef]
20. Zhu, Z.; Yu, M.; Chen, Y.; Guo, Q.; Zhang, L.; Shi, H.; Liu, L. Effects of ammonium to nitrate ratio on growth, nitrogen metabolism, photosynthetic efficiency and bioactive phytochemical production of *Prunella vulgaris*. *Pharm. Biol.* **2014**, *52*, 1518–1525. [CrossRef]
21. Prinsi, B.; Morgutti, S.; Negrini, N.; Faoro, F.; Espen, L. Insight into composition of bioactive phenolic compounds in leaves and flowers of green and purple basil. *Plants* **2019**, *9*, 22. [CrossRef] [PubMed]
22. Yang, Y.; Wang, F.; Wan, Q.; Ruan, J. Transcriptome analysis using RNA-Seq revealed the effects of nitrogen form on major secondary metabolite biosynthesis in tea (*Camellia sinensis*) plants. *Acta Physiol. Plant.* **2018**, *40*, 1–17. [CrossRef]
23. Domínguez, M.; Aparicio, P.; Lamsfus, C.; Cruz, C.; Martins Loução, M.; Moran, J. Nitrogen nutrition and antioxidant metabolism in ammonium-tolerant and -sensitive plants. *Physiol. Plant.* **2008**, *132*, 359–369. [CrossRef] [PubMed]
24. Wakasa, K.; Ishihara, A. Metabolic engineering of the tryptophan and phenylalanine biosynthetic pathways in rice. *Plant Biotech.* **2009**, *26*, 523–533. [CrossRef]
25. Li, D.; Zhang, X.; Qu, H. Delaying the biosynthesis of aromatic secondary metabolites in postharvest strawberry fruit exposed to elevated CO_2 atmosphere. *Food Chem.* **2020**, *306*, 125611. [CrossRef]
26. Palumbo, M.; Putz, F.; Talcott, S. Nitrogen fertilizer and gender effects on the secondary metabolism of yaupon, a caffeine-containing North American holly. *Oecologia* **2007**, *151*, 1–9. [CrossRef]
27. Zhong, C.; Jian, S.; Chen, D.; Huang, X.; Miao, H. Organic nitrogen sources promote andrographolide biosynthesis by reducing nitrogen metabolism and increasing carbon accumulation in *Andrographis paniculata*. *Plant Physiol. Bioch.* **2021**, *164*, 82–91. [CrossRef]
28. Cao, Y.; Pan, Y.; Wang, M.; Liu, T.; Meng, X.; Guo, S. The effects of different nitrogen forms on chlorophyll fluorescence and photosystem II in *Lonicera japonica*. *J. Plant Growth Regul.* **2023**, *42*, 4106–4117. [CrossRef]
29. Lu, Z.; Hu, W.; Ren, T.; Zhu, C.; Li, X.; Cong, R.; Lu, J. Impact of K deficiency on leaves and siliques photosynthesis via metabolomics in *Brassica napus*. *Environ. Exp. Bot.* **2019**, *158*, 89–98. [CrossRef]
30. Wang, P.; Wang, Z.; Pan, Q.; Sun, X.; Chen, H.; Chen, F.; Mi, G. Increased biomass accumulation in maize grown in mixed nitrogen supply is mediated by auxin synthesis. *J. Exp. Bot.* **2019**, *70*, 1859–1873. [CrossRef]
31. Turner, M.; Heuberger, A.; Kirkwood, J.; Collins, C.; Wolfrum, E.; Broeckling, C.; Jahn, C. Non-targeted metabolomics in diverse sorghum breeding lines indicates primary and secondary metabolite profiles are associated with plant biomass accumulation and photosynthesis. *Front. Plant Sci.* **2016**, *7*, 953. [CrossRef] [PubMed]
32. Xu, L.; Liu, H.; Wu, J.; Xu, C. Paclobutrazol improves leaf carbon-use efficiency by increasing mesophyll conductance rate, while abscisic acid antagonizes this increased rate. *Photosynthetica* **2020**, *58*, 762–768. [CrossRef]
33. Hanson, A.; Gage, D.; Shachar-Hill, Y. Plant one-carbon metabolism and its engineering. *Trends Plant Sci.* **2000**, *5*, 206–213. [CrossRef]
34. Rubio-Wilhelmi, M.; Sanchez-Rodriguez, E.; Leyva, R.; Blasco, B.; Romero, L.; Ruiz, M. Response of carbon and nitrogen-rich metabolites to nitrogen deficiency in PSARK::IPT tobacco plants. *Plant Physiol. Biochem.* **2012**, *57*, 231–237. [CrossRef]
35. Tegeder, M.; Masclaux-Daubresse, C. Source and sink mechanisms of nitrogen transport and use. *New Phytol.* **2018**, *217*, 35–53. [CrossRef]
36. Li, X.; Yan, X.; Liu, J. Effects of nitrogen forms on nitrogen metabolism-related enzymes and growth of *Phellodendron amurense* seedlings. *Chin. Bull. Bot.* **2006**, *23*, 255.
37. Foyer, C.; Parry, M.; Noctor, G. Markers and signals associated with nitrogen assimilation in higher plants. *J. Exp. Bot.* **2003**, *54*, 585–593. [CrossRef]
38. Poucet, T.; González-Moro, M.; Cabasson, C.; Beauvoit, B.; Gibon, Y.; Dieuaide-Noubhani, M.; Marino, D. Ammonium supply induces differential metabolic adaptive responses in tomato according to leaf phenological stage. *J. Exp. Bot.* **2021**, *72*, 3185–3199. [CrossRef]
39. Izui, K.; Matsumura, H.; Furumoto, T. Phosphoenolpyruvate carboxylase: A new era of structural biology. *Annu. Rev. Plant Biol.* **2004**, *55*, 69–84. [CrossRef]
40. Masumoto, C.; Gantt, E. Phosphoenolpyruvate carboxylase intrinsically located in the chloroplast of rice plays a crucial role in ammonium assimilation. *Proc. Natl. Acad. Sci. USA* **2010**, *107*, 5226–5231. [CrossRef]
41. Ruan, J.; Haerdter, R.; Gerendás, J. Impact of nitrogen supply on carbon/nitrogen allocation: A case study on amino acids and catechins in green tea [*Camellia sinensis* (L.) O. Kuntze] plants. *Plant Biol.* **2010**, *12*, 724–734. [CrossRef] [PubMed]
42. Krapp, A. Plant nitrogen assimilation and its regulation: A complex puzzle with missing pieces. *Curr. Opin. Plant Biol.* **2015**, *25*, 115–122. [CrossRef] [PubMed]
43. Liu, C.; Zhou, G.; Qin, H.; Guan, Y.; Wang, T.; Ni, W.; Ge, S. Metabolomics combined with physiology and transcriptomics reveal key metabolic pathway responses in apple plants exposure to different selenium concentrations. *J. Hazard. Mater.* **2024**, *464*, 132953. [CrossRef] [PubMed]

44. Moon, S.; Giglione, C.; Lee, D.; An, S.; Jeong, D.H.; Meinnel, T.; An, G. Rice peptide deformylase PDF1B is crucial for development of chloroplasts. *Plant Cell Physiol.* **2008**, *49*, 1536–1546. [CrossRef]
45. Balasundram, N.; Sundram, K.; Samman, S. Phenolic compounds in plants and agri-industrial by-products: Antioxidant activity, occurrence, and potential uses. *Food Chem.* **2006**, *99*, 191–203. [CrossRef]
46. Kong, D.; Li, Y.; Wang, M.; Bai, M.; Zou, R.; Tang, H.; Wu, H. Effects of light intensity on leaf photosynthetic characteristics, chloroplast structure, and alkaloid content of *Mahonia bodinieri* (Gagnep.) Laferr. *Acta Physiol. Plant.* **2016**, *38*, 120. [CrossRef]
47. Matt, P.; Krapp, A.; Haake, V.; Mock, H.; Stitt, M. Decreased Rubisco activity leads to dramatic changes of nitrate metabolism, amino acid metabolism and the levels of phenylpropanoids and nicotine in tobacco antisense RBCS transformants. *Plant J.* **2002**, *30*, 663–677. [CrossRef]
48. Montanari, M.; Degl Innocenti, E.; Maggini, R.; Pacifici, S.; Pardossi, A.; Guidi, L. Effect of nitrate fertilization and saline stress on the contents of active constituents of *Echinacea angustifolia* DC. *Food Chem.* **2008**, *107*, 1461–1466. [CrossRef]
49. Fritz, C.; Palacios-Rojas, N.; Feil, R.; Stitt, M. Regulation of secondary metabolism by the carbon-nitrogen status in tobacco: Nitrate inhibits large sectors of phenylpropanoid metabolism. *Plant J.* **2006**, *46*, 533–548. [CrossRef]
50. Moura, J.; Bonine, C.; Fernandes Viana, J.; Dornelas, M.; Mazzafera, P. Abiotic and biotic stresses and changes in the lignin content and composition in plants. *J. Integr. Plant Biol.* **2010**, *52*, 360–376. [CrossRef]
51. Liu, Q.; Zheng, L.; He, F.; Zhao, F.; Shen, Z.; Zheng, L. Transcriptional and physiological analyses identify a regulatory role for hydrogen peroxide in the lignin biosynthesis of copper-stressed rice roots. *Plant Soil* **2015**, *387*, 323–336. [CrossRef]
52. Panche, A.; Diwan, A.; Chandra, S. Flavonoids: An overview. *J. Nutr. Sci.* **2016**, *5*, e47. [CrossRef] [PubMed]
53. Duan, Y.; Yang, H.; Wei, Z.; Yang, H.; Fan, S.; Li, W. Effects of Different Nitrogen Forms on Blackberry Fruit Quality. *Foods* **2023**, *12*, 2318. [CrossRef] [PubMed]
54. Fallovo, C.; Schreiner, M.; Schwarz, D.; Colla, G.; Krumbein, A. Phytochemical changes induced by different nitrogen supply forms and radiation levels in two leafy Brassica species. *J. Agric. Food Chem.* **2011**, *59*, 4198–4207. [CrossRef]
55. Chen, Y.; Bai, Y.; Zhang, Z.; Zhang, Y.; Jiang, Y.; Sun, Z. Transcriptomics and metabolomics reveal the primary and secondary metabolism changes in *Glycyrrhiza uralensis* with different forms of nitrogen utilization. *Front. Plant Sci.* **2023**, *14*, 1229253. [CrossRef]
56. Tang, D.; Liu, M.; Zhang, Q.; Ma, L.; Shi, Y.; Ruan, J. Preferential assimilation of NH_4^+ over NO_3^- in tea plant associated with genes involved in nitrogen transportation, utilization and catechins biosynthesis. *Plant Sci.* **2020**, *291*, 110369. [CrossRef]
57. Wang, Y.; Wang, Y.; Lu, Y.; Qiu, Q.; Fan, D.; Wang, X.; Zheng, X. Influence of different nitrogen sources on carbon and nitrogen metabolism and gene expression in tea plants (*Camellia sinensis* L.). *Plant Physiol. Biochem.* **2021**, *167*, 561–566. [CrossRef]
58. Han, J.; Gu, L.; Warren, J.M.; Guha, A.; Mclennan, D.; Zhang, W.; Zhang, Y. The roles of photochemical and non-photochemical quenching in regulating photosynthesis depend on the phases of fluctuating light conditions. *Tree Physiol.* **2022**, *42*, 848–861. [CrossRef]
59. Sun, Y.; Li, Y.; Wang, B.; Li, Y.; Ding, L.; Wang, M.; Guo, S. Leaf nitrate accumulation influences the photorespiration of rice (*Oryza sativa* L.) seedlings. *Plant Soil.* **2020**, *456*, 323–338. [CrossRef]

International Journal of
Molecular Sciences

MDPI

Article

The Role of Phosphorus Fertilization in Antioxidant Responses of Drought-Stressed Common Beech and Sessile Oak Provenances

Antonia Vukmirović [1], Željko Škvorc [1], Saša Bogdan [1], Daniel Krstonošić [1], Ida Katičić Bogdan [1], Tomislav Karažija [2], Marko Bačurin [1], Magdalena Brener [1] and Krunoslav Sever [1,*]

[1] Faculty of Forestry and Wood Technology, University of Zagreb, Svetošimunska Cesta 23, HR-10000 Zagreb, Croatia; avukmiro@sumfak.unizg.hr (A.V.); zskvorc@sumfak.unizg.hr (Ž.Š.); sbogdan@sumfak.unizg.hr (S.B.); dkrstonosic@sumfak.unizg.hr (D.K.); ikaticic@sumfak.unizg.hr (I.K.B.); mbacurin@sumfak.unizg.hr (M.B.); mbrener@sumfak.unizg.hr (M.B.)

[2] Faculty of Agriculture, University of Zagreb, Svetošimunska Cesta 25, HR-10000 Zagreb, Croatia; tkarazija@agr.hr

* Correspondence: ksever@sumfak.unizg.hr; Tel.: +385-1-235-2521

Abstract: During drought, a major abiotic stressor for European forests, excessive reactive oxygen species (ROS) are produced, causing oxidative damage that affects structural and metabolic tree functions. This research examines the effects of drought, phosphorus (P) fertilization, and provenance on photosynthetic pigments, malondialdehyde (MDA) concentrations, and antioxidant enzyme activities in common beech (*Fagus sylvatica* L.) and sessile oak (*Quercus petraea* (Matt.) Liebl.) saplings from two provenances. In a common garden experiment, four treatments were applied: regular watering with (+PW) and without P fertilization (−PW), and drought with (+PD) and without (−PD) P fertilization. Results showed that drought increased both MDA concentrations and antioxidant enzyme activity, particularly superoxide dismutase (SOD), catalase (CAT), peroxidase (POD), and ascorbate peroxidase (APX), which are responsible for ROS scavenging. Additionally, chlorophyll *a* + *b* concentrations were lower in drought-exposed plants. Phosphorus fertilization minimally affected MDA levels but enhanced antioxidant responses, particularly APX and CAT activities in oak during drought. Provenance differences were notable, with oak and beech from the drier provenance showing better adaptation, reflected in lower MDA levels and higher enzyme activities. This study underscores the importance of antioxidant defenses in coping with drought stress, with phosphorus fertilization and provenance shaping the species' adaptive capacity.

Keywords: *Fagus sylvatica*; *Quercus petraea*; drought; phosphorus fertilization; antioxidant enzyme activity; provenance-specific antioxidant response

1. Introduction

Drought, a stressful abiotic factor that negatively impacts various traits in forest trees, including physiological, morphological, and biochemical processes, has become a common subject of research in recent years [1–8]. During drought, photosynthesis is limited: one of the first defense mechanisms is stomata closure to prevent excessive water loss through transpiration. However, this also inhibits CO_2 uptake, leading to an accumulation of excess energy as plants consume less light energy for photosynthetic carbon fixation [9]. This excess energy is redirected, leading to an increased flow of electrons to oxygen (O_2), which results in cellular damage due to the overproduction of reactive oxygen species (ROS) [8].

These free radicals, such as hydrogen peroxide (H_2O_2), superoxide radical ($O2^{\bullet-}$), singlet oxygen ($^1O_2$), and hydroxyl radicals ($OH^{\bullet}$), are highly reactive and can damage cellular structures, impair normal functions, and disturb the redox balance [10]. They are highly damaging to all organelles, including chloroplasts, where photosynthesis takes place and one of the typical symptoms of oxidative stress is a reduction in photosynthetic pigment content [5,11]. Excessive ROS also damage the cell membrane, which is vital for sensing and responding to structural and metabolic disturbances [12]. The accumulation of ROS triggers lipid peroxidation, leading to oxidative damage to the membrane. The degree of lipid peroxidation is commonly assessed by measuring the concentration of malondialdehyde (MDA), a by-product of this process [11].

When oxidative stress occurs, plants employ various mechanisms to protect themselves and minimize damage. One such mechanism is the activation of ROS-scavenging antioxidant enzymes. Their activity is crucial not only during acute drought stress but also plays a key role in recovery [13]. Superoxide dismutase (SOD) is often regarded as a "first line of defense" against oxidative stress, as it scavenges hydrogen peroxide (H_2O_2) and superoxide ions ($O2^{\bullet-}$) to maintain redox balance [10,11]. In addition to SOD, several other antioxidant enzymes, such as catalase (CAT), peroxidase (POD), and ascorbate peroxidase (APX), also help defend against oxidative stress caused by drought. Catalase plays a crucial role in preventing cellular oxidative damage by efficiently breaking down H_2O_2 into water and oxygen [14]. Similarly, APX reduces H_2O_2 to water by using ascorbate (ASC) as an electron donor in various subcellular compartments [15]. Peroxidases (POD) are also commonly found in plants, where they help protect cells from oxidative stress and damage caused by H_2O_2 [16]. The activity of these antioxidant enzymes depends on the plant species, as well as the intensity and duration of the stress. Some antioxidants act as effective ROS scavengers during moderate stress, while their activity decreases under high stress intensity, and for others, the opposite is true. For example, the activation of CAT and APX is stronger in drought-tolerant species compared to sensitive ones [13]. A high antioxidant capacity to neutralize toxic ROS has been associated with increased plant tolerance to environmental stresses [17].

Previous studies have shown that fertilization can reduce the negative effects of drought on overall plant growth and development, partly by positively influencing antioxidant enzyme activities [11]. In our study, we focused on phosphorus (P) fertilization, a macronutrient that regulates various metabolic processes in plants, which in turn affects their physiological functions and growth [18]. Research on the impact of P fertilization on the antioxidant capacity of plants, particularly forest trees, is relatively limited. Phosphorus addition may help regulate photo-assimilates and maintain membrane integrity, thereby reducing the negative impact of drought stress [19]. Furthermore, Iqbal et al. [8] emphasized that P fertilizer could improve plant tolerance to stress through positive adjustments in morphological, physiological, and biochemical processes, while Ullah et al. [20] showed a positive effect of P through the reduction in MDA concentration, i.e., the rate of lipid peroxidation. Tariq et al. [19,21] studied the impact of P fertilization on the antioxidant system of three species of forest trees, concluding that P positively influences antioxidant capacity and reduces ROS production during drought stress.

Common beech (*Fagus sylvatica* L.) is the most widespread tree species in Central Europe, often growing alongside sessile oak (*Quercus petraea* (Matt.) Liebl.) [22]. Numerous studies have shown that beech is more sensitive to drought than oak [1,23,24]. Additionally, it has been established that different provenances of various species exhibit varying phenotypic plasticity in their response to drought [2,25]. Therefore, it is plausible that provenances also differ in their ability to mitigate oxidative stress caused by drought. Previous studies on sessile oak and pedunculate oak (*Quercus robur* L.) have shown that offspring from humid old-growth stands accumulate more antioxidants under favorable conditions but deplete them during drought, whereas saplings from arid old-growth stands maintain their antioxidant capacity even under harsher environmental conditions [26]. Similarly, in the Asian oak species *Quercus brantii* (Lindl.), drought stress increased the activity of antioxidant enzymes like POD, SOD, CAT, and APX more in plants from arid provenances compared to those from wetter ones, while the concentration of malondialdehyde (MDA) rose more in plants from wetter provenances [27]. In another Asian tree species, plants from arid provenances exhibited higher SOD and POD activity under moderate to severe drought stress compared to those from wetter regions, suggesting enhanced drought tolerance through increased antioxidant enzyme activity [28]. Our study was conducted on saplings from two provenances. To the best of our knowledge, no published research has examined whether phosphorus (P) fertilization can enhance drought tolerance in common beech and sessile oak by affecting antioxidant enzyme activity, or whether there are differences between provenances in this response. Therefore, the aims of this research were as follows:

1. To investigate the impact of drought, phosphorus fertilization, and provenance on the antioxidative capacity of common beech and sessile oak;
2. To assess the impact of phosphorus fertilization under drought conditions on the antioxidative capacity in common beech and sessile oak;
3. To investigate the differences in antioxidative capacity in common beech and sessile oak provenances.

2. Results

2.1. Water Availability in Soil and Leaf Water Potential

During the growing season, the volumetric water content (VWC) in the regularly watered treatments ranged from 18.3% to 40.6%. In contrast, the drought treatments experienced a continuous decline in VWC, reaching low levels of 8.0% and 10.1% by the end of August, marking the peak of the drought (Figure 1). Regular watering had a significant positive impact on the pre-dawn leaf water potential (Ψ_{PD}) of both common beech and sessile oak. In the regularly watered treatments, Ψ_{PD} remained above −0.4 MPa throughout the season. However, in the drought treatments, Ψ_{PD} decreased steadily, hitting its lowest value on 29 August, with beech at −3.1 MPa and oak at −3.3 MPa (Figure 1).

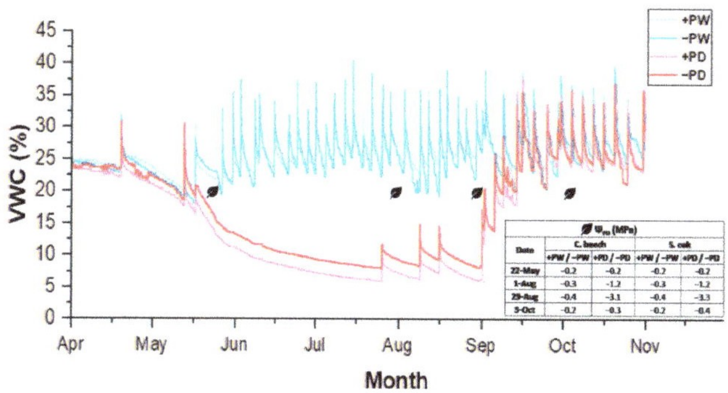

Figure 1. Patterns of substrate volumetric water content (VWC) in four treatments: regularly watered with phosphorus (+PW), regularly watered without phosphorus (−PW), drought with phosphorus (+PD), and drought without phosphorus (−PD), with mean pre-dawn leaf water potential (Ψ_{PD}) in regularly watered (+PW and −PW) and drought-treated (+PD and −PD) common beech (C. beech) and sessile oak (S. oak) saplings, measured on 22 May, 1 and 29 August as well as on 3 October.

2.2. Leaf Phosphorus Concentrations

Phosphorus fertilization significantly impacted leaf phosphorus concentrations (P_{leaves}) in both beech and oak (Figure 2a,b). For beech fertilized with P, P_{leaves} ranged from 1.79 to 1.97 mg P g^{-1} DW, reflecting a nutritional status between upper normal and surplus (Figure 2a). In contrast, oak had P_{leaves} ranging from 1.92 to 1.99 mg P g^{-1} DW, indicating upper normal nutrition (Figure 2b). In non-fertilized beech, P_{leaves} ranged from 1.49 to 1.69 mg P g^{-1} DW, suggesting a mid-normal nutritional status. Meanwhile, non-fertilized oak had P_{leaves} between 1.41 and 1.55 mg P g^{-1} DW, indicating a lower normal nutrition level.

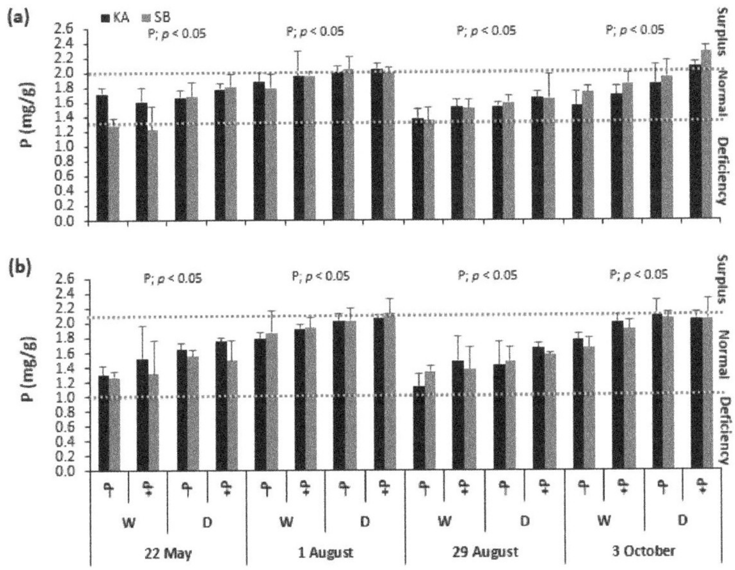

Figure 2. Mean values ±SE of phosphorus (P) concentrations in the leaves in response to drought (D) and regularly watered (W) treatments, as well as to phosphorus fertilization (+P) and non-fertilization (−P) treatments, for common beech (**a**) and sessile oak (**b**) from the Karlovac (KA) and Slavonski Brod (SB) provenances, measured on 22 May, 1 and 29 August, and 3 October. The figures show a significant effect of phosphorus fertilization (P) on P concentrations in the leaves, as determined by factorial ANOVA ($p < 0.05$). Red horizontal dashed lines represent the critical phosphorus concentrations in the leaves of common beech and sessile oak saplings [29].

2.3. Photosynthetic Pigments

Chlorophyll $a + b$ (Chl $a + b$) concentrations in beech leaves peaked on 22 May and declined through the growing season (Figure 3a). Drought significantly reduced Chl $a + b$ on 1 August and 3 October, with lower concentrations in drought (+PD and −PD) compared to regularly watered (+PW and −PW) treatments (Figure 3a). On 1 August, a drought ×P fertilization interaction showed significantly lower Chl $a + b$ in the −PD treatment, which was not observed later in the season. On 3 October, the same interaction revealed lower Chl $a + b$ in P-fertilized treatments (+PW and +PD) compared to non-fertilized treatments (−PW and −PD) (Figure 3a).

Carotenoid (Car) concentrations in beech leaves were highest on 22 May and decreased towards the season's end. However, Car concentrations were higher on 29 August, during the peak of drought, compared to 1 August and 3 October (Figure 3b). Drought significantly affected Car concentrations at the end of the season, with regularly watered treatments (+PW and −PW) having higher Car than drought treatments (+PD and −PD) (Figure 3b).

A significant drought × P fertilization interaction also showed higher Car concentrations in +PW compared to +PD treatments (Figure 3b).

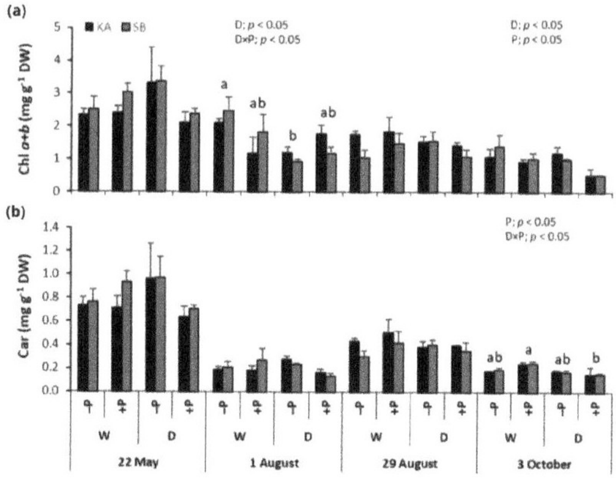

Figure 3. Mean values ±SE of (**a**) chlorophyll $a + b$ (Chl $a + b$) and (**b**) carotenoid (Car) concentrations in response to drought (D) and regularly watered (W) treatments, as well as to phosphorus fertilization (+P) and non-fertilization (−P) treatments, for common beech from the Karlovac (KA) and Slavonski Brod (SB) provenances, measured on 22 May, 1 and 29 August, and 3 October. The figures show significant effects of drought (D), phosphorus fertilization (P), provenance (Pr), and the drought × phosphorus fertilization interaction (D × P) on Chl $a + b$ and Car, as determined by factorial ANOVA ($p < 0.05$). Lowercase letters indicate significant differences among treatments.

The drought had a significant impact on Chl $a + b$ concentrations in oak, particularly on 1 and 29 August when the regularly watered treatments (+PW and −PW) showed higher concentrations compared to the drought treatments (+PD and −PD) (Figure 4a). On 1 August, a significant drought × P fertilization interaction revealed that the −PW treatment had the highest Chl $a + b$ concentrations compared to the other treatments. On 3 October, at the end of the growing season, oaks in the non-fertilized P treatments (−PW and −PD) had significantly higher Chl $a + b$ concentrations than those in the P-fertilized treatments (+PW and +PD) (Figure 4a).

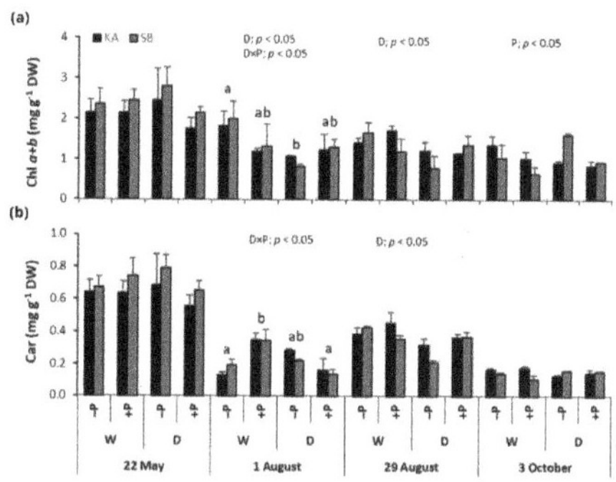

Figure 4. Mean values ±SE of (**a**) chlorophyll $a + b$ (Chl $a + b$) and (**b**) carotenoid (Car) concentrations in response to drought (D) and regularly watered (W) treatments, as well as to phosphorus fertilization (+P) and non-fertilization (−P) treatments, for sessile oak from the Karlovac (KA) and Slavonski Brod (SB) provenances, measured on 22 May, 1 and 29 August, and 3 October. The figures show significant effects of drought (D), phosphorus fertilization (P), provenance (Pr), and the drought × phosphorus fertilization interaction (D × P) on Chl $a + b$ and Car, as determined by factorial ANOVA ($p < 0.05$). Lowercase letters indicate significant differences among treatments.

Similar to beech, Car concentrations in oak leaves were highest on 22 May and decreased toward the end of the season. Additionally, Car concentrations were higher on 29 August, during the peak of the drought, compared to both 1 August and 3 October (Figure 4b). On 1 August, a significant drought × P fertilization interaction showed the highest Car concentrations in the +PW treatment. Drought also significantly affected Car concentrations on 29 August, when the regularly watered treatments (+PW and −PW) had higher concentrations compared to the drought treatments (+PD and −PD) (Figure 4b).

2.4. Malondialdehyde Concentrations

In beech, MDA concentrations generally increased from the beginning to the end of the growing season. At the peak of the drought period on 29 August, drought significantly increased MDA concentrations, with drought treatments (+PD and −PD) showing significantly higher levels compared to regularly watered (+PW and −PW) treatments (Figure 5a). Additionally, on 29 August, significant drought × P fertilization interaction revealed the lowest MDA concentration in −PW treatment compared to other treatments. By the end of the growing season, on 3 October, P fertilization led to higher MDA concentrations in P fertilized (+PW and +PD) compared to P non-fertilized treatments (−PW and −PD) (Figure 5a).

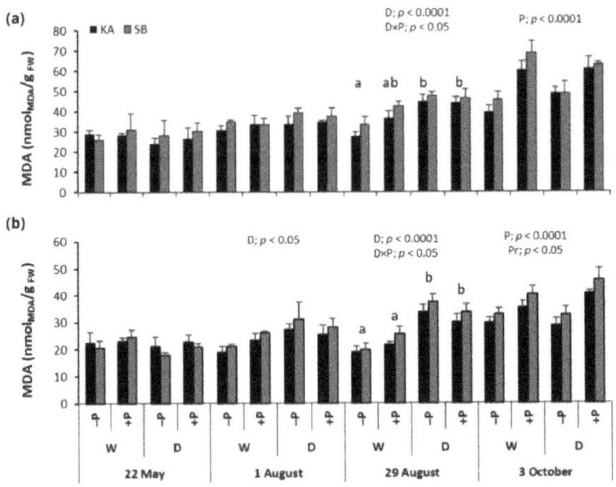

Figure 5. Mean values ±SE of malondialdehyde (MDA) concentrations in response to drought (D) and regularly watered (W) treatments, as well as to phosphorus fertilization (+P) and non-fertilization (−P) treatments, for common beech (**a**) and sessile oak (**b**) from the Karlovac (KA) and Slavonski Brod (SB) provenances, measured on 22 May, 1 and 29 August, and 3 October. The figures show significant effects of drought (D), phosphorus fertilization (P), provenance (Pr), and the drought × phosphorus fertilization interaction (D × P) on MDA, as determined by factorial ANOVA ($p < 0.05$). Lowercase letters indicate significant differences among treatments.

Malondialdehyde concentrations in oak followed the same pattern throughout the vegetation period as those in beech. However, drought significantly raised MDA concentrations on both 1 August and 29 August, with drought-exposed treatments (+PD and −PD) showing significantly higher MDA levels compared to regularly watered treatments (+PW and −PW) (Figure 5b). At the peak of the drought on 29 August, a significant interaction between drought and phosphorus fertilization was observed, with the −PW treatment exhibiting the lowest MDA concentrations. By the end of the growing season on 3 October, P fertilization led to higher MDA concentrations in P-fertilized treatments (+PW and +PD) compared to P non-fertilized treatments (−PW and −PD) (Figure 5b).

Additionally, on 3 October, saplings from the SB provenance showed significantly higher MDA concentrations than those from the KA provenance.

2.5. Antioxidant Enzyme Activities in Common Beech

During the drought period, SOD activity was significantly higher in drought treatments (+PD and −PD) compared to regularly watered treatments (+PW and −PW). After the drought was released on 29 August, SOD activity decreased, with no significant difference between the drought and regularly watered treatments. Throughout the growing season, SOD activity was significantly higher in non-P-fertilized treatments (−PD and −PW) compared to P-fertilized treatments (+PD and +PW). A significant interaction between drought and P fertilization at the peak of the drought on 29 August revealed significantly higher SOD activity in the −PD treatment compared to the other treatments. SOD activity was similar across both provenances throughout the growing season (Figure 6a).

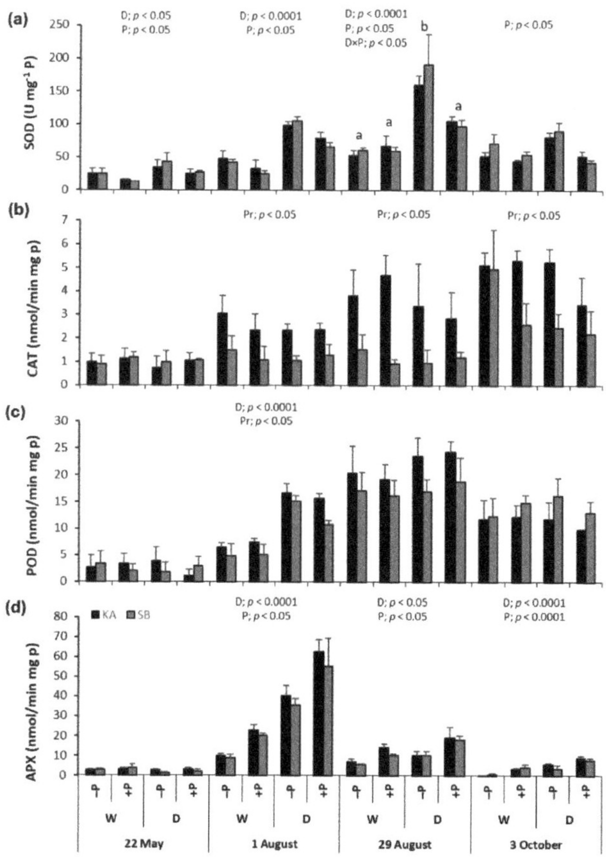

Figure 6. The mean values ± SE of (**a**) superoxide dismutase (SOD), (**b**) catalase (CAT), (**c**) peroxidase (POD), and (**d**) ascorbate peroxidase (APX) activity in response to drought (D) and regularly watered (W) treatments, as well as to phosphorus fertilization (+P) and non-fertilization (−P) treatments, for common beech from the Karlovac (KA) and Slavonski Brod (SB) provenances, measured on 22 May, 1 and 29 August, and 3 October. The figures show significant effects of drought (D), phosphorus fertilization (P), provenance (Pr), and the drought × phosphorus fertilization interaction (D × P) on SOD, CAT, POD, or APX activities, as determined by factorial ANOVA ($p < 0.05$). Lowercase letters indicate significant differences among treatments.

Drought and P fertilization did not significantly affect CAT activity during the growing season (Figure 6b). However, CAT activity was higher in the KA provenance compared to the SB provenance for almost the entire growing season, except at the very beginning (22 May). The highest CAT activities across all treatments and provenances were recorded on 3 October, at the end of the growing season (Figure 6b).

In all treatments and both provenances, POD activity increased until 29 August, the peak of the drought, after which it declined. On 1 August, POD activity was significantly higher in the drought treatments (+PD and −PD) compared to the regularly watered treatments (+PW and −PW) and in the KA provenance compared to the SB provenance (Figure 6c). Similar to CAT, P fertilization did not affect POD activity throughout the growing season.

APX activity increased sharply until 1 August, after which it declined sharply. Throughout almost the entire growing season, except at the very beginning (22 May), APX activity was significantly higher in drought treatments (+PD and −PD) compared to regularly watered treatments (+PW and −PW), and in P-fertilized treatments (+PW and +PD) compared to non-P-fertilized treatments (−PW and −PD). There were no significant differences between the provenances (Figure 6d).

2.6. Antioxidant Enzyme Activities in Sessile Oak

During the drought period (1 August and 29 August), SOD activity in the drought treatments (+PD and −PD) was significantly higher than in the regularly watered treatments (+PW and −PW) (Figure 7a). SOD activity was consistently lower in the P-fertilized treatments (+PD and +PW), compared to the non-P-fertilized treatments (−PW and −PD) throughout most of the season, except at the peak of the drought (29 August). On 1 August, a significant interaction between drought and P fertilization revealed significantly higher SOD activity in the −PD treatment compared to the other treatments. In all treatments, the highest SOD activity was recorded at the peak of the drought (29 August). Provenances showed significant differences only at the end of the season (3 October) when oaks from the KA provenance exhibited lower SOD activity compared to saplings from the SB provenance.

Generally, the highest CAT activity was recorded at the end of the growing season (3 October) (Figure 7b). On 29 August, at the peak of the drought, CAT activity was significantly higher in P-fertilized treatments (+PD and +PW) than in non-P-fertilized treatments (−PW and −PD) (Figure 7b). Provenances also differed significantly at this time, with saplings from the KA provenance showing higher CAT activity than those from the SB provenance.

POD activity was affected solely by drought, and only at the peak of the drought (29 August), when POD activity was significantly higher in the drought treatments (+PD and −PD) than in the regularly watered treatments (+PW and −PW) (Figure 7c). In all treatments, the highest POD activity occurred at the peak of the drought (29 August).

On 1 and 29 August, APX activity was higher in the drought treatments (+PD and −PD) compared to the regularly watered treatments (+PW and −PW) (Figure 7d). Throughout August and at the end of the season in October, APX activity was higher in the P-fertilized treatments (+PD and +PW) compared to non-P-fertilized treatments (−PW and −PD). A significant interaction between drought and P fertilization on 1 August and 29 August revealed significantly higher APX activity in the +PD treatment compared to the other treatments.

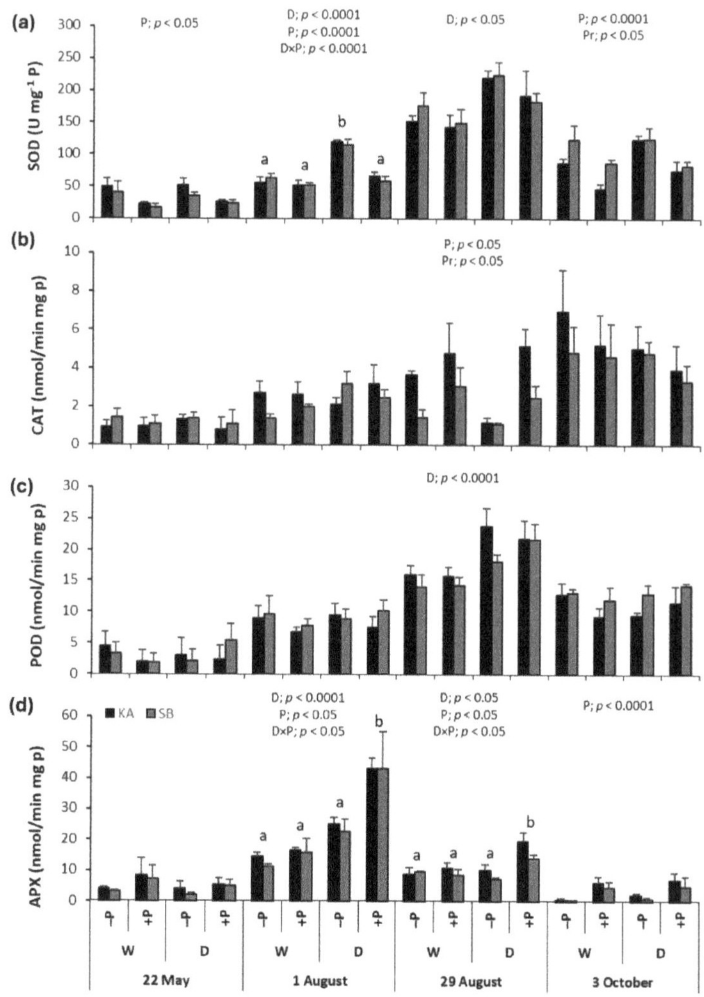

Figure 7. The mean values ± SE of (**a**) superoxide dismutase (SOD), (**b**) catalase (CAT), (**c**) peroxidase (POD), and (**d**) ascorbate peroxidase (APX) activity in response to drought (D) and regularly watered (W) treatments, as well as to phosphorus fertilization (+P) and non-fertilization (−P) treatments, for sessile oak from the Karlovac (KA) and Slavonski Brod (SB) provenances, measured on 22 May, 1 and 29 August, and 3 October. The figures show significant effects of drought (D), phosphorus fertilization (P), provenance (Pr), and the drought × phosphorus fertilization interaction (D × P) on SOD, CAT, POD, or APX activities, as determined by factorial ANOVA ($p < 0.05$). Lowercase letters indicate significant differences among treatments.

3. Discussion

3.1. Effect of Drought on Malondialdehyde Concentrations, Photosynthetic Pigments and Antioxidant Enzyme Activities

In both species, plants subjected to drought exhibited significantly higher malondialdehyde (MDA) concentrations at the peak of drought compared to regularly watered ones. This response is typical for plants under drought stress, indicating elevated rates of lipid peroxidation, as previously documented in both crops and forest trees [17,21,30,31]. These findings suggest that drought stress leads to increased oxidative stress and hydrogen peroxide (H_2O_2) accumulation, resulting in damage to cell membranes—often the primary targets of various biotic and abiotic stresses [12]. Consequently, this damage compromises plasma membrane integrity, causing significant harm to the cell membranes [10,11], thereby exacerbating the negative impacts of drought on the plants.

This oxidative damage is reflected in the significant reduction in photosynthetic pigment concentrations observed under drought stress. Chlorophyll *a* + *b* concentrations were notably lower in drought-exposed plants compared to regularly watered ones, although this was somewhat more pronounced in oak. Additionally, in beech, carotenoid concentrations were higher in regularly watered plants than in drought-exposed ones, a trend also observed in oak during the peak drought period in August. The reduction in photosynthetic pigments, including chlorophylls and carotenoids, impairs photosynthesis and severely limits primary production, as documented in several studies [5,10,19–21,32]. Such declines are typically linked to pigment photo-oxidation and chlorophyll degradation, with increasing reactive oxygen species (ROS) production during drought exacerbating these processes and impairing chlorophyll biosynthesis [5,33].

During drought conditions, the increased production of reactive oxygen species (ROS) leads to oxidative stress, making it essential for plants to have effective antioxidant defense mechanisms to mitigate the negative effects of drought. In this study, we focused on antioxidant enzymes responsible for scavenging excess ROS. Superoxide dismutase (SOD) activity significantly increased in both beech and oak under drought stress compared to regularly watered plants, a result consistent with previous studies on various crop and tree species [19–21,30,32–34]. While Chatterjee et al. [33] showed that SOD activity increases with light intensity but not with increasing drought in rice, our results indicate that both beech and oak exhibited the highest SOD activity precisely during peak drought stress. This suggests effective regulation of the oxidative stress defense mechanism, as SOD activity was maintained even under severe drought conditions.

Interestingly, catalase (CAT) activity did not show a significant change in response to drought in either species, which contrasts with findings from other studies where CAT activity was found to increase under drought stress in various crops and trees [10,20,21,30]. However, research on two oak species found that CAT activity significantly decreased during drought stress [35], while moderate water stress induced CAT activity in rice, and higher levels of water stress led to a decrease in activity [33]. Furthermore, in *Arabidopsis thaliana*, some mutants exhibited reduced CAT activity due to excessive lipid peroxidation and peroxisome destruction during drought stress [34]. In our study, the significant increase in MDA levels in drought-exposed plants suggests that severe lipid peroxidation may have caused too much damage, thereby limiting CAT's role in mitigating oxidative stress.

Peroxidase (POD) activity, however, did respond to drought stress in both species. Higher POD levels were observed in drought-treated saplings compared to regularly watered ones in beech at the beginning of August and in oak at the peak of drought at the end of August. Similar findings have been reported in other studies [19,20,30,32], underscoring the important role of POD as an antioxidant defense mechanism in both beech and oak. However, in beech, no significant difference was found between drought-treated and regularly watered plants during the peak of drought stress, suggesting that POD's effectiveness diminished under extreme drought conditions. Similar results have been reported in peanuts [36]. These results indicate that some antioxidant enzymes, like POD, may be more effective under moderate drought conditions but less so under severe stress, highlighting the variability of antioxidant responses depending on plant species and stress intensity.

As for ascorbate peroxidase (APX), we found that drought treatment throughout most of the growing season significantly increased APX activity in beech compared to regularly watered plants. In oak, higher APX activity was observed during moderate and peak drought periods in August. These results are consistent with findings from Chatterjee et al. in rice [33]. The highest APX activity occurred in early August, during moderate drought, when APX was most effective in defending against oxidative damage. However, by the peak of drought stress, APX activity decreased significantly in both species, suggesting that APX becomes less effective as an antioxidant under extreme drought conditions. This is in contrast to findings in peanuts, where APX activity remained high even under extreme drought stress [36], but mirrors the patterns observed for CAT and aligns with our hypothesis that severe lipid peroxidation during peak drought may inhibit APX's protective role.

3.2. *Effect of Phosphorus Fertilization and Drought × Phosphorus Fertilization Interaction on Malondialdehyde Concentrations, Photosynthetic Pigments and Antioxidant Enzyme Activities*

It is hypothesized that fertilization with nitrogen (N) and phosphorus (P) could help plants cope with drought stress by regulating membrane integrity [37], thereby reducing lipid peroxidation rates and MDA concentrations. Previous research has shown that P fertilization positively affects MDA levels during drought, with lower MDA rates observed in P-fertilized plants compared to non-fertilized ones [19,21,30]. However, in our study, P fertilization did not significantly reduce MDA levels in either species. While oak plants from the −PD treatment had higher MDA concentrations at the peak of drought than those from the +PD treatment, the difference was not statistically significant. Additionally, at the end of the season, both species displayed significantly higher MDA concentrations in P-fertilized plants compared to non-fertilized ones. This was observed in early October, and we conclude that P fertilization did not benefit the plants; rather, it exacerbated membrane damage and accelerated leaf senescence as the growing season ended. This is also evident from the lower chlorophyll concentrations in both species in October in the P-fertilized treatments.

Our study could not conclude that P fertilization improved chlorophyll $a + b$ concentrations in either beech or oak. A study on the tree species *Alnus cremastogyne* found no significant differences in photosynthetic pigment concentrations due to P application [19]. However, in *Eucalyptus grandis*, P application significantly increased chlorophyll concentrations in drought-exposed plants [21]. In our study, plants fertilized with P (+PD treatment) exhibited higher chlorophyll $a + b$ levels compared to non-fertilized plants (−PD treatment) during periods of advanced drought (1 August, as shown in Figures 3a and 4a), but this difference was not statistically significant. By the end of the season, chlorophyll $a + b$ concentrations were lower in the P-fertilized plants than in the non-fertilized ones. It is possible that the severe drought conditions in our study masked any potential positive effects of P on drought tolerance in these species regarding photosynthetic pigment concentrations.

Interestingly, P fertilization did not enhance SOD activity in either species; in fact, it had the opposite effect. In August, the −PD treatment showed the highest SOD activity (at the beginning of August for oak and at the end for beech), while the +PD treatment did not differ from the regularly watered treatments. Thus, P fertilization seemed to decrease SOD activity, negatively impacting the antioxidant defense during drought. This finding contrasts with previous studies that reported positive effects of P on antioxidant metabolism in various tree species [19,21]. We believe our results arise from the fact that, although our saplings were not in P surplus, they were near that threshold (Figure 2). Takagi et al. [38] noted that high inorganic P accumulation can cause an over-reduction in the photosynthetic electron transport chain, leading to additional ROS production already excessive due to drought. Under such conditions, homeostasis may be disrupted, and the antioxidant system becomes ineffective, which is reflected in the reduced SOD activity observed in our research.

Fertilization did not affect CAT activity in beech, but in oak, P-fertilized plants in both drought and regularly watered treatments showed significantly higher CAT activity than non-fertilized ones during the drought peak. Previous studies have reported mixed results regarding P fertilization: in some cases, it decreased CAT activity in drought-stressed plants compared to non-fertilized ones [20,21], while in others, it increased CAT activity [19]. In our case, P fertilization was beneficial for drought-stressed oak, as it stimulated greater CAT activity, enhancing the plants' defense against oxidative damage from drought. As for POD, P fertilization did not affect its activity in either species. The results of previous studies are quite varied: some studies found no effect of P on POD activity during drought, as in our case [30], while others reported increased POD activity in drought-stressed plants due to P fertilization [19], and some found that P reduced POD activity during drought [21]. Therefore, based on our findings and previous studies, we cannot conclude whether P fertilization improves plant responses to oxidative stress in terms of POD activity.

Phosphorus fertilization did, however, significantly enhance APX activity in both species throughout most of the growing season, except for the early stages, as evidenced by the higher APX activity in plants from fertilized treatments compared to those from non-fertilized treatments. In oak, the positive effect of P on APX activity during drought stress was most pronounced in August, when the +PD treatment showed the highest APX activity, significantly differing from all other treatments. Laxa et al. highlighted that APX activation is stronger in drought-tolerant species compared to less tolerant ones [13], a finding that aligns with our results, as oak is generally regarded as a drought-tolerant species [6].

3.3. *Effect of Provenance on the Malondialdehyde Concentrations, Photosynthetic Pigments, and Antioxidant Enzyme Activities*

Previous research has shown that provenances vary in response to drought in the context of adaptive antioxidant defense mechanisms. For example, in one oak species, MDA levels increased more in plants from wetter provenances than those from drier provenances [27]. In our study, oaks from the KA provenance exhibited significantly lower MDA levels than those from the SB provenance at the end of the growing season. This may indicate better adaptation or recovery from drought stress, consistent with our previous findings that oak from the KA provenance demonstrated better photosynthetic performance, as indicated by higher net photosynthesis rates (A_{net}) and photosynthetic performance index (PI_{abs}) compared to those from the SB provenance [39]. We believe this is due to the drier growing conditions in the natural habitats of KA provenance (2016–2020), which may have enhanced their resilience to drought, showcasing a significant degree of intra-specific diversity. Additionally, previous studies on two tree species have shown that plants from drier provenances had higher antioxidant enzyme activity compared to plants from wetter provenances [27,28]. In our research, beech from the KA provenance exhibited higher CAT activity compared to the SB provenance throughout most of the season; a similar trend was observed in oak during the peak of the drought. This suggests that KA provenance plants are better adapted to oxidative stress caused by drought due to their enhanced CAT activity. Beech from the KA provenance also exhibited higher POD activity compared to the SB provenance during the advanced drought period (early August). We believe this is due to the same factors observed in previous cases, indicating that saplings from KA are better adapted to drought stress in various ways than those from SB. These results suggest that in our study, provenances that experienced lower precipitation during the growth period of the investigated saplings showed better drought resistance, reflected by higher antioxidant enzyme activity and lower MDA concentrations compared to provenances with higher precipitation.

4. Materials and Methods

4.1. Plant Material and Provenance Habitat Conditions

This study focused on four-year-old saplings of common beech (*Fagus sylvatica* L.) and sessile oak (*Quercus petraea* (Matt.) Liebl.) from two mature mixed stands in the continental region of the Republic of Croatia. One provenance was located in the north-western part of Croatia near Karlovac (KA provenance, 15.524041° E, 45.466135° N 170 m a.s.l.), with an approximate age of 100 years, while the second provenance was near Slavonski Brod in the eastern part of the country (SB provenance, 17.973173° E, 45.273451° N, 245 m a.s.l.), with an approximate age of 105 years (Figure 8). Both sites were dominated by these species.

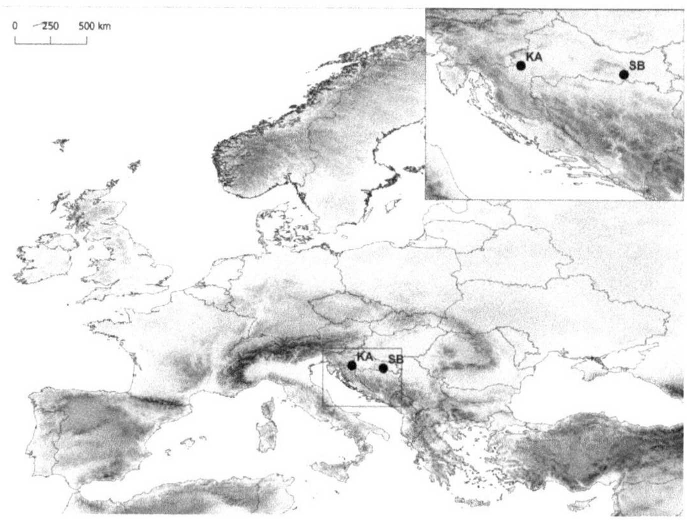

Figure 8. The geographic locations of the Slavonski Brod (SB) and Karlovac (KA) provenances.

In early March 2021, four-year-old saplings were carefully excavated from beneath mature trees, ensuring minimal disturbance to their root systems. The trees were spaced at least 100 m apart. Further details on the habitat characteristics of the studied provenances, including phytosociological, geomorphological, climatological, meteorological, and soil properties, can be found in [40]. Notably, the mean annual precipitation from 1949 to 2019 was higher in the KA (1111.8 mm) compared to the SB provenance (770.3 mm). However, between 2016 and 2020, during the growth period of the saplings, the KA provenance experienced 17 months of moderately to extremely dry months (9 of which occurred during the growing season), while the SB provenance experienced only 9 such months (4 during the growing season). During the experiment, the average daily temperature was 18.8 °C during the vegetation season (April–October), while the average daily relative humidity was 66.7%, and the average daily insolation was 7.6 h. Soil characteristics were largely similar between the two provenances. Phosphorus soil concentrations were low in both provenances with the KA provenance having a P_2O_5 concentration of 0.50 ± 0.32 mg/100 g of soil and the SB provenance having 0.64 ± 0.21 mg/100 g of soil, at a depth of 0–30 cm.

4.2. Experimental Design and Growth Conditions

The four-year-old saplings, averaging 36.6 ± 8.01 cm in height, were relocated to the garden of the Faculty of Forestry and Wood Technology at the University of Zagreb (45.82065 N, 16.02303 E, 120 m a.s.l.), where a common garden experiment was set up. The saplings were transplanted into four wooden boxes (155 × 275 × 80 cm, with a total volume of 3.41 m^3) filled with 3800 L of Klasmann TS 3 substrate, which had a P_2O_5 concentration of 160 mg/L. To raise the phosphorus concentration to 300 mg/L, two of the four boxes received 1182 g of triple superphosphate (Triplex), a fertilizer containing 45% P_2O_5, while the other two boxes remained unfertilized, maintaining a moderate P_2O_5 level of 160 mg/L.

Each box was planted with 100 saplings, consisting of 50 common beech and 50 sessile oak saplings, or 25 saplings per species from each provenance (KA and SB), for a total of 400 saplings in the experiment. The saplings were arranged randomly with a spacing of 20 × 18 cm. During the 2021 growing season, the saplings were exposed to natural weather conditions and were regularly watered in the summer to aid acclimatization and survival. In 2022, a transparent PVC roof covered all boxes to prevent natural precipitation, and four treatments were applied. Two boxes, one fertilized and one unfertilized with P, were watered manually with 40 L of water every four days. The other two boxes, also one fertilized and one unfertilized with P, were subjected to drought from 15 May to 1 September 2022. During this period, they were watered minimally (20 L per box) only when wilting leaves indicated drought stress, occurring three times: late July, early August, and mid-August.

The four experimental treatments were as follows: regular watering with P fertilization (+PW), regular watering without P fertilization (−PW), drought with P fertilization (+PD), and drought without P fertilization (−PD).

4.3. Soil Volumetric Water Content and Leaf Water Potential

The seasonal water dynamics in the substrates of all treatments were monitored using a data logger and sensors to measure the volumetric water content (VWC) in the soil (Spectrum Technologies, Inc., Aurora, CO, USA). Four sensors were installed at a depth of 5–20 cm in each treatment. Pre-dawn leaf water potential (Ψ_{PD}) was measured on three randomly selected saplings of common beech and sessile oak per provenance in each treatment (12 saplings per treatment) using a portable pressure chamber (Model 600, PMS Instrument Company, Albany, OR, USA). The first measurements were taken on May 26 (onset of the drought period), followed by early August (moderate drought), late August (drought peak), and October (at the end of the growing season).

4.4. Leaf Sampling for Chemical Analysis

During the growing season, we collected three composite samples per species and provenance within each treatment. For phosphorus concentrations, concentrations of photosynthetic pigments, total soluble proteins, malondialdehyde, and antioxidant enzymes, we sampled in May, early August, late August, and October. On each occasion, the composite sample consisted of 10 leaves, which were then analyzed. In total, there were three composite samples for beech SB, three for beech KA, three for oak KA, and three for oak SB in each treatment, for a total of forty-eight samples per date.

4.5. Leaf Phosphorus Concentration

Three composite leaf samples were collected in May, early August, late August and October to analyze phosphorus concentration. The leaves were then dried, ground, and homogenized before undergoing chemical analysis. The phosphorus concentration was determined using a spectrophotometric method, following standardized international protocols [41].

4.6. Chlorophyll and Carotenoid Concentrations

The leaves were ground and homogenized in a mortar with liquid nitrogen. From 0.1 g of the prepared powder, Eppendorf cuvettes of 0.2 mL volume were filled with the addition of cold acetone (80%). The tissue was then extracted for 15 min at 4 °C, after which it was centrifuged at 1000 rpm for 10 min. The supernatant was decanted into plastic test tubes, while the green-colored solid residue was re-extracted with acetone until the tissue lost its green color. The volume of each sample was measured using a graduated cylinder.

The prepared samples were used for spectrophotometric measurements on a UV/VIS spectrophotometer (Specord 40, Analytic Jena, Jena, Germany). Absorbance was measured at wavelengths of 661.6 nm, 664.8 nm, and 470 nm. Based on the known leaf mass (m), volume (V), and absorbance (A) at each wavelength, the concentrations of photosynthetic pigments (chlorophyll *a*, chlorophyll *b*, chlorophyll *a* + *b*, and carotenoids) were calculated using the following formulas according to [42]:

$$\text{Chlorophyll } a = (11.24 \times A\ (661.6\ \text{nm}) - 2.04 \times A\ (664.8\ \text{nm})) \times V/(m \times 1000),$$

$$\text{Chlorophyll } b = (20.13 \times A\ (664.8\ \text{nm}) - 4.19 \times A\ (661.6\ \text{nm})) \times V/(m \times 1000),$$

$$\text{Chlorophyll } a + b = (7.05 \times A\ (661.6\ \text{nm}) + 18.09 \times A\ (664.8\ \text{nm})) \times V/(m \times 1000),$$

$$\text{Carotenoids} = (1000 \times A\ (470\ \text{nm}) - 1.90 \times (11.24 \times A\ (661.6\ \text{nm}) - 2.04 \times A\ (664.8\ \text{nm})) - 63.14 \times (20.13 \times A\ (664.8\ \text{nm}) - 4.19 \times A\ (661.6\ \text{nm}))) \times V/(214 \times m \times 1000)$$

The measured concentrations of photosynthetic pigments were expressed in mg/g fresh weight (FW) of the sample.

4.7. Determination of Malondialdehyde Concentration

To determine the concentration of lipid peroxidation, leaf samples were extracted and homogenized in a mortar. A solution of trichloroacetic acid (TCA) in water was added to the homogenate, after which the sample was centrifuged (100,000 rpm, 5 min). The supernatant was used to measure lipid peroxidation concentration, based on the quantification of malondialdehyde (MDA), the final product of lipid peroxidation. To 1 mL of supernatant, 1 mL of 0.5% thiobarbituric acid (TBA) in 20% TCA was added. The mixture was heated for 30 min in a water bath at 90 °C, then rapidly cooled and centrifuged (10,000 rpm, 10 min). Absorbance was measured at 390 nm, and MDA was expressed as nmol per gram of fresh tissue (nmol g^{-1} fresh weight).

4.8. Determination of Protein Concentration and Antioxidant Enzyme Activity

Plant leaves were homogenized on ice using a mortar and pestle. For extraction, plant material was first weighed, then homogenized with the addition of 2 mL of extraction buffer (pH = 7.0) and polyvinylpolypyrrolidone (PVP, a protein stabilizer). After homogenization, the extract was centrifuged at 10,000 rpm, decanted, and stored at −20 °C until further analysis. The obtained extract was used for determining the total soluble protein concentration by the Bradford method and for measuring the activity of ascorbate peroxidase (APX), catalase (CAT), peroxidases (POD), and superoxide dismutase (SOD).

4.8.1. Determination of Total Soluble Protein Concentration by Bradford Method

The concentration of total soluble proteins was determined using the Bradford method [43]. A 96-well plate was prepared by adding 10 µL of the sample to each well, followed by the addition of 225 µL of Bradford reagent. The plate was incubated for 15 min in the dark, and absorbance was measured spectrophotometrically at 595 nm.

4.8.2. Measurement of Ascorbate Peroxidase Activity

The activity of ascorbate peroxidase (APX) was measured according to the protocol of Ambriović Ristov et al. [44]. In a quartz cuvette, 970 µL of APOX assay buffer was added, followed by 10 µL of sodium ascorbate, 10 µL of the extract, and 10 µL of hydrogen peroxide, initiating the reaction. The reaction progress was monitored spectrophotometrically at

290 nm every 15 s for 1 min. Ascorbate peroxidase activity was determined by the reduction in ascorbate, with an extinction coefficient of 2.8 mM^{-1} cm^{-1}, and was expressed as nmol per minute per mg of total protein (nmol min^{-1} mg^{-1} P).

4.8.3. Measurement of Catalase Activity

Catalase (CAT) activity was measured spectrophotometrically at 290 nm every 5 s for 1 min. The measurement was initiated by adding 5 μL of the diluted sample to 995 μL of catalase assay buffer [44]. The extinction coefficient is 40 mM^{-1} cm^{-1}, and catalase activity was expressed as nmol per minute per mg of total protein (nmol min^{-1} mg^{-1} P).

4.8.4. Measurement of Peroxidase Activity

Peroxidase (POD) activity was measured in a 96-well plate by adding 190 μL of POD assay buffer to each well, followed by the addition of 10 μL of the extract. Absorbance was measured every 15 s for 5 min at 470 nm. During the reaction, tetraguaiacol was formed, which has a molar extinction coefficient of 26.6 mM^{-1} cm^{-1}. POD activity was expressed as mmol per minute per mg of total protein (nmol min^{-1} mg^{-1} P).

4.8.5. Measurement of Superoxide Dismutase Activity

For the measurement of superoxide dismutase (SOD) activity, various dilutions of the extract were prepared by mixing specific volumes of the extract and protein extraction buffer (pH = 7.0) to a final volume of 33.3 μL in 96-well plates. To each well, 300 μL of SOD assay buffer (pH = 7.8) and 0.67 μL of 2 μM riboflavin were added, and the reaction was initiated by exposing the samples to light from a 36 W light source. The reaction lasted for 10 min, after which it was stopped by darkening the samples. SOD activity was measured spectrophotometrically at 560 nm 5–10 min after darkening. Enzyme activity was expressed as substrate reduction inhibition. Inhibition was calculated using the formula:

$$\text{NBT } (\%) = [(A - B)/A] \cdot 100,$$

where A is the absorbance measured after the reaction without enzyme (maximum absorbance value) and B is the absorbance measured after the reaction with enzyme (reduced absorbance). The enzyme activity curve was constructed by plotting the NBT reduction inhibition (%) on the y-axis and the volume of the extract used to prepare the dilutions on the x-axis. Based on this curve, the volume of extract required for 50% NBT inhibition in the reaction mixture was used to calculate the specific SOD activity, which was expressed in enzyme activity units per milligram of total protein (U mg^{-1} P).

4.9. Statistical Analysis

All statistical analyses were performed using SAS 15.1 software (SAS Institute Inc., Cary, NC, USA). The assumptions of residual normality and homogeneity of variances were assessed using the Shapiro-Wilk test and Levene's test, conducted through the GLM and UNIVARIATE procedures in SAS. Residuals were plotted against fitted values to check for homogeneity of variance, and the residual distribution was also examined. A factorial ANOVA was used to assess the fixed effects of drought, phosphorus fertilization, provenance, and their interactions on leaf water potential, phosphorus concentrations in leaves, total chlorophyll and carotenoid concentrations, protein concentrations, malondialdehyde concentrations, and antioxidant enzyme activities (APX, CAT, POD, SOD) for each species individually. In all cases, a Tukey post hoc test was performed to determine the significance of differences ($p < 0.05$) between the levels of the factors studied.

5. Conclusions

Our study highlights the complex interactions between drought stress, phosphorus fertilization, and provenance in shaping the antioxidative capacity of common beech and sessile oak. Both species exhibited increased malondialdehyde (MDA) concentrations, a marker of lipid peroxidation and membrane damage, along with reduced photosynthetic pigments under drought conditions. While antioxidant enzyme activity increased under drought conditions, the effectiveness of these enzymes varied depending on the species, provenance, and the severity of the drought stress. Phosphorus fertilization did not consistently enhance drought tolerance, but it did affect antioxidant activity, particularly in oak, where increased catalase (CAT) and ascorbate peroxidase (APX) activities were observed during the peak of the drought. The results also revealed notable differences between provenances, with plants from the provenance that experienced drier conditions during the growth of the studied saplings in natural stands showing better adaptation to drought stress, as reflected in higher antioxidant enzyme activity and lower oxidative damage. Overall, the findings suggest that phosphorus fertilization may not always alleviate the negative effects of drought and that provenance-specific differences in antioxidant capacity can influence species' resilience to drought stress. These findings underscore the complexity of plant responses to environmental stress and the potential role of provenance in shaping drought tolerance.

Author Contributions: Conceptualization, A.V. and K.S.; methodology, A.V., K.S., S.B. and Ž.Š.; validation, K.S.; formal analysis, A.V. and K.S.; investigation, A.V., K.S., S.B., Ž.Š., D.K., I.K.B., T.K., M.B. (Marko Bačurin) and M.B. (Magdalena Brener); resources, K.S.; data curation, A.V., K.S. and S.B.; writing—original draft preparation, A.V.; writing—review and editing, K.S., S.B., Ž.Š., D.K., I.K.B., T.K., M.B. (Marko Bačurin) and M.B. (Magdalena Brener); visualization, A.V., K.S. and Ž.Š.; supervision, K.S.; project administration, K.S. and A.V.; funding acquisition, K.S. All authors have read and agreed to the published version of the manuscript.

Funding: This research was funded by the Croatian Science Foundation, grant number IP-2020-02 5204; project title: "Phenotypic response of common beech and sessile oak provenances to long-lasting drought in interaction with different phosphorus concentrations in the soil".

Institutional Review Board Statement: Not applicable.

Informed Consent Statement: Not applicable.

Data Availability Statement: The data in this study are available from the corresponding author upon request.

Conflicts of Interest: The authors declare no conflicts of interest.

References

1. Arend, M.; Brem, A.; Kuster, T.M.; Günthardt-Goerg, M.S. Seasonal Photosynthetic Responses of European Oaks to Drought and Elevated Daytime Temperature. *Plant Biol.* **2013**, *15*, 169–176. [CrossRef] [PubMed]
2. Cocozza, C.; De Miguel, M.; Pšidová, E.; Ditmarová, L.; Marino, S.; Maiuro, L.; Alvino, A.; Czajkowski, T.; Bolte, A.; Tognetti, R. Variation in Ecophysiological Traits and Drought Tolerance of Beech (*Fagus sylvatica* L.) Seedlings from Different Populations. *Front. Plant Sci.* **2016**, *7*, 886. [CrossRef]
3. Gessler, A.; Bottero, A.; Marshall, J.; Arend, M. The Way Back: Recovery of Trees from Drought and Its Implication for Acclimation. *New Phytol.* **2020**, *228*, 1704–1709. [CrossRef]
4. Leuschner, C. Drought Response of European Beech (*Fagus sylvatica* L.)—A Review. *Perspect. Plant Ecol. Evol. Syst.* **2020**, *47*, 125576. [CrossRef]
5. Talbi, S.; Rojas, J.A.; Sahrawy, M.; Rodríguez-Serrano, M.; Cárdenas, K.E.; Debouba, M.; Sandalio, L.M. Effect of Drought on Growth, Photosynthesis and Total Antioxidant Capacity of the Saharan Plant Oudeneya Africana. *Environ. Exp. Bot.* **2020**, *176*, 104099. [CrossRef]

6. Bose, A.K.; Scherrer, D.; Camarero, J.J.; Ziche, D.; Babst, F.; Bigler, C.; Bolte, A.; Dorado-Liñán, I.; Etzold, S.; Fonti, P.; et al. Climate Sensitivity and Drought Seasonality Determine Post-Drought Growth Recovery of Quercus Petraea and Quercus Robur in Europe. *Sci. Total Environ.* **2021**, *784*, 147222. [CrossRef] [PubMed]
7. Zhong, Z.; He, B.; Chen, Y.; Yuan, W.; Huang, L.; Guo, L.; Zhang, Y.; Xie, X. Higher Sensitivity of Planted Forests' Productivity Than Natural Forests to Droughts in China. *JGR Biogeosci.* **2021**, *126*, e2021JG006306. [CrossRef]
8. Iqbal, A.; Huiping, G.; Qiang, D.; Xiangru, W.; Hengheng, Z.; Xiling, Z.; Meizhen, S. Differential Responses of Contrasting Low Phosphorus Tolerant Cotton Genotypes under Low Phosphorus and Drought Stress. *BMC Plant Biol* **2023**, *23*, 168. [CrossRef]
9. Waraich, E.A.; Ahmad, R.; Ashraf, M.Y. Ehsanullah Role of Mineral Nutrition in Alleviation of Drought Stress in Plants. *Aust. J. Crop Sci.* **2011**, *5*, 764–777.
10. Choudhury, S.; Moulick, D.; Ghosh, D.; Soliman, M.; Alkhedaide, A.; Gaber, A.; Hossain, A. Drought-Induced Oxidative Stress in Pearl Millet (*Cenchrus americanus* L.) at Seedling Stage: Survival Mechanisms through Alteration of Morphophysiological and Antioxidants Activity. *Life* **2022**, *12*, 1171. [CrossRef]
11. Zhou, R.; Kong, L.; Yu, X.; Ottosen, C.-O.; Zhao, T.; Jiang, F.; Wu, Z. Oxidative Damage and Antioxidant Mechanism in Tomatoes Responding to Drought and Heat Stress. *Acta Physiol. Plant.* **2019**, *41*, 20. [CrossRef]
12. Du, B.; Haensch, R.; Alfarraj, S.; Rennenberg, H. Strategies of Plants to Overcome Abiotic and Biotic Stresses. *Biol. Rev.* **2024**, *99*, 1524–1536. [CrossRef] [PubMed]
13. Laxa, M.; Liebthal, M.; Telman, W.; Chibani, K.; Dietz, K.-J. The Role of the Plant Antioxidant System in Drought Tolerance. *Antioxidants* **2019**, *8*, 94. [CrossRef]
14. Sharma, I.; Ahmad, P. Catalase. In *Oxidative Damage to Plants*; Elsevier: Amsterdam, The Netherlands, 2014; pp. 131–148. ISBN 978-0-12-799963-0.
15. Li, S. Novel Insight into Functions of Ascorbate Peroxidase in Higher Plants: More than a Simple Antioxidant Enzyme. *Redox Biol.* **2023**, *64*, 102789. [CrossRef]
16. Twala, P.; Mitema, A.; Baburam, C.; Aliye Feto, N. Breakthroughs in the Discovery and Use of Different Peroxidase Isoforms of Microbial Origin. *AIMS Microbiol.* **2020**, *6*, 330–349. [CrossRef]
17. Sharma, P.; Jha, A.B.; Dubey, R.S.; Pessarakli, M. Reactive Oxygen Species, Oxidative Damage, and Antioxidative Defense Mechanism in Plants under Stressful Conditions. *J. Bot.* **2012**, *2012*, 217037. [CrossRef]
18. Malhotra, H.; Vandana; Sharma, S.; Pandey, R. Phosphorus Nutrition: Plant Growth in Response to Deficiency and Excess. In *Plant Nutrients and Abiotic Stress Tolerance*; Hasanuzzaman, M., Fujita, M., Oku, H., Nahar, K., Hawrylak-Nowak, B., Eds.; Springer: Singapore, 2018; pp. 171–190. ISBN 978-981-10-9043-1.
19. Tariq, A.; Pan, K.; Olatunji, O.A.; Graciano, C.; Li, Z.; Sun, F.; Zhang, L.; Wu, X.; Chen, W.; Song, D.; et al. Phosphorous Fertilization Alleviates Drought Effects on Alnus Cremastogyne by Regulating Its Antioxidant and Osmotic Potential. *Sci. Rep.* **2018**, *8*, 5644. [CrossRef]
20. Ullah, A.; Tariq, A.; Zeng, F.; Sardans, J.; Graciano, C.; Ullah, S.; Chai, X.; Zhang, Z.; Keyimu, M.; Asghar, M.A.; et al. Phosphorous Supplementation Alleviates Drought-Induced Physio-Biochemical Damages in *Calligonum mongolicum*. *Plants* **2022**, *11*, 3054. [CrossRef]
21. Tariq, A.; Pan, K.; Olatunji, O.A.; Graciano, C.; Li, Z.; Li, N.; Song, D.; Sun, F.; Wu, X.; Dakhil, M.A.; et al. Impact of Phosphorus Application on Drought Resistant Responses of *Eucalyptus grandis* Seedlings. *Physiol. Plant.* **2019**, *166*, 894–908. [CrossRef]
22. Yang, F.; Magh, R.-K.; Ivanković, M.; Lanšćak, M.; Haberstroh, S.; Du, B.; Dannenmann, M.; Rennenberg, H.; Herschbach, C. Foliar P Nutrition of European Beech (*Fagus sylvatica* L.) Depends on the Season but Remains Unaffected by Co-Cultivation with Silver Fir (*Abies alba* Mill.). *Eur. J. For. Res.* **2020**, *139*, 853–868. [CrossRef]
23. Kuster, T.M.; Arend, M.; Günthardt-Goerg, M.S.; Schulin, R. Root Growth of Different Oak Provenances in Two Soils under Drought Stress and Air Warming Conditions. *Plant Soil* **2013**, *369*, 61–71. [CrossRef]
24. Pretzsch, H.; Bielak, K.; Block, J.; Bruchwald, A.; Dieler, J.; Ehrhart, H.-P.; Kohnle, U.; Nagel, J.; Spellmann, H.; Zasada, M.; et al. Productivity of Mixed versus Pure Stands of Oak (*Quercus petraea* (Matt.) Liebl. and *Quercus robur* L.) and European Beech (*Fagus sylvatica* L.) along an Ecological Gradient. *Eur. J. For. Res.* **2013**, *132*, 263–280. [CrossRef]
25. Wang, F.; Israel, D.; Ramírez-Valiente, J.-A.; Sánchez-Gómez, D.; Aranda, I.; Aphalo, P.J.; Robson, T.M. Seedlings from Marginal and Core Populations of European Beech (*Fagus sylvatica* L.) Respond Differently to Imposed Drought and Shade. *Trees* **2021**, *35*, 53–67. [CrossRef]
26. Schlosser, F.; Böhm, S.; Horder, N.; Seegmüller, S. Antioxidative Capacity of Oak Trees—Tannins as a Substantial Prerequisite for a Provenance-Specific Response to Drought. *Austrian J. For. Sci.* **2022**, *139*, 73–94.
27. Jafarnia, S.; Akbarinia, M.; Hosseinpour, B.; Modarres Sanavi, S.; Salami, S. Effect of Drought Stress on Some Growth, Morphological, Physiological, and Biochemical Parameters of Two Different Populations of *Quercus brantii*. *iForest* **2018**, *11*, 212–220. [CrossRef]
28. Ying, Y.Q.; Song, L.L.; Jacobs, D.F.; Mei, L.; Liu, P.; Jin, S.H.; Wu, J.S. Physiological Response to Drought Stress in *Camptotheca acuminata* Seedlings from Two Provenances. *Front. Plant Sci.* **2015**, *6*, 361. [CrossRef]

29. Mellert, K.H.; Göttlein, A. Comparison of New Foliar Nutrient Thresholds Derived from van Den Burg's Literature Compilation with Established Central European References. *Eur. J. For. Res* **2012**, *131*, 1461–1472. [CrossRef]
30. Tariq, A.; Pan, K.; Olatunji, O.A.; Graciano, C.; Li, Z.; Sun, F.; Sun, X.; Song, D.; Chen, W.; Zhang, A.; et al. Phosphorous Application Improves Drought Tolerance of Phoebe Zhennan. *Front. Plant Sci.* **2017**, *8*, 1561. [CrossRef]
31. Hasanuzzaman, M.; Bhuyan, M.H.M.; Zulfiqar, F.; Raza, A.; Mohsin, S.; Mahmud, J.; Fujita, M.; Fotopoulos, V. Reactive Oxygen Species and Antioxidant Defense in Plants under Abiotic Stress: Revisiting the Crucial Role of a Universal Defense Regulator. *Antioxidants* **2020**, *9*, 681. [CrossRef]
32. Ullah, A.; Tariq, A.; Zeng, F.; Asghar, M.A.; Sardans, J.; Graciano, C.; Ali, I.; Peñuelas, J. Drought Priming Improves Tolerance of *Alhagi sparsifolia* to Subsequent Drought: A Coordinated Interplay of Phytohormones, Osmolytes, and Antioxidant Potential. *Plant Stress* **2024**, *12*, 100469. [CrossRef]
33. Chatterjee, A.; Dey, T.; Galiba, G.; Kocsy, G.K.; Dey, N.; Kar, R.K. Effect of Combination of Light and Drought Stress on Physiology and Oxidative Metabolism of Rice Plants. *Plant Sci. Today* **2021**, *8*, 762–777. [CrossRef]
34. Niu, Y.; Wang, Y.; Li, P.; Zhang, F.; Liu, H.; Zheng, G. Drought Stress Induces Oxidative Stress and the Antioxidant Defense System in Ascorbate-Deficient Vtc1 Mutants of Arabidopsis Thaliana. *Acta Physiol Plant* **2013**, *35*, 1189–1200. [CrossRef]
35. Özden, H.; Bayçu Kahyaoğlu, G. Physiological, Oxidative, and Antioxidative Responses of Quercus Vulcanica Boiss. and Quercus Aucheri Jaub. & Spach. under Drought Stress Conditions. *İstanb. J. Pharm.* **2024**, *54*, 69–79. [CrossRef]
36. Celikkol Akcay, U.; Ercan, O.; Kavas, M.; Yildiz, L.; Yilmaz, C.; Oktem, H.A.; Yucel, M. Drought-Induced Oxidative Damage and Antioxidant Responses in Peanut (*Arachis hypogaea* L.) Seedlings. *Plant Growth Regul.* **2010**, *61*, 21–28. [CrossRef]
37. Qiao, M.; Hong, C.; Jiao, Y.; Hou, S.; Gao, H. Impacts of Drought on Photosynthesis in Major Food Crops and the Related Mechanisms of Plant Responses to Drought. *Plants* **2024**, *13*, 1808. [CrossRef] [PubMed]
38. Takagi, D.; Miyagi, A.; Tazoe, Y.; Suganami, M.; Kawai-Yamada, M.; Ueda, A.; Suzuki, Y.; Noguchi, K.; Hirotsu, N.; Makino, A. Phosphorus Toxicity Disrupts Rubisco Activation and Reactive Oxygen Species Defence Systems by Phytic Acid Accumulation in Leaves. *Plant Cell Environ.* **2020**, *43*, 2033–2053. [CrossRef] [PubMed]
39. Vukmirović, A.; Škvorc, Ž.; Bogdan, S.; Krstonošić, D.; Bogdan, I.K.; Karažija, T.; Bačurin, M.; Brener, M.; Sever, K. Photosynthetic Response to Phosphorus Fertilization in Drought-Stressed Common Beech and Sessile Oak from Different Provenances. *Plants* **2024**, *13*, 2270. [CrossRef] [PubMed]
40. Sever, K.; Vukmirović, A.; Hodak, L.; Bogdan, S.; Katičić Bogdan, I.; Krstonošić, D.; Karažija, T.; Franjić, J.; Škvorc, Ž. Funkcionalna Prilagodba Prirodnog Pomlatka Hrasta Kitnjaka i Obične Bukve na Različite Stanišne Prilike. *Šumar. List Online* **2022**, *146*, 293–307. [CrossRef]
41. AOAC. *Official Methods of Analysis*, 17th ed.; Association of Official Analytical Chemists: Washington, DC, USA, 2000.
42. Lichtenthaler, H.K. Chlorophylls and Carotenoids: Pigments of Photosynthetic Biomembranes. In *Methods in Enzymology*; Elsevier: Amsterdam, The Netherlands, 1987; Volume 148, pp. 350–382. ISBN 978-0-12-182048-0.
43. Bradford, M.M. A Rapid and Sensitive Method for the Quantitation of Microgram Quantities of Protein Utilizing the Principle of Protein-Dye Binding. *Anal. Biochem.* **1976**, *72*, 248–254. [CrossRef]
44. Ambriović Ristov, A.; Brozović, A.; Bruvo Mađarić, B.; Ćetković, H.; Herak Bosnar, M.; Hranilović, D.; Katušić Hećimović, S.; Mihaljević, S.; Slade, N.; Vujaklija, D. *Metode u Molekularnoj Biologiji;* Institut Ruđer Bošković: Zagreb, Croatia, 2007; ISBN 978-953-6690-72-5.

International Journal of
Molecular Sciences

MDPI

Article

The Overexpression of *SlPLATZ17* Can Increase the Tolerance of Tomatoes to Drought and Salt Stress

Xueli Jia [1,2], Xinyue Pang [1,2], Min Xu [1,2], Congmin Wang [1,2], Huixin Wei [1,2], Jing Liu [1,2], He Zhang [1,2], Dalong Li [1,2], Xiangyang Xu [1,2,*] and Tingting Zhao [1,2,*]

1 Tomato Research Institute, College of Horticulture and Landscape Architecture, Northeast Agricultural University, Harbin 150030, China; jiaxueli2023@163.com (X.J.); pxinyue1201@163.com (X.P.); 15024995257@163.com (M.X.); 15954153922@163.com (C.W.); weihuixin011014@163.com (H.W.); 18334804395@163.com (J.L.); esculentum@neau.edu.cn (H.Z.); lidalong@neau.edu.cn (D.L.)

2 Key Laboratory of Biology and Genetic Improvement of Horticultural Crops (Northeast Region), Ministry of Agriculture and Rural Affffairs, Northeast Agricultural University, Harbin 150030, China

* Correspondence: xuxy@neau.edu.cn (X.X.); ttzhao2016@neau.edu.cn (T.Z.); Tel.: +86-130-0986-9477 (X.X.); +86-136-4457-2690 (T.Z.)

Abstract: PLATZ, a novel zinc finger DNA-binding protein, plays crucial regulatory roles in the growth and development of various plants and in the modulation of abiotic stress responses. However, knowledge of the functions of PLATZ genes in tomato (*Solanum lycopersicum*) is lacking. In our study, we established stable *SlPLATZ17* gene overexpression and knockout lines to further analyze the regulatory functions of *SlPLATZ17* under drought and salt stress. It was found that the overexpression lines presented increased drought and salt tolerance under stress. Transcriptome sequencing and screening for interacting proteins revealed that SlPLATZ17 may exert its effects by interacting with POR1, thereby influencing glutathione metabolism and arginine and proline metabolism. In conclusion, the findings lay the groundwork for a deeper understanding of the regulatory mechanisms underlying the role of *SlPLATZ17* in stress resistance, as well as facilitating the application of *SlPLATZ17* in the breeding of stress-resistant tomato varieties.

Keywords: tomato; *SlPLATZ17*; gene editing; drought and salt stresses; protein interactions

1. Introduction

The growth and development of tomatoes (*Solanum lycopersicum*) are significantly influenced by environmental factors such as light, air, water, and temperature. When tomato cells are subjected to abiotic stress, the stability of the cell membrane is compromised, leading to alterations in membrane protein structure and increased permeability [1]. Abiotic stress can lead to the excessive accumulation of proline in plants, resulting in the production of reactive oxygen species (ROS), which are byproducts of physiological metabolism [2,3]. This process is precisely regulated by both enzymatic and nonenzymatic antioxidant defence systems [4–6]. Superoxide dismutase (SOD) disproportionates superoxide anions ($O_2^{-\bullet}$) into hydrogen peroxide (H_2O_2) and O_2 [7]. Peroxidase (POD) plays distinct roles during various stages of ageing and stress, influencing processes both before and after these events [8]. The physiological changes induced by stress can significantly impact tomato quality and yield. Therefore, it is crucial to investigate the stress resistance mechanisms in tomatoes, explore genetic resources for stress tolerance, and cultivate and select varieties that are resilient to adverse conditions.

PLATZ is a novel type of zinc finger DNA-binding protein. The nomenclature of zinc finger transcription factors is derived from their distinctive finger-like structure and their ability to form specific interactions with zinc ions [9]. It can activate or inhibit gene transcription through three distinct mechanisms [10]. Zinc finger proteins are classified into nine major categories. Zinc finger proteins play crucial roles in the expression and regulation of plant genes, as well as in growth and development, antibacterial immunity, and responses to both abiotic and biotic stresses [11–13]. PLATZ1 possesses two distinct structural domains: C-x2-H-x11-C-x2-C-x(4-5)-C-x2-C-x(3-7)-H-x2-H and C-x2-C-x(10-11)-C-x3-C [14].

Numerous reports have confirmed that PLATZ transcription factors play a significant role in plant growth and development and are involved in the response to various abiotic stresses [15,16]. In rice, the *SG6* gene encodes a PLATZ protein. The overexpression of the *G6* gene results in a significant increase in grain size and weight, along with a notable height increase in the plants. This observation is consistent with the high rate of spikelet hull cell division associated with *SG6* [17]. Transgenic Arabidopsis and soybean hairy roots overexpressing *GmPLATZ17* exhibited drought sensitivity [18]. *PLATZ4* interacts with *AITR6* to increase ABA sensitivity and drought tolerance in Arabidopsis by upregulating the expression of PR1, ABI3, ABI4, and ABI5 and suppressing the expression of PIP2;8 and related genes [19]. In vitro and in vivo experiment have shown that in *A. thaliana*, *AtPLATZ2* directly binds to A/T-rich sequences in the CBL4/SOS3 and CBL10/SCaBP8 promoters and represses the activity of the CBL4/SOS3 promoter in the leaves of the plant, thereby inhibiting the salt tolerance of the plant [20]. However, few stress-related studies on these PLATZ family genes have been conducted on tomatoes, and the molecular mechanisms by which tomato PLATZ genes regulate abiotic stress and plant growth and development are unknown.

In a previous study, we examined the *SlPLATZ17* gene, which exhibited a significant response under various stress conditions. Using virus-induced gene silencing (VIGS), we initially demonstrated that *SlPLATZ17* is involved in regulating the tomato plant's resistance to drought and salt stress. [21]. However, given the limitations of VIGS technology, the function of this gene needs to be further validated. Based on preliminary research findings and the investigation of the PLATZ gene's role in stress resistance across various species, we can hypothesize that *SlPLATA17* positively regulates the drought and salt tolerance of tomatoes.

In this study, we constructed a stable T2 overexpression strain and a double allelic mutant T2 gene-edited strain of the *SlPLATZ17* gene through genetic transformation, analyzed its response to drought and salt stress, screened for intercalating proteins via yeast two-hybrid technology, and analyzed the metabolic pathways involved in *SlPLATZ17* and its possible involvement in functional modules by combining the transcriptomic data of different strains. We also analyzed the metabolic pathways in which *SlPLATZ17* and its interacting proteins are involved and explored the functional modules in which *SlPLATZ17* might be involved. The key mechanism by which *SlPLATZ17* regulates tomato resistance was ultimately elucidated, laying the foundation for the breeding of resistant tomato plants.

2. Results

2.1. Expression Pattern of SlPLATZ17 in Different Tissues of Tomato and Its Subcellular Localization

To elucidate the potential function of *SlPLATZ17*, we first analyzed the expression of *SlPLATZ17* in the roots, stems, leaves, flowers, green fruit, and red fruit of AC. The qRT-PCR results revealed that the expression of *SlPLATZ17* was extremely high in flowers and roots, 10–20 times higher than that in red fruit, whereas its expression was relatively low in leaves, stems, and green fruit (Figure 1A). To determine the subcellular localization of the SlPLATZ17 protein, the pCAMBIA2300-*SlPLATZ17* construct was generated and introduced into tobacco leaves. As shown in Figure 1B, compared with that of the pCAMBIA2300-eGFP control, the fluorescence signal of pCAMBIA2300-*SlPLATZ17* was detected only in the nuclei of the cells, indicating that *SlPLATZ17* was localized in the nucleus.

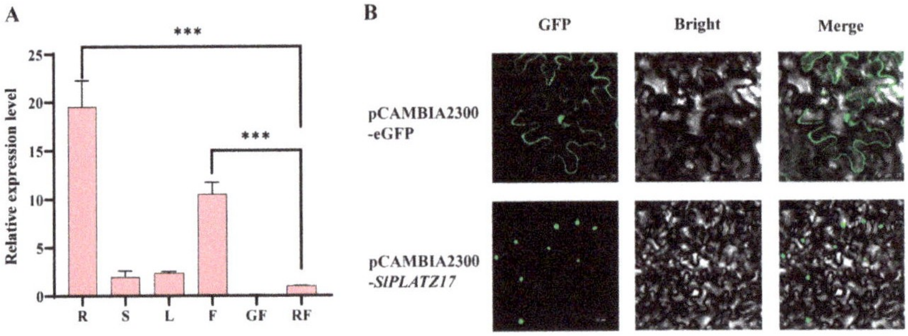

Figure 1. Expression pattern analysis of *SlPLATZ17* in tomato and its subcellular localization. (**A**) Expression of *SlPLATZ17* in R, roots; S, stems; L, leaves; F, flowers; GF, fruit at mature green stage; RF, fruit at mature red stage in WT (***; $p \leq 0.001$). (**B**) Subcellular localization of SlPLATZ17 protein in tobacco leaves. Bars: 25 μm; 60 μm.

2.2. Construction of SlPLATZ17 Overexpression and Knockout Lines

In this study, we used *SlPLATZ17*-overexpressing T2 generation strains obtained in a previous study and performed fluorescence quantitative PCR for *SlPLATZ17*. Compared with that in the control AC plants, gene expression in all four strains was significantly upregulated, and the OE-2 and OE-4 strains are chosen, which presented greater expression, for the subsequent experiments (Figure 2A). The successful *pYLCRISPR/Cas9-SlPLATZ17* vector was subsequently transformed into the tomato variety AC via Agrobacterium, and four knockout plants were obtained. The designed targets are shown in Figure 2B. The *SlPLATZ17* gene-edited plants from four T0 generation plants were identified via PCR to amplify a 342 bp fragment, and the products were subsequently sent to a sequencing company. Alignment of the sequencing results with the target fragments was performed via DNAMAN software (6.0), and target resolution was performed via the web-based tool DSDecodeM [22]. Through two consecutive generations of planting and testing, we generated two T2 generation pure knockout strains, QC-1 and QC-2 (Figure 2C), which were used for subsequent stress and other experiments.

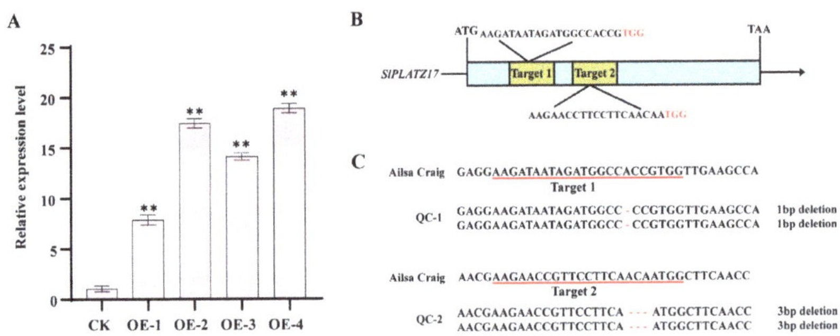

Figure 2. Identification of *SlPLATZ17*-overexpressing and *SlPLATZ17* knockout plants. (**A**) *SlPLATZ17* gene expression in CK and *SlPLATZ17*-overexpressing plants. Data are presented as means ± SEs ($n = 3$). ** ($p \leq 0.01$) indicates significant differences compared with CK (AC), as determined by one-way ANOVA. (**B**) Schematic diagram of tomato genome editing mediated by CRISPR/Cas9. Arrow indicates direction of genome, from 5′ to 3′. Yellow boxes represent sgRNA target sites, sequences of which are also shown. PAM motifs are marked in red. (**C**) Sequence near mutation site of *SlPLATZ17* gene. Lower part of sgRNA target site is marked with red line.

2.3. *SlPLATZ17* Overexpression Improves Drought Tolerance in Tomatoes

Three groups of tomato seedlings were subjected to 15% PEG6000 treatment. As shown in Figure 3A, the leaf blades of the three groups of plants started to wilt, and the stems became bent and drooped at 1.5 h. The degree of stem bending and leaf wilting were most severe in the knockout plants, whereas the overexpression plants performed the best, exhibiting only slightly bent stems; at 3 h, the three groups of plants further wilted, but the overexpression plants performed slightly better, and the remaining two groups performed similarly. The knockout plants presented the most severe degree of wilting, whereas the overexpression plants presented the best performance, with only slightly bent stems. At 6 h, the three groups of plants exhibited further wilting, but the overexpression plants presented slightly better performance, and the remaining two groups presented similar performance. In addition to performing phenotypic observations, we measured the related physiological and biochemical indices, and the results revealed that the four indices, Superoxide dismutase (SOD), Phenylalanine ammonia lyase (PAL), and Peroxidase (POD) activities and proline (Pro) content, were elevated in the overexpression plants but reduced in the knockout plants compared with the wild-type plants overall, and the decrease was significant at some time points (Figure 3B–E). Taken together, these findings indicate that drought-related physiological indices were increased in the overexpression plants but reduced in the knockout plants overall. Measurement of the malondialdehyde (MDA) content and relative conductivity in the three groups of plants revealed that both indicators tended to increase, and the increase was most rapid in the knockout plants, peaking at 12 h. The increase in the overexpression plants was slower than that in the wild-type, with smaller values overall (Figure 3F,G). Overall, the analysis revealed that the cell membranes of the overexpression plants presented the least damage under drought stress, whereas those of the knockout plants presented the most severe damage under drought stress.

To further characterize the damage to tomato leaves under stress conditions, we subsequently quantified the ROS content. As shown in Figure 4, with prolonged drought stress, the staining area of the plant leaves gradually expanded, and the colour deepened; however, the degree of staining of the leaves from the overexpression plants was clearly the lightest, and all three groups of leaves were almost completely stained at 12 h. The degree of staining of the knockout plants was the strongest, exhibiting a reddish-brown and dark blue colour, whereas the degree of staining of the overexpression plants was lower than that of the wild-type plants overall, with the smallest staining area and the lightest colour. Under drought stress, *SlPLATZ17* overexpression inhibited the production of ROS in leaves, reduced the damage caused by stress to leaves, and improved the drought tolerance of tomato plants.

2.4. *SlPLATZ17* Overexpression Improves Salt Tolerance in Tomatoes

Three groups of tomato seedlings were subjected to 200 mM NaCl salt treatment. As shown in Figure 5A, the stems of all three groups of plants became bent and drooped at 1.5 h of treatment, with the stems of the overexpression plants bending the least and those of the knockout plants bending the most severely; from 3 to 6 h, the stems of the wild-type plants and the knockout plants were completely bent and drooping, whereas the changes in the overexpression plants were not obvious, and the stems of the overexpression plants did not completely droop until 12 h. The stems of the wild-type and knockout plants were completely bent and drooping, whereas the stems of the overexpression plants did not completely droop. The phenotype of the stems of the overexpression plants was not obvious until 12 h, when complete drooping was observed. Moreover, the physiological indices related to the salt resistance of the overexpression plants increased strongly, whereas those of the knockout plants decreased (Figure 5B–E). The MDA content and conductivity of the gene-edited plants were consistently greater than those of the other two groups, indicating that knockdown of the *SlPLATZ17* gene resulted in a high degree of cellular plasma peroxidation, severe damage to the cellular membrane, and severe injury caused by salt stress in the plants (Figure 5F,G). The leaves from the different treatment timepoints were stained; the whole leaves of the knockout plants were reddish-brown or dark blue at 12 h of treatment, whereas those of the overexpression plants showed staining in only the upper part of the leaves, and the degree of staining of the overexpression plants was the lightest among the three groups of plants (Figure 6). Taken together, these results indicate that the overexpression of *SlPLATZ17* reduced the degree of damage caused by salt stress to tomato plants and improves their salt tolerance.

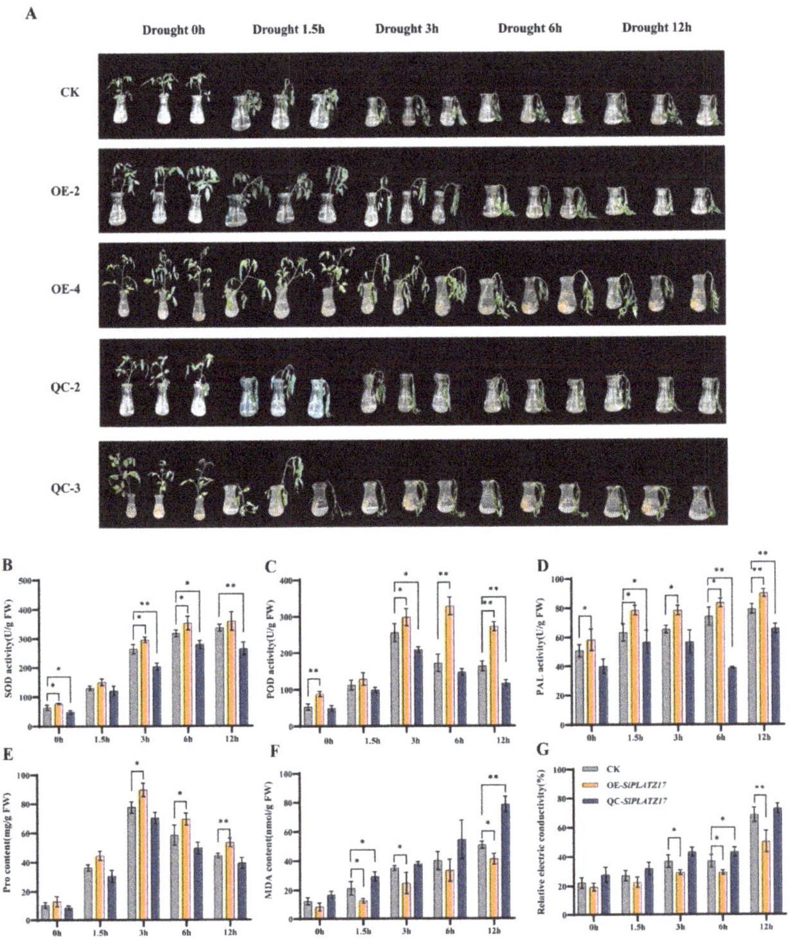

Figure 3. Overexpression of *SlPLATZ17* improved drought tolerance in tomato. (**A**) Changes in phenotypes of different plant lines under drought stress. Analysis of physiological indicators of different plant lines under drought stress: (**B**) SOD activity; (**C**) POD activity; (**D**) PAL activity; (**E**) Pro content; (**F**) MDA content; (**G**) relative electrical conductivity. All treatments were performed in triplicate with biological replicates. * ($p \leq 0.05$) and ** ($p \leq 0.01$) indicate significant differences compared with CK at each time point, as analyzed by one-way ANOVA.

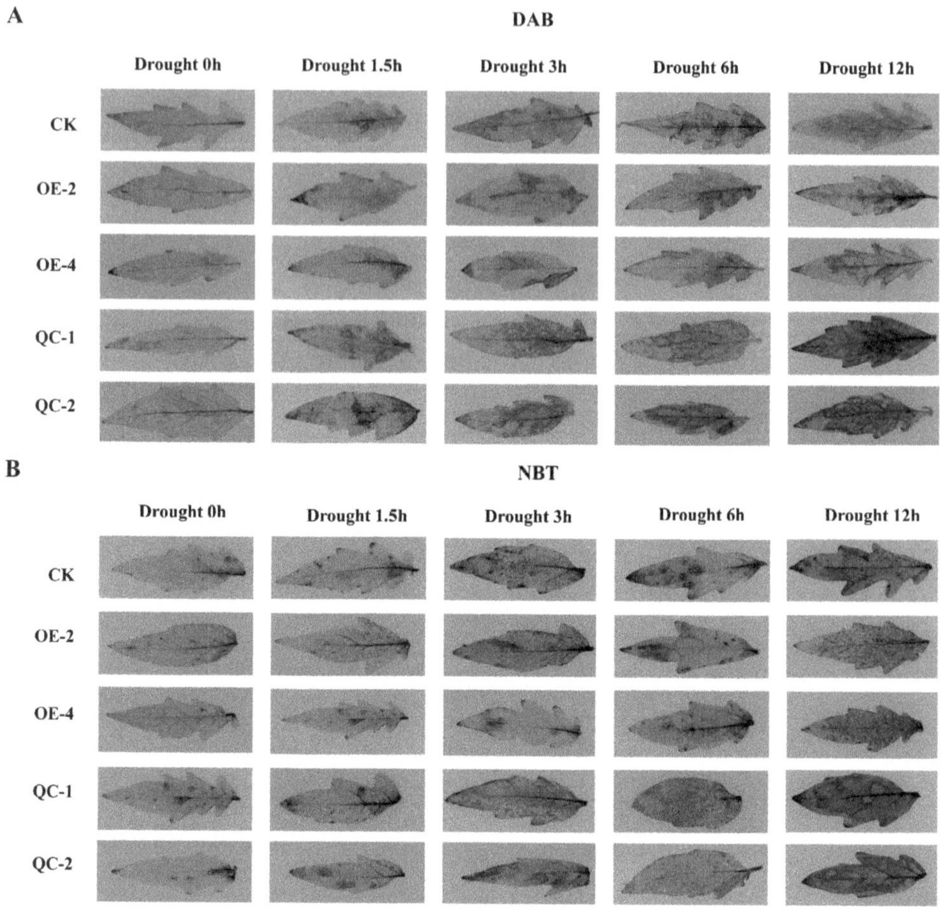

Figure 4. Reactive oxygen species staining in different plant lines under drought stress. (**A**) DAB staining of different lines under drought treatment, with brown representing H_2O_2. (**B**) NBT staining of different lines under drought treatment, with blue representing O^{2-}.

2.5. Sequencing Analysis of SlPLATZ17 Transcriptome

We performed transcriptome sequencing analysis for three biological replicates of three groups of *SlPLATZ17* gene-overexpressing plants (OE), knockout plants (QC), and AC plants (CK), with a total of nine samples, and statistical analysis showed that the total number of bases in the individual samples in the filtered data was greater than 6 Gb, the percentage of filtered sequenced bases with a quality level of Q20 or above was greater than 97%, and the average percentage of Q30 bases was greater than 92% (Table S2). In the reference genome comparison analysis, among the nine groups of samples, the number of reads and the proportion of valid reads that were localized to the reference genome were greater than 95%, whereas the number of reads and the proportion of valid reads that were unique compared to the reference genome were greater than 90%, which is a high comparison rate in transcriptome sequencing (Table S3). These results indicated that the transcriptomic data were of high quality and could be subsequently analyzed.

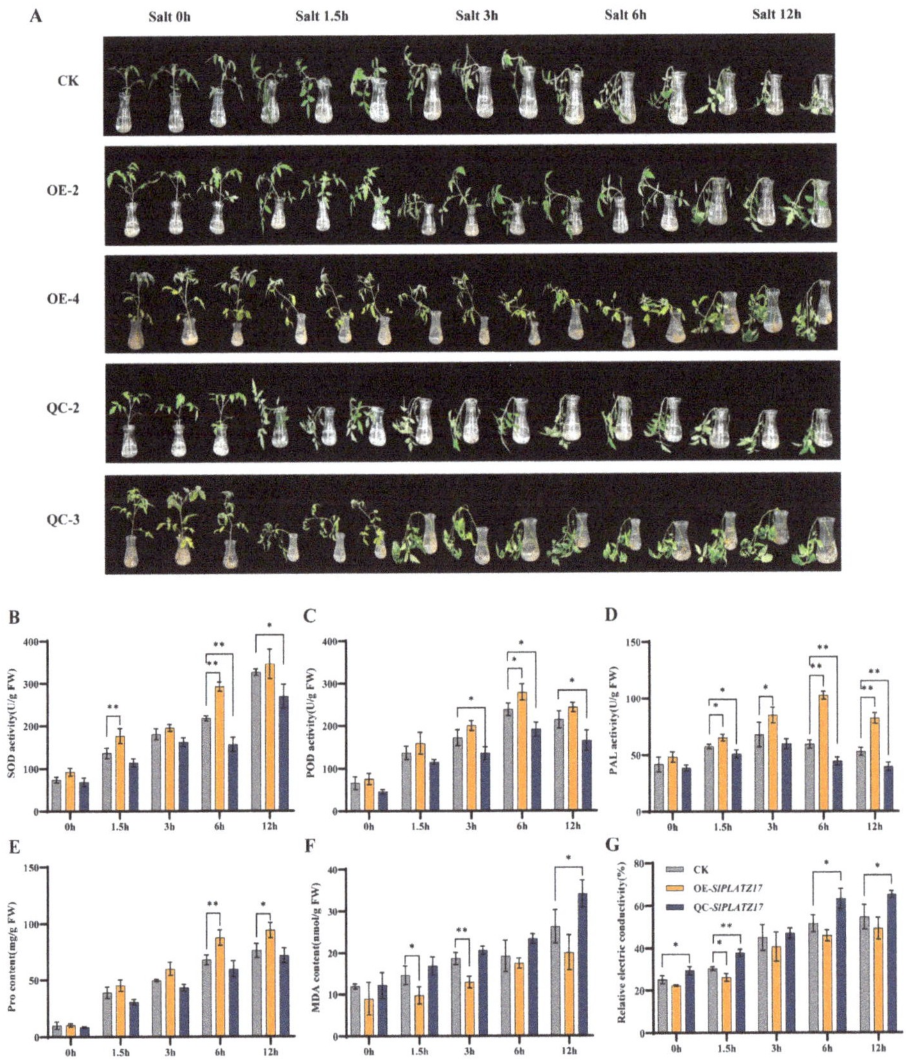

Figure 5. Overexpression of *SlPLATZ17* improved salt tolerance in tomatoes. (**A**) Changes in phenotypes of different plant lines under salt stress. Analysis of physiological indicators of different plant lines under drought stress: (**B**) SOD activity; (**C**) POD activity; (**D**) PAL activity; (**E**) Pro content; (**F**) MDA content; (**G**) relative electrical conductivity. All treatments were performed in triplicate with biological replicates. * ($p \leq 0.05$) and ** ($p \leq 0.01$) indicate significant differences compared with CK at each time point, as analyzed by one-way ANOVA.

We compared AC plants, OE plants, and gene-edited plants in a pairwise manner on the basis of their gene expression levels in the 0 h transcriptome; compared with OE plants, CK plants presented a lower number of upregulated genes; compared with QC plants, CK plants presented a greater number of upregulated genes (227); and compared with QC plants, OE plants presented 600 upregulated genes and 247 downregulated genes (Figure 7A). On the basis of the significantly differentially expressed genes in the comparison group, we performed volcano plot analysis (Figure 7B). The distribution of upregulated and downregulated is visualized in the figure for a more intuitive depiction. The volcano plot for the OE plants compared with the QC revealed the presence of significantly more upregulated genes and a wider distribution at both ends, indicating a greater degree of genetic differentiation. GO analysis of the three groups of materials, CK, OE, and QC, revealed that the three groups of materials presented the most variation in the number of differentially expressed genes in the biological process category, and in the comparison of two groups of materials, OE and QC, the number of upregulated genes related to biological processes was also the greatest, including genes involved in metabolic processes and cellular processes. The cellular component terms with the greatest number of upregulated genes were the cellular, cellular components, etc.; the molecular function terms with the highest number of upregulated genes were functions such as catalytic activity and nucleic acid binding transcription factor activity (Figure 7C–E). KEGG analysis of the differentially expressed genes was performed on the three groups of materials, and pairwise comparison revealed that the most enriched metabolic pathways in the CK vs. OE group included glutathione metabolism, steroid biosynthesis, arginine and proline metabolism, and secondary metabolism biosynthesis; the differentially abundant metabolites in the CK vs. QC group were enriched mainly in glycerophospholipid metabolism, arginine and proline metabolism, sulphur metabolism, and glutathione metabolism; the pathways with the most significant differences in the OE vs. QC group were secondary metabolism biosynthesis, plant–pathogen interactions, and steroid biosynthesis (Figure 7F).

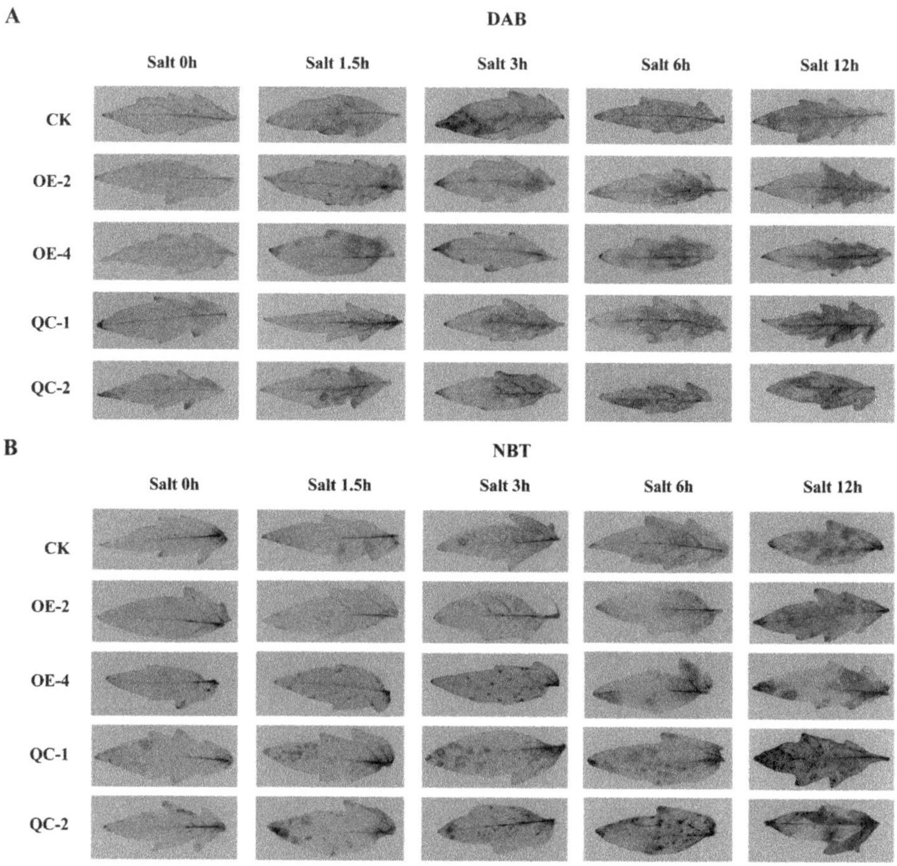

Figure 6. Reactive oxygen species staining in different plant lines under salt stress. (**A**) DAB staining of different lines under salt treatment, with brown representing H_2O_2. (**B**) NBT staining of different lines under salt treatment, with blue representing O_2^-.

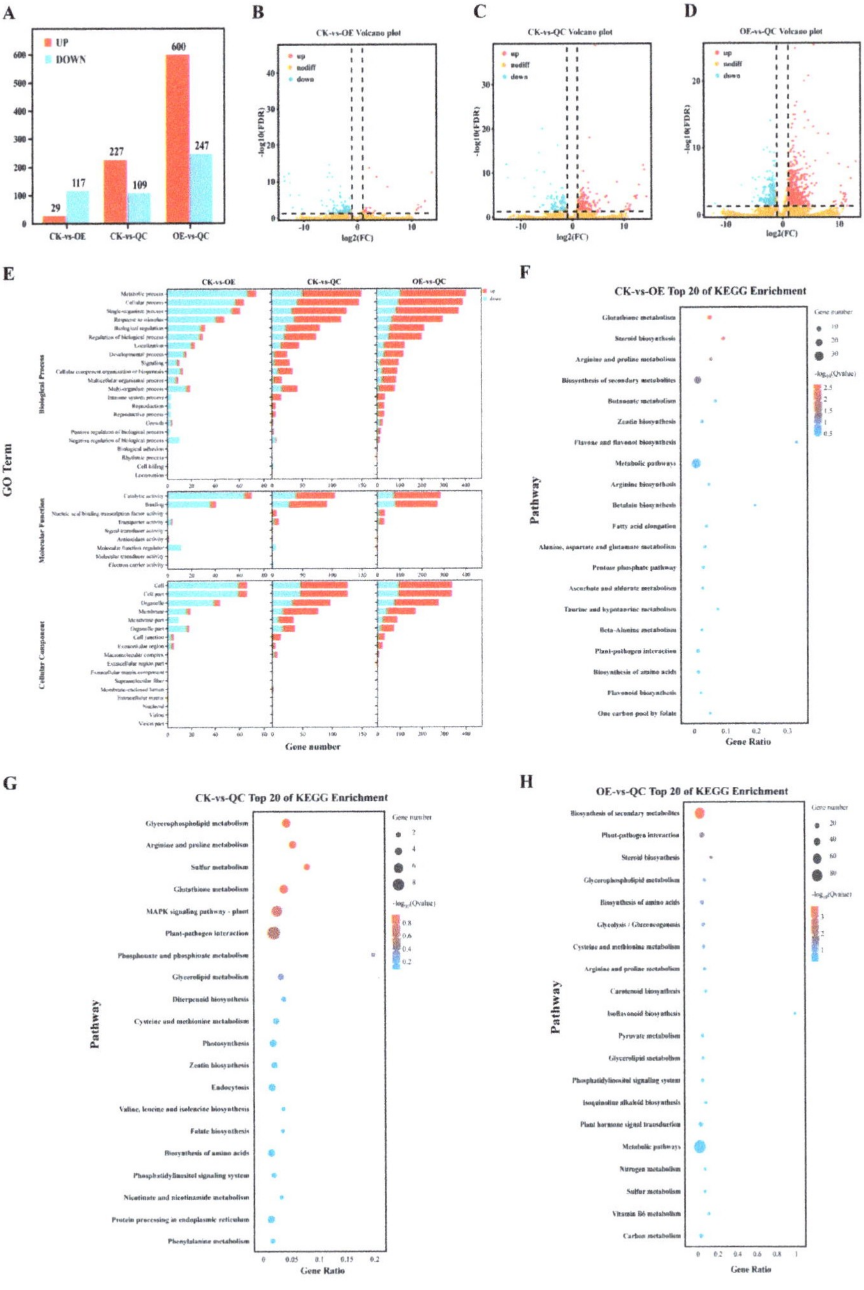

Figure 7. Transcriptome sequencing analysis. (**A**) Statistics of number of differentially expressed genes. (**B–D**) Volcano plot of differentially expressed genes between groups. Red, upregulated expression; blue, downregulated expression; orange, no difference in expression. (**E**) GO analysis of differentially expressed genes. (**F–H**) Scatter plot of enriched KEGG pathways of differentially expressed genes.

2.6. SlPLATZ17 and POR1 Proteins Have a Reciprocal Relationship

To screen for proteins that interact with SlPLATZ17, we performed a yeast two-hybrid library screen. First, the toxicity and self-activation of the pGBKT7-SlPLATZ17 bait gene were examined, and it was found that all the transformants grew on SD medium lacking Leu and Trp (SD/Trp-Leu), whereas their growth on SD medium lacking Leu, Trp, His and Ade (SD/Trp-Leu-His-Ade) was effectively inhibited, indicating that *SlPLATZ17* did not have toxic effects and that self-activation did not occur (Figure 8A,B). After screening and analysis, we ultimately selected seven proteins that may have relevant functions for interaction validation (Table S4). Among these, only pGBKT7-SlPLATZ17 and pGADT7-POR1 showed normal growth in SD/Trp-Leu and SD/Trp-Leu-His-Ade + X-α-gal medium, and the clones presented blue colouration (Figure 8C), whereas the clones with the remaining six proteins presented no blue colouration on the quadruple dropout medium. These findings indicated that SlPLATZ17 interacted with the POR1 protein but not with the remaining six proteins. These interactions were further verified via a bimolecular fluorescence complementation (BiFC) assay, which revealed that SlPLATZ17 interacted with POR1 proteins in the nucleus (Figure 8D).

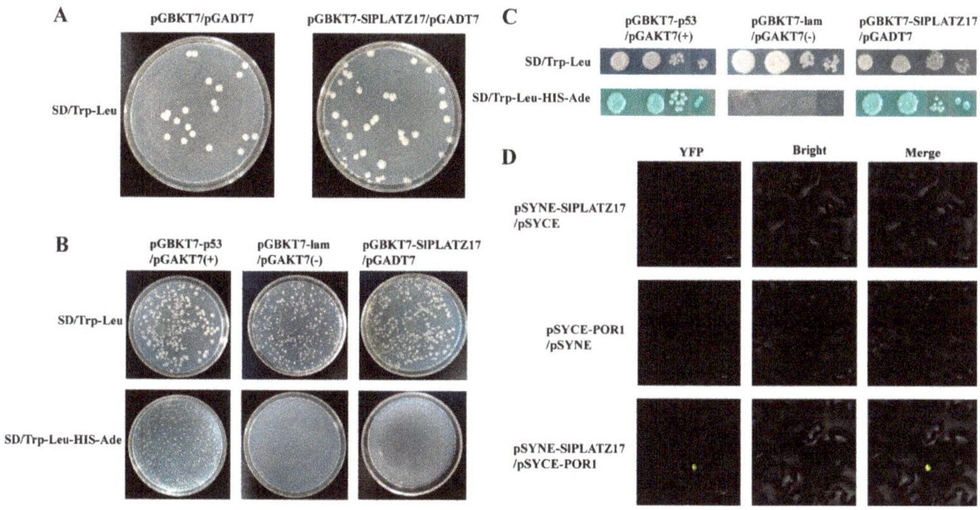

Figure 8. POR1 is an important SlPLATZ17-interacting protein. (**A**) SlPLATZ17 genotoxicity testing. (**B**) *SlPLATZ17* gene autoactivation test. (**C**) Yeast two-hybrid validation of the SlPLATZ17-interacting protein. (**D**) The identification of the SlPLATZ17-interacting proteins via BiFC. Bars: 20 μm.

2.7. Expression Pattern Analysis and Subcellular Localization of SlPLATZ17-Interacting Protein POR1

We examined the expression of *POR1* in various tissue fractions of tomatoes and found that *POR1* expression was high in the roots and flowers of AC plants, 5- to 8-fold higher than that in red fruit (control) (Figure 9A). Further analysis of its expression pattern under stress conditions revealed that *POR1* expression was low from 0 to 6 h of drought stress, and by 12 h, *POR1* expression increased rapidly in both the AC and OE plants, even exhibiting a 20-fold increase in the OE plants (Figure 9C). Under salt stress, *POR1* expression tended to decrease overall in the three groups of plants, but the change in value was small (Figure 9D). Subcellular localization revealed that POR1 was localized to the nucleus (Figure 9B), which was consistent with the expression profile of *SlPLATZ17*.

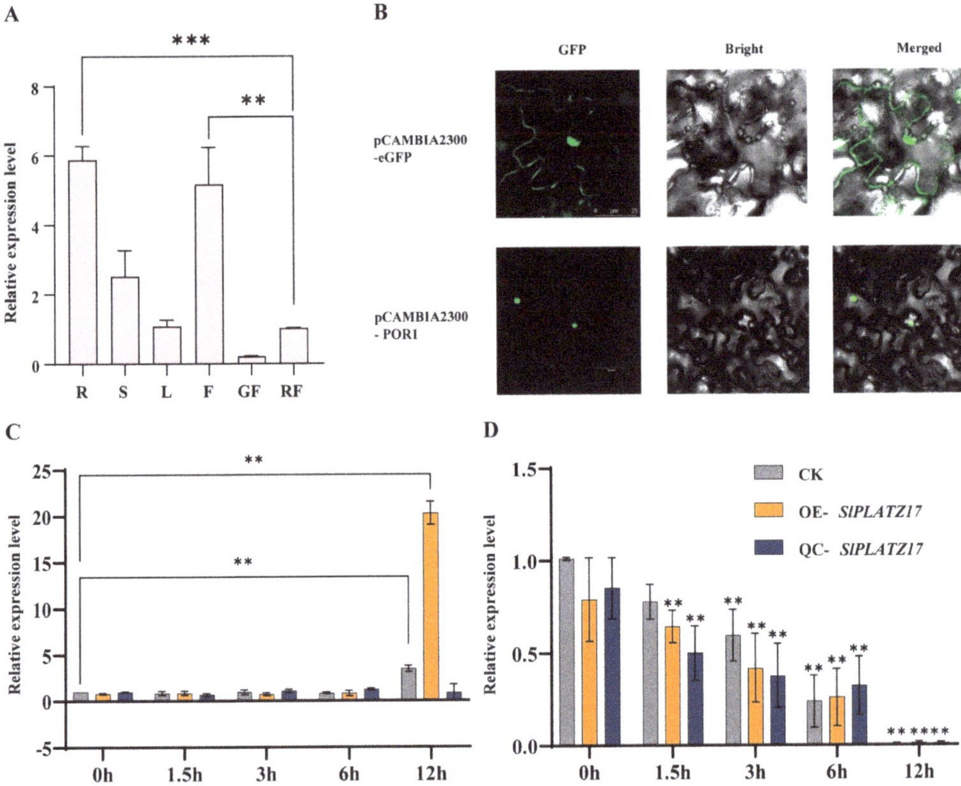

Figure 9. Expression pattern analysis of *POR1* in tomato and its subcellular localization. (**A**) Expression of POR1 in R, roots; S, stems; L, leaves; F, flowers; GF, fruit at mature green stage; RF, fruit at mature red stage in WT. (**B**) Subcellular localization of POR1 protein in tobacco leaves. (**C**) Analysis of expression pattern of POR1 under drought stress. Bars: 25 μm; 60 μm. (**D**) Analysis of expression pattern of POR1 under salt stress. Data are presented as means (±SEs) of three independent biological replicates (**, $p \leq 0.01$; ***, $p \leq 0.001$).

2.8. Functional Prediction for the SlPLATZ17-Interacting Protein POR1 on the Basis of Transcriptome Sequencing Data

To further explore the regulatory pathways in which POR1 may be involved, we established a network of interacting genes in the transcriptome database for the protein POR1, as shown in Figure 10A. The set of POR1 protein-interacting genes was subjected to GO and KEGG analyses (Figure 10B,C). The genes that interact with the POR1 protein were enriched mainly in cellular processes, individual tissue processes, metabolic processes, and stimulus responses in the GO analysis, and meticulous delineation revealed them to be involved in small molecule binding, purine-ribonucleoside binding, ribonucleoside binding, purine-ribonucleoside binding, nucleoside binding, carbohydrate derivative binding, cytoplasmic fractions, and the cytoplasm. The genes interacting with POR1 proteins identified via KEGG analyses were enriched in protein processing in the endoplasmic reticulum, glutathione metabolism, glycolysis/gluconeogenesis, and carbon metabolism.

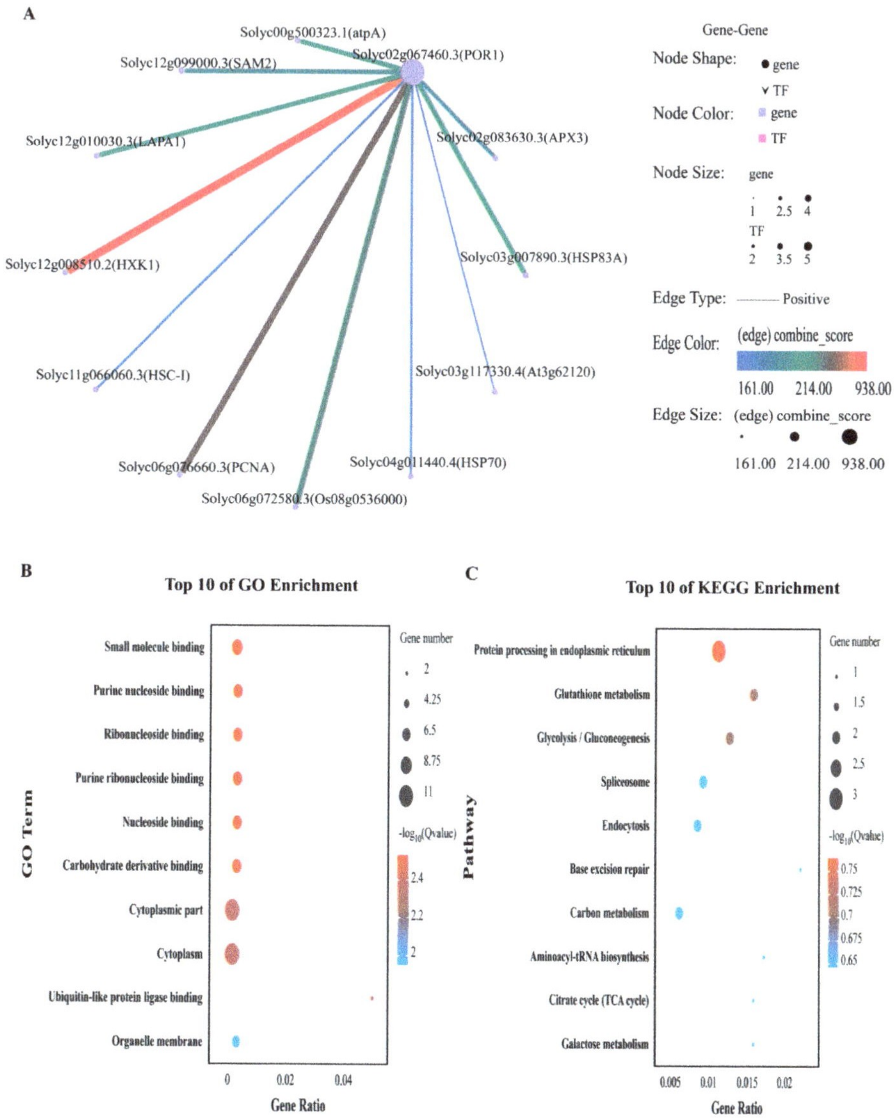

Figure 10. POR1 protein-interacting gene set. (**A**) POR1 protein-interacting gene set. (**B**) GO analysis. (**C**) KEGG analysis.

We analyzed the correlations of the coexpression of the reciprocal protein POR1 and found that most of the genes were highly correlated with POR1 (Figure 11A). These genes were subsequently analyzed via GO, and most of the genes were enriched in the processes of carboxylic acid metabolism, membrane lipid biosynthesis, oxyacid metabolism, organic acid metabolism, small molecule metabolism, polyamine metabolism, membrane lipid metabolism, and lipid biosynthesis (Figure 11B). KEGG analysis revealed that most of the differentially expressed genes were associated with the biosynthesis of secondary metabolites, steroid biosynthesis, metabolic pathways, arginine and proline metabolism, glutathione metabolism, and other metabolic processes (Figure 11C).

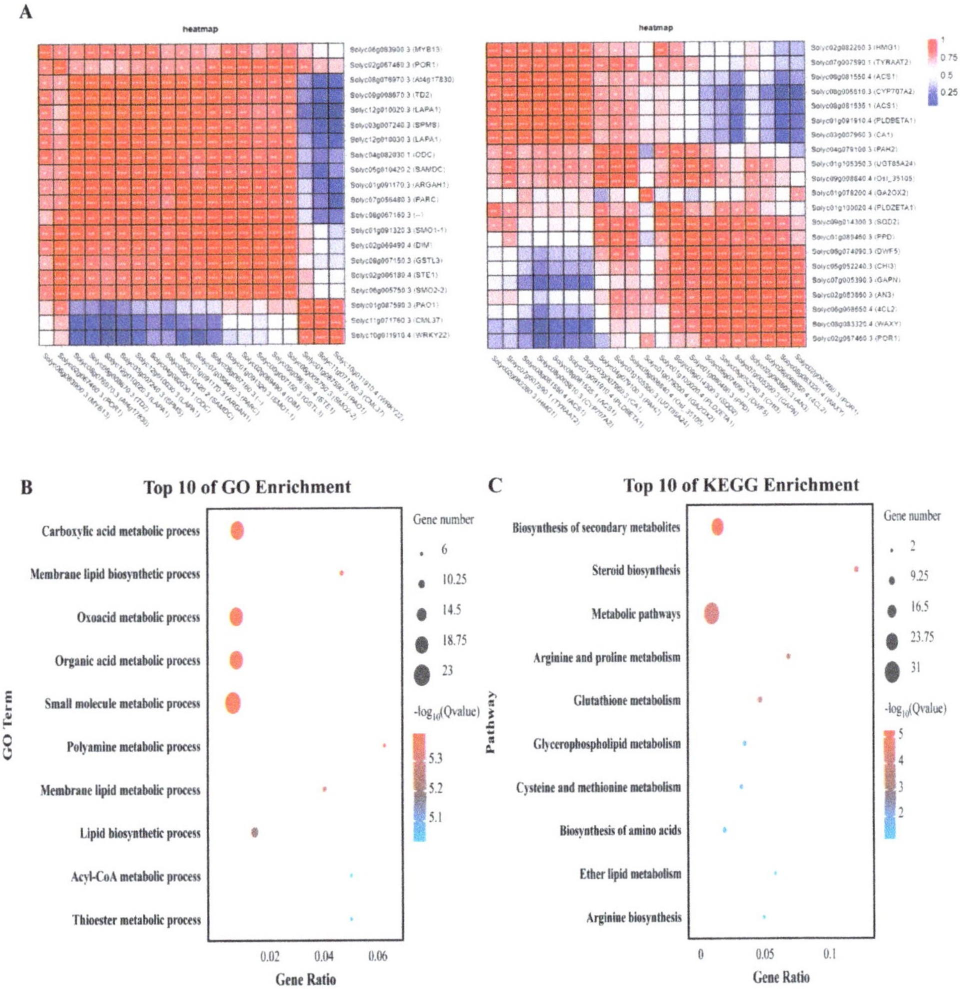

Figure 11. Set of genes coexpressed with POR1 protein. (**A**) Genes coexpressed with POR1 protein. (**B**) GO analysis. (**C**) KEGG analysis (*, $p \leq 0.05$; **, $p \leq 0.01$; ***, $p \leq 0.001$).

2.9. Validation of qRT-PCR Expression Patterns of Important Differentially Expressed Genes

For the important differentially expressed pathways, such as glutathione metabolism and arginine and proline metabolism, screened in the above analysis, we selected nine differentially expressed genes for further expression pattern analysis and validation, as shown in Figure 12. These genes showed obvious expression changes in OE or QC, indicating that the expression of these genes was indeed affected by changes in the expression of the *SlPLATZ17* gene, which may further affect the metabolic pathways in which these genes are located. In addition, the expression patterns in the qRT-PCR results showed the same trend of change as the transcriptome sequencing results, indicating that our RNA-Seq data were accurate and reliable.

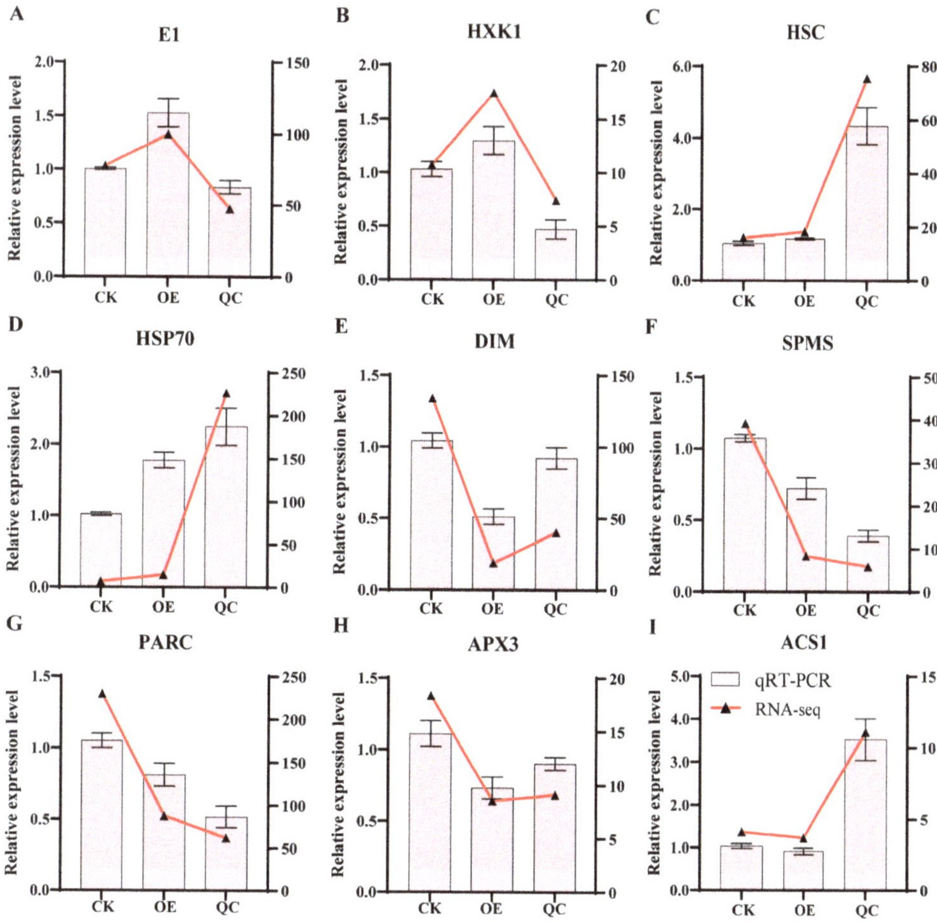

Figure 12. qRT-PCR validation of RNA-Seq results. (**A**,**B**) Relative expression of genes related to glycolysis. (**C**,**D**) Relative expression of genes related to protein processing in endoplasmic reticulum. (**E**) Relative expression of genes related to steroid biosynthesis. (**F**) Relative expression of genes related to arginine and proline metabolism. (**G**,**H**) Relative expression of genes related to glutathione metabolism. (**I**) Relative expression of genes related to secondary metabolism. Vertical coordinate on left represents qRT-PCR results for differentially expressed genes, while vertical coordinate on right represents RNA-seq results. Data are presented as means (±SEs) of three independent biological replicates.

3. Discussion

3.1. SlPLAT17 Can Regulate the Salt and Drought Tolerance of Tomatoes by Reducing the Accumulation of ROS

In the normal growth state of a plant, a relative balance is maintained between the production of ROS and the antioxidant system in its body, which protects the plant from damage [23,24]. However, when a plant is subjected to stress, it is not possible to maintain this balance, and a large amount of ROS accumulates in the plant cells, disrupting the membrane structure of the cells and damaging the corresponding functions [25,26]. In previous research [21], we found that the reduced expression of *SlPLATZ17* decreased the drought and salt tolerance of tomato plants, and we hypothesized that the downregulation of *SlPLATZ17* might lead to the accumulation of free radicals and ROS, thus weakening plant stress tolerance. Here, we used stable overexpression and knockout to further analyze the function of *SlPLATZ17* in stress tolerance and found that the behaviour of the *SlPLATZ17* knockout plants was consistent with that of the gene-silenced plants and that the *SlPLATZ17*-overexpressing plants presented increased drought and salt tolerance, suggesting that the *SlPLATZ17* gene positively regulates the stress tolerance of these plants (Figures 3 and 5). H_2O_2 and O_2^- levels can be detected by nitroblue tetrazolium (DAB) and 3,3′-diaminobenzidine (NBT) staining, which allows for the measurement of the amount of ROS produced by plants after stress injury [27]. We stained the leaves of the drought- and salt-treated plants at different times and confirmed that the accumulation of ROS was lowest in the overexpression plants and highest in the knockout plants, which exhibited the most severe damage. In conclusion, these results suggest that *SlPLATZ17* can reduce the production and accumulation of ROS by increasing the free radical scavenging and antioxidant capacity of plants, which in turn modulates the drought and salt tolerance of tomatoes.

3.2. SlPLAT17 May Be Involved in the Regulation of Root and Flower Development in Tomato Plants

We screened the *SlPLATZ17* gene in a tomato stress response-related yeast library via yeast two-hybridization and obtained one unreported reciprocal protein in tomato, POR1, after double validation via yeast two-hybrid one-to-one and BiFC assays. An analysis of tissue-specific expression patterns (Figures 1A and 9A) revealed that *SlPLATZ17* and *POR1* presented the highest expression in roots and flowers, whereas the expression changes in other tissues were not obvious, and a previous experimental study revealed that the development of roots and flowers of *SlPLATZ17*-overexpressing plants differed from that of wild-type plants [21]. Therefore, the *SlPLATZ17* gene and POR1 protein may be jointly involved in the regulation of root and flower development in tomato plants. In wheat, the PLATZ transcription factor *GL6* positively regulates grain length by altering cell division and affecting cell proliferation to control the number of cells in the seed grain during spikelet development [28]. In poplar, *PtrPLATZ14* regulates leaf development by affecting the cell size control genes *PtrGRF/GIF* and *PtrTCP* [29]. Therefore, POR1 may also play a regulatory role in the expansion or elongation of plant organs together with *SlPLATZ17*, but the specific function of POR1 needs to be investigated via further experiments.

3.3. The Expression Pattern of POR1 Under Salt Stress Is Different from That of SlPLATZ17

Both SlPLATZ17 and POR1 were localized in the nucleus (Figures 1B and 9B). BiFC showed that the interaction sites were also located in the nucleus (Figure 8D). The specific nuclear localization of *SlPLATZ17* suggests that it has a specific role as a transcription factor in transcriptional regulation and a synergistic regulatory role with POR1. However, the expression pattern of POR1 under stress was different from that of *SlPLATZ17*, especially under salt treatment, and the expression of *POR1* decreased with increasing treatment time (Figure 9D), whereas the expression of *SlPLATZ17* was significantly greater under stress than under normal conditions. The different individual response periods of the two to stress suggest that the interaction between SlPLATZ17 and POR1 may occur at the later stage of the stress response, and there may be a more complex mode of action between the two, which should be investigated in depth with respect to the specific function of POR1 and the specific mode of regulation with *SlPLATZ17*.

3.4. *SlPLATZ17* Mainly Exerts Its Functions Through the Glutathione Metabolism and the Arginine and Proline Metabolism Pathways

In further clarifying the mechanism by which the *SlPLATZ17* gene is involved in resistance to abiotic stress in tomato, by transcriptome sequencing, we found that the differentially expressed genes in the *SlPLATZ17*-overexpressing, wild-type, and gene-edited plants were enriched in different metabolic pathways. Kyoto Encyclopedia of Genes and Genomes (KEGG) analysis revealed alterations in several pathways, and the metabolic pathways associated with both the *SlPLATZ17*-overexpressing and the *SlPLATZ17* knockout materials compared with the wild-type plants were related to glutathione metabolism and arginine and proline metabolism (Figure 7). We further established a set of reciprocal genes and genes coexpressed with the reciprocal protein POR1 in the transcriptomic database (Figures 10 and 11) and found that the metabolic pathway involved in plant resistance to stress overlapped with the metabolic pathway via which *SlPLATZ17* functions in glutathione metabolism and arginine and proline metabolism. These results suggest that *SlPLATZ17* most likely functions through the glutathione metabolic pathway as well as the arginine and proline metabolic pathways.

Glutathione (GSH) is a ubiquitous, abundant, and indispensable thiol involved in a variety of biological processes, such as ROS scavenging, redox signalling, sulphur storage and transport, detoxification of hazardous substances, and metabolism of many compounds [30]. The overexpression of the *SbGST* (Tau-like glutathione transferase) gene in transgenic tobacco promotes seed germination and growth under salt stress, and *GST* can be genetically engineered to increase abiotic stress tolerance in plants [31]. The application of exogenous Pro or the transformation of genes involved in Pro metabolism enhances plant tolerance to oxidative stress [2]. Strawberry *FaNAC2* improves tolerance to abiotic stress by regulating proline metabolism [32]. Plant researchers agree that proline accumulation is beneficial to plants, especially after recovery from stress [33]. The maize protein arginine methyltransferase gene plays an important role in flowering regulation and abiotic stress tolerance in Arabidopsis [34]. Therefore, we hypothesized that *SlPLATZ17* is involved in glutathione metabolism pathway when plants are subjected to abiotic stress, which in turn improves the stress tolerance of tomato plants by controlling the accumulation of ROS and other activities involving glutathione and other substances [35]. Simultaneously, the metabolism of arginine and proline maintains the osmotic balance within and outside the cells while scavenging free radicals, thereby enhancing the drought resistance of plants and reducing damage from salt stress. However, in this study, we did not measure the contents of glutathione and arginine and could not determine the specific metabolites through which *SlPLATZ17* and POR1 play regulatory roles; thus, the downstream regulatory mechanisms should be further investigated in detail in subsequent studies.

In this study, we demonstrated the role of the *SlPLATZ17* gene in regulating abiotic stress tolerance in tomatoes by analyzing expression, phenotypic, physiological, and biochemical data. By integrating physiological assessments, molecular techniques, and transcriptomics, we have provided new insights into the biological function of *SlPLAZT17* in the regulation of drought and salt tolerance, and the phenotypes observed in transgenic plants are expected to be used to improve drought and salt tolerance in crops. In future work, we will conduct a more detailed investigation into the downstream regulatory mechanisms of the *SlPLATZ17* gene and hope to widely apply the *SlPLAZT17* gene in the breeding of salt- and drought-tolerant crop varieties.

4. Materials and Methods

4.1. Generation of SlPLATZ17 Overexpression and Knockout Strains

In this study, tomato *SlPLATZ17*-overexpressing T2 generation lines obtained in our previous study were used [21]. *SlPLATZ17* knockout plants were obtained via CRISPR/Cas9 technology. First, the target design programme on the website http://skl.scau.edu.cn/ (accessed on 15 January 2023) was used to design the target. Then, primers for constructing expression boxes were generated via the "primer Design V" programme (http://skl.scau.edu.cn/primerdesign/vector/, accessed on 15 January 2023) (Table S1). The pYLCRISPR/Cas9-SlPLATZ17 vector was constructed and transformed into the DH10B colonic receptor for the Agrobacterium-mediated genetic transformation of tomatoes. The editing results were verified via PCR and Sanger sequencing, and the T0 knockout-positive seedlings were naturally grown for two generations to obtain the T2 generation with genetically stable double allelic mutations for testing.

4.2. Plant Stress Treatments

The T2 generation of *SlPLATZ17*-overexpressing, T2 generation of gene knockout, and wild-type AC (Ailsa Craig) tomato seedlings were placed in a light incubator for cultivation (light period, 16 h; light intensity, 35,000 lx; temperature, 20–25 °C; dark period, 8 h; temperature, 13–15 °C; and relative humidity, 45%). When the seedlings grew to the 4-leaf and 1-heart stage, uniformly growing plants were selected and cultured in Hoagland nutrient solution for 12 h. Afterwards, the plants were divided into two groups and subjected to drought treatment with 15% PEG6000 and salt treatment with 200 mM NaCl. Each treatment was replicated three times, and the phenotypic changes in the plants were observed at 0, 1.5, 3, 6, and 12 h after the treatment. The samples were obtained, snap-frozen in liquid nitrogen, and stored at −80 °C until used for the determination of physiological indices.

4.3. Quantitative Real-Time PCR (qRT-PCR) Analysis

Total RNA was extracted from samples using Trizol. cDNA synthesis was performed using the M-MLVRTase Kit (TaKaRa, Dalian, China). qRT-PCR was performed using the iQ5 system. For each qRT-PCR, the mixture consisted of 10 μL SYBR® Green Master Mix, 0.5 μL each of forward primer and reverse primer, 1 μL of diluted cDNA, and 8 μL of ddH_2O in a 20 μL reaction. The reaction was carried out as follows: 95 °C for 5 min, followed by 40 cycles of 94 °C for 5 s, 60 °C for 15 s, and 72 °C for 10 s. The $2^{-\Delta\Delta CT}$ method was used for data analysis, and *Actin-7* was used as a reference gene. All specific primers used for qRT-PCR are listed in Table S1. Each qRT-PCR assay included three biological replicates.

4.4. Subcellular Localization

The coding sequence (CDS) of the target gene without a stop codon was amplified and fused with the *pCAMBIA2300-eGFP* vector to obtain the *pCAMBIA2300*-target gene construct, which was subsequently introduced into *Agrobacterium rhizogenes* GV3101. The recombinant Agrobacterium cells were subsequently transformed into tobacco leaves. After two days of incubation in the dark, fluorescence signals were detected with a laser confocal microscope (LSM800, Zeiss, Jena, Germany). The specific primers used for vector construction are listed in Table S1.

4.5. Determination of Physiological Indicators

SOD activity was determined following the methods of Beyer and Fridovich [36], and POD activity was determined according to the methods of Chance and Maehly [37]. The proline content in leaves was estimated via the methods of Claussen [38], and the MDA content was estimated via the methods of Cervilla et al. [39]. ROS were detected using via nitroblue tetrazolium (NBT) and 3,3′-diaminobenzidine (DAB) staining methods [40,41]. The blade conductivity was measured with an FE38-ST conductivity metre.

4.6. Yeast Two-Hybrid Screening

To identify *SlPLATZ17*-interacting proteins, we used a yeast two-hybrid library screening system. The CDS of *SlPLATZ17* was inserted into pGBKT7 (BD) to construct the pGBKT7-SlPLATZ17 bait plasmid. In addition, a positive control (pGBKT7-p53/pGADT7-T), a negative control (pGBKT7-lam/pGADT7-T), and an empty plasmid control (pGBKT7/pGADT7-T) were used. First, the bait gene pGBKT7-SlPLATZ17 was tested for toxicity and self-activation. The pGBKT7-SlPLATZ17 plasmid and the two-hybrid library plasmid pGADT7 (AD) were then integrated into the Y2HGold yeast strain via cotransformation. The transformants were screened with selective quadruple dropout SD medium (SD/-Trp-Leu-His-Ade) and X-α-gal. After 3–5 days of incubation at 30 °C, single colonies that turned blue on the medium were subjected to Sanger sequencing to obtain sequences for homology analysis. In addition, candidate interacting proteins were screened via yeast two-hybrid one-to-one validation to identify proteins that interact with SlPLATZ17.

4.7. Bimolecular Fluorescence Complementation (BiFC) Experiment

The CDSs of *SlPLATZ17* and *POR1* without a terminator were constructed in the pSYNE-YFP vector and pSYCE-YFP vector, respectively, followed by infiltration in tobacco, and yellow fluorescent signals were captured under a laser confocal microscope.

4.8. Transcriptomic Analysis

Young leaves of *SlPLATZ17* gene-overexpressing plants, *SlPLATZ17* knockout plants, and CK (AC) plants at the 4-leaf–1-heart stage were collected in three physiological replicates, with a total of nine samples, and stored at −80 °C after quick freezing in liquid nitrogen. After the RNA was extracted, it was sent to Gene Denovo Biotechnology for subsequent library construction and sequencing. To ensure data quality, we removed various types of low-quality reads and then used HISAT2 for alignment analysis on the basis of the reference genome [42]. On the basis of the results of the HISAT2 comparison, the transcripts were then reconstructed via Stringtie, the expression of all the genes in each sample was calculated via RSEM, and GO analysis, KEGG analysis, etc., were performed on the differentially expressed genes (DEGs). We then established the protein POR1 interacting gene set and co-expressed gene set to find metabolic pathways.

4.9. Statistical Analysis

Univariate analysis of variance (ANOVA) was performed on the data using SPSS (26.0) software. Post hoc multiple comparisons were conducted using the least significant difference (LSD) and Duncan's tests. Asterisks (*, **, and ***) are used to denote significant differences ($p \leq 0.05$, $p \leq 0.01$, and $p \leq 0.001$, respectively). The data in this report were obtained from a minimum of three independent experiments, each comprising three technical replicates.

5. Conclusions

We report that the PLATZ transcription factor *SlPLATZ17* is involved in regulating tomato resistance to drought and salt stress. The overexpression of *SlPLATZ17* increased seedling resistance to PEG6000 and NaCl, whereas knockout of the *SlPLATZ17* gene exacerbated the damage caused by these stressors to the plants. Furthermore, our study revealed that SlPLATZ17 interacts with POR1 in response to stress and may regulate the accumulation of ROS through pathways involving glutathione metabolism and arginine and proline metabolism. In summary, this research identified the role of *SlPLATZ17* in the response to salt and drought stress, expanding our understanding of the biological functions of PLATZ transcription factors in plants and providing new targets for future studies aimed at enhancing plant resistance.

Supplementary Materials: The following supporting information can be downloaded at: https://www.mdpi.com/article/10.3390/ijms26051836/s1.

Author Contributions: Conceptualization, X.J., X.P., X.X. and T.Z.; methodology, X.J., X.P. and M.X.; software, X.J., X.P., M.X. and C.W.; validation, X.P., C.W., H.W. and J.L.; formal analysis, X.J., M.X. and H.W.; investigation, X.J., X.P. and M.X.; resources, X.J., M.X. and C.W.; data curation, X.P., H.W. and J.L.; writing—original draft preparation, X.J. and X.P.; writing—review and editing, X.J., X.X. and T.Z.; visualization, X.J. and X.P.; supervision, H.Z., D.L., X.X. and T.Z.; project administration, X.X. and T.Z.; funding acquisition, X.X. and T.Z. All authors have read and agreed to the published version of the manuscript.

Funding: This research was funded by the Key research and development plan of Heilongjiang Province (SC2022ZX02C0202), the National Natural Science Foundation of China (Grant No. 32102390), and supported by the earmarked fund for CARS (CARS-23-A11).

Institutional Review Board Statement: Not applicable.

Informed Consent Statement: Not applicable.

Data Availability Statement: All data are displayed in the manuscript and Supplementary Files.

Acknowledgments: We acknowledgment the Heilongjiang Province Modern Agricultural Industry Technology Collaborative Innovation Promotion System.

Conflicts of Interest: The authors declare no conflicts of interest.

References

1. Shakespear, S.; Sivaji, M.; Kumar, V.; Arumugam Pillai, M.; Wani, S.H.; Penna, S.; Yasin, J.K. Navigating Through Harsh Conditions: Coordinated Networks of Plant Adaptation to Abiotic Stress. *J. Plant Growth Regul.* **2024**, 1–19. [CrossRef]
2. Zulfiqar, F.; Ashraf, M. Proline alleviates abiotic stress induced oxidative stress in plants. *J. Plant Growth Regul.* **2023**, *42*, 4629–4651. [CrossRef]
3. Naing, A.H.; Kim, C.K. Abiotic stress-induced anthocyanins in plants: Their role in tolerance to abiotic stresses. *Physiol. Plant.* **2021**, *172*, 1711–1723. [CrossRef] [PubMed]
4. Tavanti, T.R.; Melo, A.A.R.; Moreira, L.D.K.; Sanchez, D.E.J.; Silva, R.D.S.; Silva, R.M.D.; Reis, A.R.D. Micronutrient fertilization enhances ROS scavenging system for alleviation of abiotic stresses in plants. *Plant Physiol. Biochem.* **2021**, *160*, 386–396. [CrossRef]
5. You, J.; Chan, Z. ROS Regulation During Abiotic Stress Responses in Crop Plants. *Front. Plant Sci.* **2015**, *6*, 1092. [CrossRef]
6. Guo, Y.; Jiang, Q.; Hu, Z.; Sun, X.; Fan, S.; Zhang, H. Function of the auxin-responsive gene *TaSAUR75* under salt and drought stress. *Crop J.* **2018**, *6*, 181–190. [CrossRef]
7. Yadav, S.; Gill, S.S.; Passricha, N.; Gill, R.; Badhwar, P.; Anjum, N.A.; Francisco, J.-B.J.; Tuteja, N. Genome-wide analysis and transcriptional expression pattern-assessment of superoxide dismutase (SOD) in rice and Arabidopsis under abiotic stresses. *Plant Gene* **2019**, *17*, 100165. [CrossRef]
8. Xu, X.; Ye, X.; Xing, A.; Wu, Z.; Li, X.; Shu, Z.; Wang, Y. Camellia sinensis small GTPase gene (CsRAC1) involves in response to salt stress, drought stress and ABA signaling pathway. *Gene* **2022**, *821*, 146318. [CrossRef]
9. Kiełbowicz-Matuk, A. Involvement of plant C_2H_2-type zinc finger transcription factors in stress responses. *Plant Sci.* **2012**, *185*, 78–85. [CrossRef] [PubMed]
10. Gourcilleau, D.; Lenne, C.; Armenise, C.; Moulia, B.; Julien, J.L.; Bronner, G.; Leblanc-Fournier, N. Phylogenetic study of plant Q-type C_2H_2 zinc finger proteins and expression analysis of poplar genes in response to osmotic, cold and mechanical stresses. *DNA Res.* **2011**, *18*, 77–92. [CrossRef] [PubMed]
11. Mysore, K.S.; Crasta, O.R.; Tuori, R.P.; Folkerts, O.; Swirsky, P.B.; Martin, G.B. Comprehensive transcript profiling of Pto- and Prf-mediated host defense responses to infection by Pseudomonas syringae pv. tomato. *Plant J.* **2002**, *32*, 299–315. [CrossRef] [PubMed]

12. Guo, Y.H.; Yu, Y.P.; Wang, D.; Wu, C.A.; Yang, G.D.; Huang, J.G.; Zheng, C.C. GhZFP1, a novel CCCH-type zinc finger protein from cotton, enhances salt stress tolerance and fungal disease resistance in transgenic tobacco by interacting with GZIRD21A and GZIPR5. *New Phytol.* **2009**, *183*, 62–75. [CrossRef]
13. Wei, W.; Lu, L.; Bian, X.H.; Li, Q.T.; Han, J.Q.; Tao, J.J.; Yin, C.C.; Lai, Y.C.; Li, W.; Bi, Y.D. Zinc-finger protein GmZF351 improves both salt and drought stress tolerance in soybean. *J. Integr. Plant Biol.* **2023**, *65*, 1636–1650. [CrossRef] [PubMed]
14. Nagano, Y.; Furuhashi, H.; Inaba, T.; Sasaki, Y. A novel class of plant-specific zinc-dependent DNA-binding protein that binds to A/T-rich DNA sequences. *Nucleic Acids Res.* **2001**, *29*, 4097–4105. [CrossRef] [PubMed]
15. Han, X.; Rong, H.; Tian, Y.; Qu, Y.; Xu, M.; Xu, L.A. Genome-Wide Identification of PLATZ Transcription Factors in *Ginkgo biloba* L. and Their Expression Characteristics During Seed Development. *Front. Plant Sci.* **2022**, *13*, 946194. [CrossRef]
16. Zhang, X.; Lan, Y.; Wang, L.; Liu, H.; Jiang, N.; He, W.; Yan, H.; Wu, M.; Xiang, Y. Whole-genome identification and multiple abiotic stresses expression pattern profiling analysis of PLATZ transcription factor family members in Pecan (*Carya illinoensis*). *Int. J. Biol. Macromol.* **2023**, *248*, 125959. [CrossRef]
17. Zhou, S.R.; Xue, H.W. The rice PLATZ protein SHORT GRAIN6 determines grain size by regulating spikelet hull cell division. *J Integr. Plant Biol.* **2020**, *62*, 847–864. [CrossRef] [PubMed]
18. Liu, M.; Wang, C.; Ji, Z.; Lu, J.; Zhang, L.; Li, C.; Huang, J.; Yang, G.; Yan, K.; Zhang, S.; et al. Regulation of drought tolerance in *Arabidopsis* involves the PLATZ4-mediated transcriptional repression of plasma membrane aquaporin *PIP2;8*. *Plant J.* **2023**, *115*, 434–451. [CrossRef]
19. Zhao, J.; Zheng, L.; Wei, J.; Wang, Y.; Chen, J.; Zhou, Y.; Chen, M.; Wang, F.; Ma, Y.; Xu, Z.-S. The soybean PLATZ transcription factor GmPLATZ17 suppresses drought tolerance by interfering with stress-associated gene regulation of GmDREB5. *Crop J.* **2022**, *10*, 1014–1025. [CrossRef]
20. Liu, S.; Yang, R.; Liu, M.; Zhang, S.; Yan, K.; Yang, G.; Huang, J.; Zheng, C.; Wu, C. PLATZ2 negatively regulates salt tolerance in Arabidopsis seedlings by directly suppressing the expression of the *CBL4/SOS3* and *CBL10/SCaBP8* genes. *J. Exp. Bot.* **2020**, *71*, 5589–5602. [CrossRef]
21. Xu, M.; Gao, Z.; Li, D.; Zhang, C.; Zhang, Y.; He, Q.; Qi, Y.; Zhang, H.; Jiang, J.; Xu, X. Functional prediction of tomato PLATZ family members and functional verification of SlPLATZ17. *J. Integr. Agric.* **2024**, *23*, 141–154. [CrossRef]
22. Liu, W.; Xie, X.; Ma, X.; Li, J.; Chen, J.; Liu, Y.G. DSDecode: A Web-Based Tool for Decoding of Sequencing Chromatograms for Genotyping of Targeted Mutations. *Mol. Plant* **2015**, *8*, 1431–1433. [CrossRef] [PubMed]
23. Gill, S.S.; Tuteja, N. Reactive oxygen species and antioxidant machinery in abiotic stress tolerance in crop plants. *Plant Physiol. Biochem.* **2010**, *48*, 909–930. [CrossRef] [PubMed]
24. Al-Khayri, J.M.; Abdel-Haleem, M.; Khedr, E.H. Harnessing GABA Pathways to Improve Plant Resilience Against Salt Stress. *Horticulturae* **2024**, *10*, 1296. [CrossRef]
25. Khedr, E.H.; Khedr, N. Unraveling the mechanisms behind salt stress alleviation and enhanced fruit storability in pomegranate through triacontanol treatment. *Sci. Hortic.* **2024**, *325*, 112634. [CrossRef]
26. Hu, Y.; Zhao, H.; Xue, L.; Nie, N.; Zhang, H.; Zhao, N.; He, S.; Liu, Q.; Gao, S.; Zhai, H. *IbMYC2* Contributes to Salt and Drought Stress Tolerance via Modulating Anthocyanin Accumulation and ROS-Scavenging System in Sweet Potato. *Int. J. Mol. Sci.* **2024**, *25*, 2096. [CrossRef] [PubMed]
27. Chen, Y.; Feng, P.; Zhang, X.; Xie, Q.; Chen, G.; Zhou, S.; Hu, Z. Silencing of *SlMYB50* affects tolerance to drought and salt stress in tomato. *Plant Physiol. Biochem.* **2022**, *193*, 139–152. [CrossRef]
28. Wang, A.; Hou, Q.; Si, L.; Huang, X.; Luo, J.; Lu, D.; Zhu, J.; Shangguan, Y.; Miao, J.; Xie, Y.; et al. The PLATZ Transcription Factor GL6 Affects Grain Length and Number in Rice. *Plant Physiol.* **2019**, *180*, 2077–2090. [CrossRef] [PubMed]
29. Wang, X.; Yuan, W.; Yuan, X.; Jiang, C.; An, Y.; Chen, N.; Huang, L.; Lu, M.; Zhang, J. Comparative analysis of PLATZ transcription factors in six poplar species and analysis of the role of PtrPLATZ14 in leaf development. *Int. J. Biol. Macromol.* **2024**, *263*, 130471. [CrossRef]
30. Ito, T.; Ohkama-Ohtsu, N. Degradation of glutathione and glutathione conjugates in plants. *J. Exp. Bot.* **2023**, *74*, 3313–3327. [CrossRef] [PubMed]
31. Jha, B.; Sharma, A.; Mishra, A. Expression of *SbGSTU* (*tau* class glutathione S-transferase) gene isolated from *Salicornia brachiata* in tobacco for salt tolerance. *Mol. Biol. Rep.* **2011**, *38*, 4823–4832. [CrossRef] [PubMed]
32. Liang, J.; Zheng, J.; Wu, Z.; Wang, H. Strawberry *FaNAC2* Enhances Tolerance to Abiotic Stress by Regulating Proline Metabolism. *Plants* **2020**, *9*, 1417. [CrossRef] [PubMed]
33. Hosseinifard, M.; Stefaniak, S.; Ghorbani Javid, M.; Soltani, E.; Wojtyla, Ł.; Garnczarska, M. Contribution of Exogenous Proline to Abiotic Stresses Tolerance in Plants: A Review. *Int. J. Mol. Sci.* **2022**, *23*, 5186. [CrossRef]
34. Ling, Q.; Liao, J.; Liu, X.; Zhou, Y.; Qian, Y. Genome-Wide Identification of Maize Protein Arginine Methyltransferase Genes and Functional Analysis of *ZmPRMT1* Reveal Essential Roles in *Arabidopsis* Flowering Regulation and Abiotic Stress Tolerance. *Int. J. Mol. Sci.* **2022**, *23*, 12793. [CrossRef] [PubMed]

35. Dorion, S.; Ouellet, J.C.; Rivoal, J. Glutathione Metabolism in Plants under Stress: Beyond Reactive Oxygen Species Detoxification. *Metabolites* **2021**, *11*, 641. [CrossRef]
36. Beyer, W.F., Jr.; Fridovich, I. Assaying for superoxide dismutase activity: Some large consequences of minor changes in conditions. *Anal. Biochem.* **1987**, *161*, 559–566. [CrossRef] [PubMed]
37. Change, B.; Maehly, A. Assay of catalases and peroxidase. *Methods Enzym.* **1955**, *2*, 764–775.
38. Claussen, W. Proline as a measure of stress in tomato plants. *Plant Sci.* **2005**, *168*, 241–248. [CrossRef]
39. Cervilla, L.M.; Blasco, B.; Ríos, J.J.; Romero, L.; Ruiz, J.M. Oxidative stress and antioxidants in tomato (*Solanum lycopersicum*) plants subjected to boron toxicity. *Ann. Bot.* **2007**, *100*, 747–756. [CrossRef] [PubMed]
40. Bournonville, C.F.; Díaz-Ricci, J.C. Quantitative determination of superoxide in plant leaves using a modified NBT staining method. *Phytochem. Anal.* **2011**, *22*, 268–271. [CrossRef] [PubMed]
41. Ramel, F.; Sulmon, C.; Bogard, M.; Couée, I.; Gouesbet, G. Differential patterns of reactive oxygen species and antioxidative mechanisms during atrazine injury and sucrose-induced tolerance in *Arabidopsis thaliana* plantlets. *BMC Plant Biol.* **2009**, *9*, 28. [CrossRef] [PubMed]
42. Kim, D.; Langmead, B.; Salzberg, S.L. HISAT: A fast spliced aligner with low memory requirements. *Nat. Methods* **2015**, *12*, 357–360. [CrossRef] [PubMed]

International Journal of
Molecular Sciences

MDPI

Review

Resilience of Maize to Environmental Stress: Insights into Drought and Heat Tolerance

Huaijun Tang [1], Lei Zhang [1], Xiaoqing Xie [1], Yejian Wang [1], Tianyu Wang [2,*] and Cheng Liu [1,*]

[1] Institute of Crops Research, Xinjiang Academy of Agricultural Sciences, Urumqi 830091, China; tanghuaijun83@sina.com (H.T.); 18999224030@163.com (L.Z.); 13565990757@163.com (X.X.); wangyejian0815@163.com (Y.W.)

[2] Institute of Crop Sciences, Chinese Academy of Agricultural Sciences, Beijing 100081, China

* Correspondence: wangtianyu@caas.cn (T.W.); liuchengxj@126.com (C.L.)

Abstract: Maize (*Zea mays* L.) is a staple cereal crop worldwide, but its productivity is significantly affected by extreme weather conditions such as drought and heat stress. Plant growth, physiological processes, and yield potential are all affected by these conditions; as such, resilient maize crops are required to tackle these abiotic challenges. With an emphasis on morphological, physiological, and biochemical reactions, this review paper investigates the processes that underlie resistance to certain environmental challenges. Features including deep root systems, osmotic adaptations, and antioxidant enzyme activity help maize withstand drought. Activation of drought- and heat-responsive genes, accumulation of osmoregulatory compounds, and changes in membrane fluidity are all components of abiotic stress tolerance. Likewise, improved transpiration efficiency, modified photosynthetic processes, and improved heat shock proteins are used to produce heat resistance. Enhancing resilience requires progress in breeding methods, genetic engineering, and agronomic techniques, such as the use of stress-tolerant cultivars, biotechnology interventions, and climate-smart agriculture tactics. A special focus was given to cutting edge technologies like CRISPER-Cas9-mediated recent advances in heat and drought resistance. This review sheds light on recent studies and potential avenues for enhancing resilience to harsh climatic conditions, guaranteeing food security in the face of climate change.

Keywords: abiotic stress; drought resistance; heat stress; climate adaptation; physiological responses

1. Introduction

Air temperatures have increased since the turn of the 20th century and are predicted to continue to climb due to climate unpredictability. High-temperature stress (HTS), which causes major harm to plants, may be caused by these extreme conditions [1,2]. Extreme drought and heat stress severely reduce crop productivity by disrupting key physiological processes like photosynthesis, pollination, and grain filling. These stresses are major threats to global food security, especially for staple crops like maize, wheat, and rice. As climate change intensifies, the frequency and severity of such events are expected to rise, putting additional pressure on agricultural systems. Therefore, given the current agroclimatic conditions, food security has emerged as a critical challenge [3–5]. High day and night temperatures are endangering the world's agricultural production system, according to climate models [6]. Globally, this leads to a decrease in maize crop yield and productivity [7]. Maize is one of the most widely cultivated crops globally, serving diverse purposes as a staple food source. Efforts to improve its yield, quality, and stability across various growing conditions remain a primary focus [5,8]. It is the most important cereal crop in the world and is used as a raw ingredient in many culinary and feed businesses.

Heat stress is one of the growth-limiting variables that significantly affect maize's growth and nutritional content at various stages of development [9,10]. The development of better breeding practices is crucial for raising maize yield and quality, since many abiotic challenges, such as heat stress and drought stress, occur simultaneously [11]. Further research is vital for creating maize genotypes resistant to drought stress and high temperatures. Several experts consider stomatal conductance a key indirect parameter for selecting heat-tolerant crops [12]. Similarly, osmoprotectants and chaperone proteins play a vital role in maize's adaptive response to heat stress and other environmental stressors. Additionally, proteins associated with leaf senescence enhance maize's resilience to both heat and drought stress [13]. It is possible to create maize hybrids that are resistant to heat and drought stress by introducing these features into locally adapted hybrids through possible donor hybrids. Moreover, identification of donor genotypes possessing favorable traits is important in heat stress breeding programs [14]. Cereal plants of the grass family provide nearly half of humanity's total caloric intake and offer more nutrition than any other food group. Although more than a dozen cereal crops are consumed as food, for example, wheat, maize, and rice dominate human diets, accounting for 94% of total cereal consumption. Maize is the third most important cereal crop contributing to global food security [15]. Cereal consumption varies significantly across regions, with rice being the predominant crop in Central Asia, the Middle East, North and South America, and Europe. While maize, commonly known as corn, is preferred in Southern and Eastern Africa, Central America, and Mexico, rice is the predominant cereal in Asia.

Maize is a good source of fiber, several B vitamins, and essential minerals. However, it is generally low in calcium, folate, and iron, and lacks certain nutrients, including vitamin B12 and vitamin C. Certain foods or components in the diet, such as vegetables, tea (like oxalates), coffee (like polyphenols), eggs, and milk (like calcium), might limit the absorption of iron, especially the nonheme iron found in maize [16]. In countries where anemia and iron deficiency are considered moderate or severe public health problems, the fortification of maize flour and cornmeal with iron and other vitamins and minerals has been used to improve micronutrient intake and prevent iron deficiency (WHO, 2009). As maize is a vital staple food crop and contributes significantly to global food security, efforts were made to fortify maize. The development of Zn- and Fe-enriched cultivars are great examples of fortification through biotechnological approaches and conventional breeding [17,18].

The aim of this review is to provide a comprehensive understanding of the mechanisms underlying maize resistance to drought and heat stress, focusing on morphological, physiological, and biochemical responses. The objective is to highlight recent advances in breeding strategies, genetic engineering, and agronomic practices that enhance abiotic stress tolerance, with the ultimate goal of supporting the development of climate-resilient maize varieties to ensure food security under changing climatic conditions.

2. Major Challenges Posed by Abiotic Stress in Maize

The increasing ecological impacts of climate change and the rising competition for environmental resources due to population growth underscore the challenges abiotic stress poses to plant growth and development. Heat and drought stress can significantly reduce maize yields by affecting plant growth, pollination, and kernel development. These stresses often occur together and worsen each other's impact. During critical growth stages like flowering and grain filling, even short periods of high temperature or water shortage can cause substantial yield loss. Maize is particularly sensitive to water deficits during silking, leading to poor kernel set. Overall, the combination of heat and drought can lead to high yield losses, depending on the severity and timing of the stress [19–21]. Agricultural production is expected to be most affected by climate change, particularly in low-latitude regions where developing nations are concentrated [22]. The negative consequences of rising carbon dioxide and high temperatures will compel researchers to develop effective adaptation strategies [23]. These limitations on the world's food supply and environmental balance promote the study and creation of climate-smart and climate-resilient crops [24].

All research on environmental stressors or abiotic elements that can cause stress to a range of species is included in the field of plant abiotic stress [25]. Extreme and low levels of light, UV-B and UV-A radiation, temperature extremes (cold and freezing), water extremes (drought, flooding, and submersion), chemical factors (heavy metals and pH), salinity from too much Na^+, a lack or surplus of vital nutrients, gaseous pollutants (ozone and sulfur dioxide), mechanical factors, and other less common conditions are some examples of these stressors. A range of physiological interactions is expected, as multiple stresses such as heat and drought often occur simultaneously in field conditions, leading to unique effects that cannot be predicted from individual stressors alone, thereby requiring innovative, tailored solutions [26].

Plants, being deeply rooted in their environment, must continuously adapt to changing conditions influenced by various environmental factors. When these factors exceed optimal levels, they can cause abiotic stress, challenging the plant's growth and survival. Determining how plants sense various stressors, how early signals are transduced inside the plant, what variety of response pathways they trigger, and how they are genetically defined is a major challenge in abiotic stress biology [27]. Depending on a plant's genetic makeup and adaptive response, environmental stimuli, whether biotic or abiotic, might present a challenge or stress. An environment suitable for one genotype of plant may not be for another. Numerous impacts in response to the environment are provided by particular genotype $\times$ environment interaction combinations [28,29].

3. Significance of Understanding and Improving Maize Adaptability to Drought and Heat

Climate factors such as temperature, precipitation, and humidity influence crop development, with the impact of climate change on yields varying by crop type and region. Crops can experience growth suppression due to abiotic stressors including drought and severe temperatures [30]. These stressors' main characteristics include the potential to lower crop vigor, impede growth and development, and lower crop output. Extreme weather events, including heat waves and droughts, severely affect maize productivity. These elements hinder maize growth, damage crops during drought, and drastically lower output [31]. As temperatures rise and rainfall deficits grow, the frequency of droughts is expected to increase. Crop yields are most severely impacted by drought stress during the reproductive or growth stage. Plant growth, reproduction, and physiology are all negatively impacted by drought, which has a significant impact on agricultural output [32]. Field experiment data from 1980 to 2015 indicate that drought, causing approximately 40% water loss, reduced maize yields by 39.3% and wheat yields by 20.6% [33]. In both dry and non-arid areas, maize was more vulnerable to drought than wheat, particularly during the reproductive stage. Drought stress can significantly reduce kernel number, ear size, and overall plant growth. During the vegetative stage, up to V12, drought stress often affects the final maize yield. Maize is particularly vulnerable to stress during fertilization and pollination, with yields potentially decreasing by 3–4% per day if drought stress persists for two weeks before pollination, causing the plant to wilt. Yield losses might reach 8% per day, depending on the degree of stress experienced during the formation of silk and pollen. Furthermore, yield losses might amount to as much as 6% per day if drought stress continues for two weeks following silking.

In maize, pollination refers to the transfer of pollen from tassels to ears. When pollen germinates on the silk, it forms a pollen tube that delivers genetic material to each ovule, leading to kernel formation. However, extreme heat stress and drought can disrupt the synchronization of silk emergence and pollen availability, affecting fertilization (Figure 1). Additionally, it can dry out exposed silk, which prevents it from absorbing pollen grains. In the meantime, drought stress is typically accompanied with high-temperature damage during the pollination phase in maize, though it can also happen on its own. Temperatures exceeding 35 °C, combined with low relative humidity, generally dry exposed silk but have little effect on its elongation. Conversely, low relative humidity and temperatures above 32 °C increase the risk of pollen damage or loss. Studies suggest that prolonged exposure to temperatures above 32 °C can negatively affect moisture levels and grain filling [34]. Fortunately, pollen release typically occurs between early and mid-morning, ensuring a daily supply of fresh pollen until maturation and production are complete.

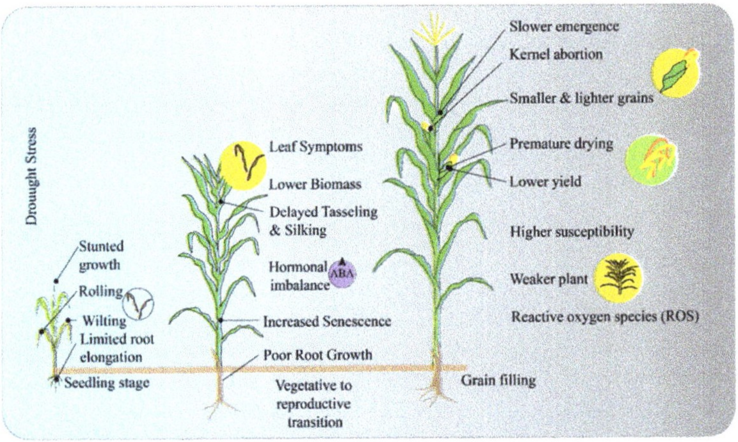

Figure 1. Maize response to drought stress at critical growth stages. The symptoms such as stunted growth, wilting, hormonal imbalance (ABA), and reduced yield. It illustrates how drought impacts plant development from seedling emergence to grain filling, leading to issues like slower emergence and kernel abortion.

Every plant has a range of ideal temperatures for growth, and temperatures outside of this range hinder the development and growth of the plant. One of the most dangerous abiotic stresses is the slow rise in the yearly mean temperature and heat. Ultimately, rising temperatures can significantly impact crop growth and development, which could influence agricultural products' output and quality [35]. In cooler climates, some crops may benefit from rising temperatures; however, overall, higher temperatures reduce yields and negatively impact crop productivity. The first application of the Crop-Environment Resource Synthesis (CERES) maize model was to estimate historical variations in maize yield over around 200 sub-Saharan African regions [36,37]. The study modeled a fictitious future in which maize yields in sub-Saharan Africa would decline more with a 2 °C rise in temperature than with a 20% drop in precipitation. Compared to other organs, reproductive organs have a much lower temperature threshold for heat stress injury [38]. The synthesis of viable pollen, its transfer to the male gamete's germ cell, and the start and maintenance of embryo and endosperm development are all necessary for maize to mature into a kernel. High temperatures are linked to lower production because they reduce grain weight and number, especially during reproductive growth. Fewer ovules will mature into kernels when fertilized at high temperatures. Furthermore, heat stress can alter the shape and physiological function of the tassels, the reproductive structures responsible for pollen production [10]. Rising temperatures and shifting precipitation patterns, driven by increased greenhouse gas emissions, pose a significant threat to crop yields and global food security. As temperatures rise, drought intensifies due to rapid moisture loss from soil surfaces and plant tissues, while high temperatures can also directly damage crops [39]. Food security depends on agriculture, which is severely impacted by heat stress and drought, either separately or in combination. Prolonged drought and extreme climate variability significantly reduce crop yields and increase losses during critical stages of maize growth and development. Maize yields have been steadily declining due to irregular rainfall patterns, while drought conditions have also led to a reduction in maize cultivation areas. Additionally, the combined effects of heat and drought during critical growth stages can significantly impact crop productivity. These abiotic stressors are widely recognized for influencing the growth and spread of weeds, insects, and diseases, often leading to the emergence of harmful pests [40]. Through its effects on photosynthesis [41], drought reduces crop growth and yield by inducing wilting, slowed growth, delayed leaf emergence, and reduced leaf area, particularly during the seedling stage [42]. In maize, yield is decided during the flowering stage due to the close association between pollen release and the growth of maize silk; further, there is a close relationship between the final yield and the ASI (Anthesis–Silking Interval). Consequently, drought stress during blooming inhibits the growth of maize silk and lowers production [43,44].

4. Drought Resistance in Maize

4.1. Root Architecture and Water Uptake Efficiency

A promising strategy for enhancing crop drought tolerance is breeding crops with root systems optimized for efficient water uptake [45]. Given the complex relationship between root traits and diverse hydrological conditions, modeling provides essential insights for trait-based selection. In one study, the effect of root architecture was examined [46]. Other studies have investigated whether a universally drought-adapted root system ideotype exists or if water uptake efficiency depends on specific hydrological conditions by integrating a root architecture model with a soil–hydrological model [47,48]. To achieve this, the transpiration of 48 root designs was modeled across sixteen drought scenarios, varying in soil texture, rainfall patterns, and initial soil moisture availability. The findings indicate that hydrological conditions directly influence the efficiency of water absorption by root structures. It is not always the case that deep and dense root systems are better at absorbing water. In a different study, findings showed that identifying root systems with optimal functionality requires more than just architectural description [49]. When rainfall is the primary source of water during root system development, root density, especially near the soil surface, becomes crucial for optimizing soil moisture uptake. Therefore, we conclude that trait-based root breeding should prioritize root systems adapted to the hydrological conditions of the target environment. For instance, some root-related parameters are presented in Figure 2. Root architecture plays a vital role in a plant's resource-foraging strategy and is a key determinant of productivity [50,51]. The key factors influencing overall water uptake are shown in Figure 2.

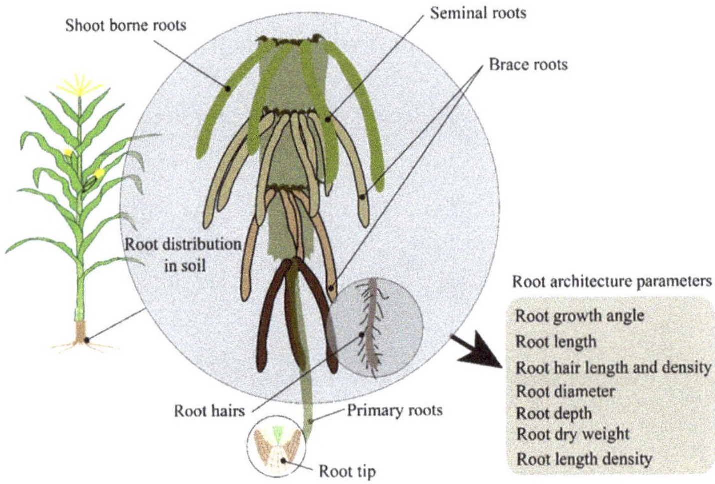

Figure 2. Root architecture parameters regulating drought tolerance in maize. The maize root system and different types of rooting in maize.

The efficiency (units of nutrient acquired per unit of resource invested by the plant) with which various root architecture forms obtain nutrients from soil have been attempted to be quantified using theoretical analysis. According to these findings, highly branched (dichotomous-like) root structures could efficiently capture immobile ions (like phosphate) from limited soil volumes. This capability paves the way for the targeted exploration of local soil volumes with the highest level of precision [52,53]. Additionally, these results indicate that herringbone-like root systems with coarser, sparser branches are better at capturing mobile nutrients over broad soil volumes (maximizing scale). The latter are better at examining spatially varied soil, but they can be costly in terms of carbon and may only be preferred when development is constrained by the soil's resource availability [53]. These analyses offer a valuable theoretical framework for studying root architecture. However, root systems were depicted as ineffective structures that were uniformly nutrient-supplied, and the conclusions were solely based on diffusion theory. Plant root systems are typically not insensitive to their surroundings, and soils are not all the same. This raises the question of whether such generalizations hold true for real-world root systems, which interact dynamically with soils that vary across space and time. Locally elevated absorption kinetics and locally accelerated root proliferation are two well-known plasticity mechanisms that allow root systems to adapt to their varied environment [48]. Individually, these two processes can play a crucial role in nutrient acquisition, helping to mitigate supply inconsistencies [54].

4.2. Osmotic Adjustment and Accumulation of Compatible Solutes

In response to drought, plants either sustain water absorption or reduce water loss (stomatal conductance). Osmotic adjustment (OA), a biochemical process that aids plants in adapting to dry and salty environments, facilitates the latter process within plant cells [55]. The quantity of osmotically active compounds in the cell increases because of OA [56]. In leaf tissue and other metabolically active cells, this increase in solutes results in a greater negative osmotic potential, which can enhance the level of cell hydration and preserve turgor. In other words, if OA happens, plants can continue their metabolic activities and live longer in drying soil. OA has improved productivity and growth of several crop cultivars under drought stress (Figure 3). OA can be caused by a variety of chemicals, such as organic acids, amino acids, carbohydrates, and inorganic cations and anions (Figure 3).

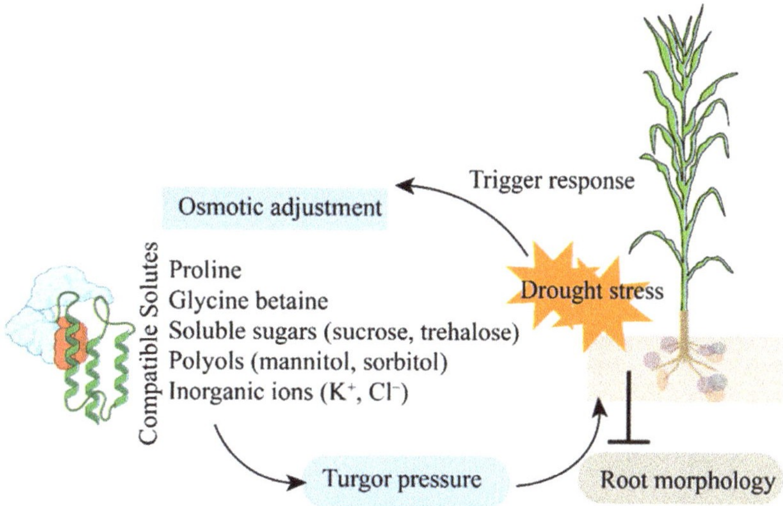

Figure 3. Illustration of the adaptive mechanisms of maize under drought stress, including osmotic adjustment, maintenance of turgor pressure, and modifications in root morphology.

The accumulation of certain solutes with protective properties is frequently linked to OA. These hydroxyls (−OH), a group rich in compatible solutes, which include sugars, cyclitols, proline, and glycine betaine, can build up in the cytoplasm and aid in preventing dehydration of cellular membranes, proteins, and enzymes [57]. However, it is crucial to understand that in many species, OA is primarily caused by the accumulation of several solutes, and that individual solutes do not significantly contribute to OA. OA is typically a gradual process that is sensitive to the timing and severity of stress since it necessitates metabolism or absorption of solutes.

There is some indication that studies have underestimated leaf relative water content (RWC) when assessing OA, which is on top of the intrinsic heterogeneity in OA expression. An accurate evaluation of the relative capacity for OA in various plants depends on the proper monitoring of plant water status, such as RWC. When threatened by desiccation due to drought or external lowering of the osmotic pressure, such as increase in soil salinity, most organisms increase the cellular concentration of osmotically active compounds, known as compatible solutes [58–60]. At high quantities, the accumulating chemicals are "compatible" with regular cellular metabolism [61]. The idea that suitable solutes could take the role of water at the surface of proteins, protein complexes, or membranes stems from the fact that they are often hydrophilic [62]. The term "compatible solute" carries a physiological connotation but does not explicitly define the functions these solutes perform. Although the biochemical processes by which suitable solutes provide protection are currently unclear, this does not necessarily mean that efforts to create transgenic plants that promote metabolite accumulation are off the table.

Many studies supported the concept of osmotic adjustment in maize to overcome the drought stress conditions. Two tropical lowland maize populations were used in a study to understand components of its genetic variance and heritability in maize [63]. The findings showed that more genetic variation was detected with data collected at the flowering stage, when water stress was more severe, than at the vegetative stage [63]. Later in a different experiment, osmotic adjustment was examined in a group of maize hybrids at different stages of growth. The results showed a positive tendency was observed between osmotic adjustment and phenotypic stability [64]. OA is used as a critical parameter in maize to measure the effect of drought on plants [65,66]. OA of cells helps to conserve the water balance of the plant, and this adjustment is generally achieved through increased amounts of various common solutes. Such evidence shows that maize crops have showed tolerance to water deficit through the mechanism of osmotic adjustment [67], and therefore genes related to OA should be mined and breeding goals can be adjusted accordingly.

4.3. Role of ABA (Abscisic Acid) in Stomatal Regulation

Abscisic acid (ABA), a plant hormone, is essential to plant viability. ABA is involved in developmental processes and rapidly accumulated in plants is observed in response to abiotic stress [68]. When plants encounter abiotic stress, they rapidly trigger the ABA signaling pathway, leading to the activation of transcription factors that respond to ABA and the subsequent expression of ABA-responsive genes [69]. Protein kinases and phosphatases play pivotal roles in ABA transduction in plants. During stress conditions, ABA accumulation starts and these molecules bind to resistance/regulatory components of ABA receptor (RCARs). Mutant plants that lack components of ABA signal transduction or ABA production exhibit extreme drought susceptibility. Our knowledge of ABA signal transduction has improved over the past ten years since the ABA receptor genes were discovered [70,71]. Numerous investigations using genetic, physiological, biochemical, chemical biology, and evolutionary methodologies were used to advance research about role of phytohormones in stomatal regulation. It has been well documented that abscisic acid-induced stomatal closure during abiotic stress [72]. ABA regulates stomatal closure by a chemical signal that can activate the metabolic process through a series of signaling cascades [73]. Furthermore, recent research has demonstrated that ABA signaling, also known as basal ABA signal transduction, plays a crucial role in regulating stomatal control, plant development, and metabolic pathways in stress condition [27,74]. Ion efflux from guard cells mediates stomatal closure triggered by abscisic acid. Stomatal closure is the outcome of ion efflux, which also causes osmotic water efflux and a decrease in the guard cells' volume and turgor. S-type and R-type anion channel activation are key factors in stomatal closure [75,76]. Anion efflux from guard cells is mediated by these channels, and they also depolarize the plasma membrane. These channels triggers depolarization-activated potassium (K) efflux channels. Such channels have an important role in regulation of anion efflux from guard cells and depolarize the plasma membrane. Finally it activates potassium efflux channels responsive to depolarization [76]. The activity of both potassium (K) efflux channels and S-type anion channels facilitates efficient solute release from guard cells and enables stomatal closure. ABA is a plant hormone that can regulate the physiology metabolism under drought conditions in maize. For instance, exogenous ABA was found effective remedies for maize ear dysplasia at grain filling stage under drought stress [77]. Maize Calmodulin-like 3 Gene (*ZmCML3*) positively regulates drought resistance in maize. The gene was targeted after being induced by ABA [78]. Therefore, the ABA pathway is extensively targeted in different crops to understand the drought mechanism. The genes and pathways related to ABA synthesis and signaling are hot topics for research related to the effects of drought on maize.

4.4. Molecular and Genetic Approaches for Drought Tolerance

Understanding the molecular mechanisms of drought-resistant genes in maize is crucial. This knowledge helps develop resilient varieties. Such varieties can address challenges caused by drought. This entails identifying potential genes linked to drought resistance, as well as examining how these genes express themselves during drought stress. Numerous techniques, like genome-wide association studies (GWAS), QTL mapping, comparative genomics, and gene expression profiling, have made it easier to identify drought-resistant genes in maize. Several key drought-resistant genes play a crucial role in drought tolerance by regulating osmotic balance, enhancing water-use efficiency, and activating stress-induced signaling pathways [79]. Determining the function of drought-resistant genes in stress tolerance requires an understanding of how these genes are expressed during drought. The regulation of these genes during dehydration and their role in preserving cellular homeostasis are clarified by gene expression investigations. Table 1 shows some recently identified genes and their role in drought resistance.

Quantitative PCR is a widely used technique for analyzing the expressions of specific drought-resistant genes. Researchers can determine how the expression of genes like *ZmDREB2/2.5/A* [80–82], *ZmAREB*, and *ZmP5CS* [83] change in response to drought stress by analyzing the mRNA levels of these genes. A more thorough method that enables the simultaneous profiling of thousands of genes is RNA sequencing, or RNA-Seq. In maize under drought stress, RNA-Seq has been utilized to find genes that are differently expressed and to shed light on changes in gene expression around the world. Research has demonstrated that during drought stress, genes related to ABA signaling, osmotic control, and ROS scavenging are significantly increased [84]. The expression of genes in maize that respond to drought has been examined using microarrays. According to these findings, a complex regulatory network that includes stress proteins, transporters, and transcription factors is induced during drought and aids maize plants in adjusting to water shortages.

Utilizing the natural genetic variation of plant quantitative traits linked to stress tolerance—where quantitative genetics plays a crucial role through classical and molecular breeding—is the primary genetic strategy for improving multiple-stress tolerance. Genetic areas linked to drought tolerance have been found using QTL analysis. Certain loci have been connected to important characteristics, such as leaf water potential, stomatal conductance, and root depth. Through the overexpression of genes resistant to drought, transgenic techniques have been used to improve drought tolerance. For instance, it has been demonstrated that increasing *ZmP5CS* and other genes linked to drought enhances tolerance to drought in maize plants [84].

Table 1. List of the genes and their role identified by researchers in recent years.

S. No	Gene	Details	Function/s	Reference
1	*ZmHB53*	Homeodomain-leucine zipper I (HD-Zip I) transcription factors (TFs)	ABA receptor *ZmPYL4*	[85]
2	*ZmPHR1*	Transcription factor	phosphorus homeostasis	[86]
3	*ZmTIP2;3*	Tonoplast intrinsic protein	arbuscular mycorrhiza fungi symbiosis	[87]
4	*ZmSCE1a*	E3 SUMO ligase	enhancing the stability of *ZmGCN5*	[88]
5	*ZmNAC55*	Trnacription factor	negatively regulate drought stress via increasing *ZmHOP3* expression in maize	[89]
6	*ZmMIK2-ZmC2DP1*	Kinase 2 proteins	negative regulatory module in maize drought- and salt-stress responses	[90]
7	*ZmCML3*	Calmodulin-like proteins (CMLs)	through increasing proline (Pro) content	[78]

Table 1. *Cont.*

S. No	Gene	Details	Function/s	Reference
8	*ZmGA20ox3*	Loss-of-function mutations of GA biosynthesis enzyme	significantly increased ABA, JA, and DIMBOA levels in mutants	[91]
9	*ZmEULD1b*	*Euonymus europaeus* (EUL) related lectin family,	stomatal development and promotes water-use efficiency	[92]
10	*ZmMYB39*	Transcription factor	stomatal development and promotes water-use efficiency	[93]
11	*ZmGLYI-8*	Glyoxalase I (GLYI)	Overexpressed in model plants	[94]
12	*ZmbHLH47-ZmSnRK2.9*	Transcription factor	ABA response and drought tolerance	[95]
13	*ZmAPX2*	Ascorbate peroxidase 2	reducing ROS content	[96]
14	*ZmSK1*	Glycogen synthase kinase 3 (GSK3)-like kinases	reduces drought tolerance in maize	[97]
15	*ZmDST44*	Drought and salinity tolerance (DST) gene	positive regulator of drought tolerance (*ZmmiR139* regulates *ZmDST44* by cleaving its mRNA)	[98]
16	*ZmPL1*	Phylloplanin-like	negatively regulates drought tolerance in maize (CRISPER-cas9)	[81]
17	*ZmC2H2-149*	Cys(2)/His(2) zinc-finger-proteins (C2H2-ZFPs)	repressing *ZmHSD1* in maize (negative regulator)	[99]
18	*ZmPRX1*	Peroxidase genes	promoting root development and lignification	[100]
19	*ZmSUS1*	Sucrose synthase (SUS)	regulating sucrose metabolism and increasing soluble sugar content	[101]
20	*ZmGRAS15*	GRAS transcription factor	regulating primary root length at the seedling stage	[102]
21	*ZmCYB5-1*	Cytochrome b5 proteins (CYB5s)	negative regulator of drought stress	[103]
22	*ZmHsf28*	Transcription factors	*ZmSnRK2.2-ZmHsf28-ZmJAZ14/17* module is identified to regulate drought tolerance through coordinating ABA and JA signaling	[104]
23	*miR166e/ZmATHB14*	Micro RNAs	*miR166e-ZmATHB14* module regulates drought tolerance	[105]
24	*ZmSNAC06*	NAC transcription factor family	hypersensitivity to abscisic acid (ABA)-positive regulator	[106]

5. Heat Resistance in Maize: Effects and Mechanisms

5.1. Impact of High Temperatures on Maize Physiology

High temperatures can profoundly affect maize physiology, particularly during critical growth stages such as blooming and grain filling. Maize, as a warm-season crop, thrives across a wide range of temperatures. However, when it is subjected to temperatures higher than its ideal range, which is typically between 30 °C and 35 °C, physiological processes are interfered with, which results in decreased growth and productivity. Developing methods to reduce heat stress and increase crop resilience in the face of climate change requires an understanding of how high temperatures affect maize physiology. In maize, high temperatures, particularly those above 35 °C, can inhibit seed germination and seedling establishment. Heat stress on maize plants at different stages of growth can affect growth and development (Figure 4). From seed germination to grain filling, heat stress can lead to a range of adverse effects, including poor seedling vigor, desiccation risk, wilting, chlorophyll damage, asynchronous flowering, pollen viability issues, silk desiccation, shortened grain filling, poor kernel development, kernel shrinkage, dehydration, weaker stalks, and impaired nutrient uptake (Figure 4). These factors collectively result in reduced crop yield and quality, highlighting the critical need for strategies to mitigate heat stress in agricultural practices.

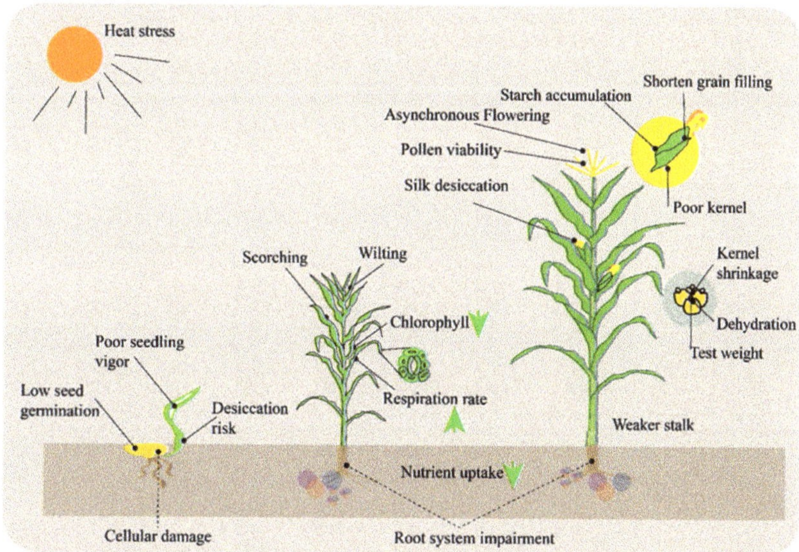

Figure 4. Comprehensive overview of heat stress on maize plants. Three different plants are depicted to show different stages of growth and development. Heat stress can lead to issues such as poor seedling vigor, desiccation risk, and cellular damage during early development. As the plant matures, it may experience chlorophyll degradation, increased respiration rates, nutrient uptake impairment, and root system damage. Later stages show effects like asynchronous flowering, pollen viability issues, silk desiccation, and ultimately poor kernel formation, kernel shrinkage, and weaker stalks, which result in shortened grain filling and reduced test weight.

Defective seed imbibition, often caused by high temperatures during germination, can hinder the activation of metabolic processes essential for seedling emergence. Additionally, high temperatures can interfere with the enzyme activity necessary for the decomposition of seed reserves, which hinders the seed's capacity to sprout [107]. Both root and shoot growth may be impeded by rising temperatures. As plant's ability to absorb water and nutrients may be limited if the root system does not grow sufficiently to support it. High temperatures can also harm the cell membranes in developing tissues, which inhibits cell elongation and division two processes essential to early plant growth [108]. Maize physiology is significantly affected by high temperatures, especially with regard to photosynthesis, reproductive development, and water relations. Reduced growth and yield are the results of oxidative stress, membrane damage, and water deprivation, particularly during crucial reproductive periods. Developing methods to reduce heat stress, such as breeding heat-tolerant cultivars and putting agronomic techniques like irrigation control and shade cover into place, requires an understanding of these physiological reactions. Enhancing maize's ability to withstand heat will be crucial to maintaining global food security as climate change continues to increase the frequency and severity of heat events.

5.2. Heat Tolerance Mechanism

The physiology of maize is negatively impacted by heat stress, particularly at crucial developmental stages like flowering and grain filling, which lowers yield. In order to survive heat stress, maize plants have developed several coping strategies that enable them to preserve cellular integrity, continue growing, and maximize reproductive success in adverse circumstances. The production of heat shock proteins (HSPs) and the preservation of cellular homeostasis are two important processes for heat stress resistance. These systems play a crucial role in protecting the plant from oxidative stress, protein denaturation, and other damage caused by high temperatures. The function of heat shock proteins (HSPs) in protein folding is related to their role in heat tolerance. The class of molecular chaperones known as heat shock proteins (HSPs) is essential for shielding cells from harm while they are under heat stress. These proteins support healthy protein folding under stressful circumstances, aid in the refolding of damaged proteins, and guard against the denaturation of cellular proteins. Heat shock factors (HSFs), which function as transcription factors to start the synthesis of HSPs, are activated in response to high temperatures, increasing the expression of HSPs. Certain HSPs are upregulated in maize in response to heat stress. For example, it has been demonstrated that HSP70 plays a crucial role in heat tolerance, helping to preserve protein homeostasis and shield cellular structures from heat [108]. Furthermore, another heat-induced chaperone, HSP101, contributes to heat tolerance by aiding in the recovery of plants from heat-induced damage, which improves their ability to survive and develop in hot conditions [109]. The stability of proteins, the integrity of cellular structures, and the general equilibrium of ions and metabolites are all at risk under heat stress. Maize plants use a variety of tactics, mediated by HSPs, antioxidants, and modifications in cellular metabolism, to preserve cellular homeostasis and survival. The lipid structures in membranes can be disrupted by high temperatures, which can impair the activity of proteins and enzymes that are membrane-bound and cause membrane fluidity. Under heat stress, maize plants modify their lipid composition by producing unsaturated fatty acids for preserving the integrity of their membranes. This modification stops cellular contents by keeping the membrane steady [110]. Recent studies for the genes and transcription factors identified were listed in Table 2.

Table 2. List of genes identified in recent years with basic studies related to drought resistance in maize.

Gene	Details	Mechanisms/Method	References
ZmHsp18	ZmHsp20 gene family	Gene family-based perfection	[111]
ZmENO1	Enolase (ENO, 2-phospho-D-glycerate hydrolyase)	Antioxidant enzyme activities and osmotic regulation	[112]
ZmDnaJ genes	HSP40s	Correlation between heat stress tolerance and the regulation of genes	[113]
ZmDnaJ-ZmNCED6	Heat shock protein	Involved in ABA signal transduction pathways	[114]
cpSRP43	CMT2 and cpSRP43	CHROMO domain family genes	[115]
ZmDnaJ96	DnaJ/HSP40 gene family	Increased antioxidant enzyme activity	[116]
ZmATL10 and AtATL27	ATL family		[49]

6. Emerging Technologies for Enhancing Drought and Heat Resistance in Maize: From a CRISPR Perspective

Several genetic engineering methods have been used to create maize cultivars that can withstand drought and heat. To maximize drought resistance, these strategies can be roughly divided into three categories: gene editing, introducing target genes, and manipulating multiple genes. The overexpression of genes resistant to drought is one of the most popular strategies for enhancing drought tolerance. Researchers can improve maize's resistance to water stress by introducing genes like *ZmDREB2*, *ZmP5CS*, and *ZmSOD* into the plant's genome. For example, it has been demonstrated that overexpressing *ZmDREB2* increases maize production during drought by triggering downstream stress-response genes. Alternatively, genes that prevent drought tolerance can be silenced via RNA interference (RNAi). For instance, prolonging the growth period of maize under drought stress can increase productivity by silencing genes linked to stress-induced premature senescence. The CRISPR-Cas9 genome-editing technique provides a more accurate method for changing genes linked to drought tolerance [117]. This method enables the targeted modification of particular genes in maize, such as *ZmABF* or *ZmDREB2* regulation, to improve the plant's resistance to drought. CRISPR can be used to eliminate genes that negatively impact water use efficiency, such as those associated with excessive transpiration or premature senescence. For example, CRISPR-Cas9-mediated editing of *ZmPL1* gene showed negative role of Phylloplanin-like 1 in drought tolerance [118]. There are QTLs linked to heat tolerance in maize, and certain loci have been linked to features like pollen viability, photosynthetic efficiency, and thermotolerance. Increased heat resistance is linked to QTLs such as qHT1 on chromosome 2. Genetic engineering aims to introduce heat-resistant genes into maize for improved resilience. For instance, the capacity to tolerate high temperatures, particularly during reproductive development, can be improved by overexpressing *HSP70* and other genes linked to heat stress [119]. Even expression analysis of the *Hsf* and *Hsp70* has showed positive regulation of stress in maize [120]. Recent advancements in omics technologies (Figure 5) have provided researchers with a comprehensive understanding of how maize responds to environmental stressors. While proteomic and metabolomic investigations shed light on changes in proteins and metabolites that occur during stress, transcriptomic analyses assist in identifying genes that are differentially expressed under drought and heat stress conditions. By focusing on important genes involved in stress responses, CRISPR-Cas9 genome-editing technology has the potential to produce maize variants with increased resistance to a variety of pressures (Figure 5) [117,121].

Stress-resilient maize varieties may be developed more quickly thanks accurate and effective methods [122]. Because of its accuracy and effectiveness, CRISPR-Cas9 is a potent tool for creating maize cultivars with specific drought-resistant characteristics without adding extraneous genes that can raise regulatory issues. In certain situations, a single gene might not offer enough resistance to drought in the field. To get around this, several genes implicated in several drought tolerance systems, including osmotic control, stress signaling, and antioxidant defense, are introduced using multigene engineering. This method can simultaneously improve multiple physiological processes, leading to more resilient drought-resistant cultivars. For example, maize plants that have been modified with both *ZmDREB2* [80,123] and *ZmP5CS* show increased drought tolerance and better osmotic adjustment.

Emerging biotechnology advancements are essential in improving maize's resilience to environmental challenges like heat, drought, and salinity as a result of climate change. Of them, the gene-editing technique known as CRISPR-Cas9 is one of the most promising means of producing crops that are more resilient to these difficulties. A groundbreaking technology in genetic engineering, CRISPR-Cas9 enables precise DNA changes at specified sites in an organism [124,125]. Because it allows for the targeted alteration of genes involved in stress-response pathways without introducing foreign DNA, this is very useful for enhancing crop resilience (Figure 5). CRISPR has several advantages over conventional genetic alteration methods, including accuracy, speed, and affordability. Through the targeting of genes that control important stress-response pathways, CRISPR-Cas9 can be used to increase maize's resistance to abiotic stress like heat, drought, and salinity. Researchers can make maize more resilient to harsh environments by suppressing or deleting genes that are sensitive to stress. CRISPR-Cas9 is one of the emerging technologies that has the potential to improve maize's resistance to environmental challenges like heat, drought, and salinity. New possibilities for developing more adapted crops to a changing environment are made possible by the capacity to precisely modify genes linked to stress-response pathways [2,126]. While CRISPR-Cas9 has revolutionized crop improvement by enabling precise genetic modifications, it is not without limitations [127]. One major concern is the potential for off-target effects, where unintended genomic changes may occur, leading to unpredictable phenotypes or undesirable traits [128]. Additionally, regulatory hurdles pose a significant challenge, as many countries classify gene-edited crops under strict GMO regulations, slowing down their adoption and commercialization. To address these issues, emerging genome-editing techniques such as prime editing and base editing offer more precision and control, minimizing off-target activity and enabling subtle, targeted modifications without inducing double-strand breaks, thereby enhancing both safety and regulatory acceptability.

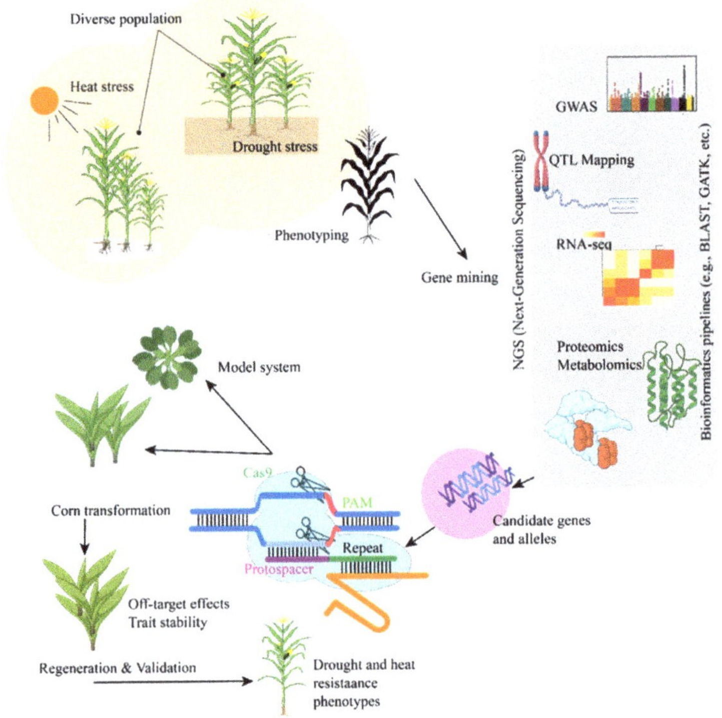

Figure 5. Model to illustrate the application of CRISPR-Cas9 technology in maize improvement to enhance resistance against heat and drought. The process involves targeting specific genes using guide RNA (gRNA) designed to bind to the protospacer region, enabling precise genome editing for stress tolerance enhancement.

7. Conclusions

Breeding programs today face the dual challenge of increasing crop yields to meet global food demands while enhancing resilience to climate-induced stresses. In crops like maize, achieving this balance is critical due to the increasing frequency of extreme weather events such as drought, heat, and flooding. Advances in biotechnology, including gene editing (CRISPR-Cas9), marker-assisted selection, and genomic tools, offer powerful strategies to accelerate the development of high-yielding, stress-tolerant varieties. However, for these innovations to have a lasting impact, their scalability across diverse agroecosystems must be carefully considered. CRISPR-edited maize lines developed under controlled conditions must undergo rigorous testing in varied environments to ensure stable performance across different climates, soils, and management practices. Local adaptation remains key; therefore, integrating edited traits into region-specific germplasm is essential for successful deployment. Furthermore, multi-trait breeding approaches and strong collaborations between researchers, breeders, and policymakers will be vital to translate laboratory breakthroughs into field-level solutions. Utilizing genetic diversity from landraces and wild relatives can further enhance the adaptability of elite lines. Ultimately, combining conventional breeding with modern biotechnological tools will enable the development of maize varieties that not only thrive under optimal conditions but also withstand the growing uncertainties posed by climate change—ensuring food security for future generations.

Author Contributions: C.L. and T.W. designed and supervised this study and revised the manuscript. H.T. conceived and wrote the manuscript; X.X., L.Z., Y.W. and C.L. revised the paper and made suggestions. All authors have read and agreed to the published version of the manuscript.

Funding: This research was funded by an open project of the Corps Key Laboratory of Efficient Utilization of Water and Fertilizer Resources (2023OWSL-04): "Introduction and selection of drought-resistant varieties of maize and research on their water consumption characteristics"; Xinjiang Uygur Autonomous Region Key Research and Development Project (2022B02015-2) "Research and Development of Key Technologies for Water Saving, Fertilizer Saving and Fertilizer Efficiency of Major Grain Crops in Yili River Valley"; Xinjiang Uygur Autonomous Region Key Research and Development Project (2023B02040-1) "Exploration of Population Characteristics and Potential of High-Yielding Maize with Close Planting"; and Xinjiang Agriculture Research System (XJARS-02).

Institutional Review Board Statement: Not applicable.

Informed Consent Statement: Not applicable.

Data Availability Statement: Not applicable.

Conflicts of Interest: The authors declare no conflicts of interest.

References

1. Farooq, M.; Bramley, H.; Palta, J.A.; Siddique, K.H. Heat Stress in Wheat during Reproductive and Grain-Filling Phases. *Crit. Rev. Plant Sci.* **2011**, *30*, 491–507. [CrossRef]
2. Molla, M.S.H.; Nakasathien, S.; Ali, M.A.; Khan, A.; Alam, M.R.; Hossain, A.; Farooq, M.; El Sabagh, A. Influence of nitrogen application on dry biomass allocation and translocation in two maize varieties under short pre-anthesis and prolonged bracketing flowering periods of drought. *Arch. Agron. Soil Sci.* **2019**, *65*, 928–944. [CrossRef]
3. Rojas-Downing, M.M.; Nejadhashemi, A.P.; Harrigan, T.; Woznicki, S.A. Climate change and livestock: Impacts, adaptation, and mitigation. *Clim. Risk Manag.* **2017**, *16*, 145–163. [CrossRef]
4. Vogel, E.; Donat, M.G.; Alexander, L.V.; Meinshausen, M.; Ray, D.K.; Karoly, D.; Meinshausen, N.; Frieler, K. The effects of climate extremes on global agricultural yields. *Environ. Res. Lett.* **2019**, *14*, 54010. [CrossRef]
5. Shiferaw, B.; Prasanna, B.M.; Hellin, J.; Bänziger, M. Crops that feed the world 6. Past successes and future challenges to the role played by maize in global food security. *Food Secur.* **2011**, *3*, 307–327. [CrossRef]
6. Khaliq, A.; Iqbal, M.A.; Zafar, M.; Gulzar, A. Appraising economic dimension of maize production under coherent fertilization in Azad Kashmir, Pakistan. *Custos E Agronegocio* **2019**, *15*, 243–253.

7. El Sabagh, A.; Barutcular, C.; Islam, M.S. Relationships between stomatal conductance and yield under deficit irrigation in maize (*Zea mays* L.). *J. Exp. Biol. Agric. Sci.* **2017**, *5*, 14–21. [CrossRef]
8. Rezaei, E.E.; Webber, H.; Asseng, S.; Boote, K.; Durand, J.L.; Ewert, F.; Martre, P.; MacCarthy, D.S. Climate change impacts on crop yields. *Nat. Rev. Earth Environ.* **2023**, *4*, 831–846. [CrossRef]
9. El-Sappah, A.H.; Rather, S.A.; Wani, S.H.; Elrys, A.S.; Bilal, M.; Huang, Q.; Dar, Z.A.; Elashtokhy, M.M.A.; Soaud, N.; Koul, M.; et al. Heat Stress-Mediated Constraints in Maize (*Zea mays*) Production: Challenges and Solutions. *Front. Plant Sci.* **2022**, *13*, 879366. [CrossRef]
10. Li, T.; Zhang, X.-p.; Liu, Q.; Liu, J.; Chen, Y.-q.; Sui, P. Yield penalty of maize (*Zea mays* L.) under heat stress in different growth stages: A review. *J. Integr. Agric.* **2022**, *21*, 2465–2476. [CrossRef]
11. Cairns, J.E.; Sonder, K.; Zaidi, P.H.; Verhulst, N.; Mahuku, G.; Babu, R.; Nair, S.K.; Das, B.; Govaerts, B.; Vinayan, M.T.; et al. Chapter one—Maize Production in a Changing Climate: Impacts, Adaptation, and Mitigation Strategies. In *Advances in Agronomy*; Sparks, D.L., Ed.; Academic Press: Cambridge, MA, USA, 2012; Volume 114, pp. 1–58.
12. Iqbal, M.A.; Iqbal, M.A.; Aslam, Z.; Maqsood, M.; Ahmad, Z.; Akbar, N.; Khan, H.Z.; Abbas, R.; Khan, R.; Abbas, G.; et al. Boosting forage yield and quality of maize (*Zea mays* L.) with multi-species bacterial inoculation in Pakistan. *Phyton* **2017**, *86*, 84.
13. Zhao, F.; Zhang, D.; Zhao, Y.; Wang, W.; Yang, H.; Tai, F.; Li, C.; Hu, X. The Difference of Physiological and Proteomic Changes in Maize Leaves Adaptation to Drought, Heat, and Combined Both Stresses. *Front. Plant Sci.* **2016**, *7*, 1471. [CrossRef] [PubMed]
14. Cairns, J.E.; Hellin, J.; Sonder, K.; Araus, J.L.; MacRobert, J.F.; Thierfelder, C.; Prasanna, B.M. Adapting maize production to climate change in sub-Saharan Africa. *Food Secur.* **2013**, *5*, 345–360. [CrossRef]
15. Djalovic, I.; Kundu, S.; Bahuguna, R.N.; Pareek, A.; Raza, A.; Singla-Pareek, S.L.; Prasad, P.V.V.; Varshney, R.K. Maize and heat stress: Physiological, genetic, and molecular insights. *Plant Genome* **2024**, *17*, e20378. [CrossRef]
16. Nuss, E.T.; Tanumihardjo, S.A. Maize: A Paramount Staple Crop in the Context of Global Nutrition. *Compr. Rev. Food Sci. Food Saf.* **2010**, *9*, 417–436. [CrossRef]
17. Chhabra, G.; Kaur, G.; Singh, K.; Kaur, J.; Praba, U.P.; Singh, R.; Karnatam, K.S.; Garg, T.; Ranjan, R.; Vikal, Y. Biofortification of Maize: A Promising Approach for Better Nutrition. In *Harnessing Crop Biofortification for Sustainable Agriculture*; Tiwari, S., Singh, B., Eds.; Springer Nature: Singapore, 2024; pp. 145–178.
18. Goredema-Matongera, N.; Ndhlela, T.; Magorokosho, C.; Kamutando, C.N.; van Biljon, A.; Labuschagne, M. Multinutrient Biofortification of Maize (*Zea mays* L.) in Africa: Current Status, Opportunities and Limitations. *Nutrients* **2021**, *13*, 1039. [CrossRef]
19. Hatfield, J.L.; Prueger, J.H. Temperature extremes: Effect on plant growth and development. *Weather Clim. Extrem.* **2015**, *10*, 4–10. [CrossRef]
20. Schlenker, W.; Roberts, M.J. Nonlinear temperature effects indicate severe damages to U.S. crop yields under climate change. *Proc. Natl. Acad. Sci. USA* **2009**, *106*, 15594–15598. [CrossRef]
21. Lobell, D.B.; Schlenker, W.; Costa-Roberts, J. Climate Trends and Global Crop Production Since 1980. *Science* **2011**, *333*, 616–620. [CrossRef]
22. Bellard, C.; Bertelsmeier, C.; Leadley, P.; Thuiller, W.; Courchamp, F. Impacts of climate change on the future of biodiversity. *Ecol. Lett.* **2012**, *15*, 365–377. [CrossRef]
23. Rosenzweig, C.; Elliott, J.; Deryng, D.; Ruane, A.C.; Müller, C.; Arneth, A.; Boote, K.J.; Folberth, C.; Glotter, M.; Khabarov, N.; et al. Assessing agricultural risks of climate change in the 21st century in a global gridded crop model intercomparison. *Proc. Natl. Acad. Sci. USA* **2014**, *111*, 3268–3273. [CrossRef] [PubMed]
24. Wheeler, T.; von Braun, J. Climate Change Impacts on Global Food Security. *Science* **2013**, *341*, 508–513. [CrossRef] [PubMed]
25. Sulmon, C.; van Baaren, J.; Cabello-Hurtado, F.; Gouesbet, G.; Hennion, F.; Mony, C.; Renault, D.; Bormans, M.; El Amrani, A.; Wiegand, C.; et al. Abiotic stressors and stress responses: What commonalities appear between species across biological organization levels? *Environ. Pollut.* **2015**, *202*, 66–77. [CrossRef] [PubMed]
26. Suzuki, N.; Rivero, R.M.; Shulaev, V.; Blumwald, E.; Mittler, R. Abiotic and biotic stress combinations. *New Phytol.* **2014**, *203*, 32–43. [CrossRef]
27. Yoshida, T.; Christmann, A.; Yamaguchi-Shinozaki, K.; Grill, E.; Fernie, A.R. Revisiting the Basal Role of ABA—Roles Outside of Stress. *Trends Plant Sci.* **2019**, *24*, 625–635. [CrossRef]
28. Des Marais, D.L.; Hernandez, K.M.; Juenger, T.E. Genotype-by-Environment Interaction and Plasticity: Exploring Genomic Responses of Plants to the Abiotic Environment. *Annu. Rev. Ecol. Evol. Syst.* **2013**, *44*, 5–29. [CrossRef]
29. Liu, H.-J.; Liu, J.; Zhai, Z.; Dai, M.; Tian, F.; Wu, Y.; Tang, J.; Lu, Y.; Wang, H.; Jackson, D.; et al. Maize2035: A decadal vision for intelligent maize breeding. *Mol. Plant* **2025**, *18*, 313–332. [CrossRef]
30. Aliniaeifard, S.; Rezayian, M.; Mousavi, S.H. Drought Stress: Involvement of Plant Hormones in Perception, Signaling, and Response. In *Plant Hormones and Climate Change*; Ahammed, G.J., Yu, J., Eds.; Springer Nature: Singapore, 2023; pp. 227–250.
31. Khalili, M.; Naghavi, M.R.; Aboughadareh, A.P.; Rad, H.N. Effects of Drought Stress on Yield and Yield Components in Maize cultivars (*Zea mays* L.). *Int. J. Agron. Plant Prod.* **2013**, *4*, 809–812.

32. Fahad, S.; Bajwa, A.A.; Nazir, U.; Anjum, S.A.; Farooq, A.; Zohaib, A.; Sadia, S.; Nasim, W.; Adkins, S.; Saud, S.; et al. Crop Production under Drought and Heat Stress: Plant Responses and Management Options. *Front. Plant Sci.* **2017**, *8*, 1147. [CrossRef]
33. Daryanto, S.; Wang, L.; Jacinthe, P.-A. Global Synthesis of Drought Effects on Maize and Wheat Production. *PLoS ONE* **2016**, *11*, e0156362. [CrossRef]
34. Sánchez, B.; Rasmussen, A.; Porter, J.R. Temperatures and the growth and development of maize and rice: A review. *Glob. Change Biol.* **2014**, *20*, 408–417. [CrossRef] [PubMed]
35. Calleja-Cabrera, J.; Boter, M.; Oñate-Sánchez, L.; Pernas, M. Root Growth Adaptation to Climate Change in Crops. *Front. Plant Sci.* **2020**, *11*, 544. [CrossRef] [PubMed]
36. Gobezie, A.; Ademe, D.; Sharma, L.K. CERES-Maize (DSSAT) Model Applications for Maize Nutrient Management Across Agroecological Zones: A Systematic Review. *Plants* **2025**, *14*, 661. [CrossRef] [PubMed]
37. Song, L.; Jin, J. Improving CERES-Maize for simulating maize growth and yield under water stress conditions. *Eur. J. Agron.* **2020**, *117*, 126072. [CrossRef]
38. Lobell, D.B.; Hammer, G.L.; McLean, G.; Messina, C.; Roberts, M.J.; Schlenker, W. The critical role of extreme heat for maize production in the United States. *Nat. Clim. Change* **2013**, *3*, 497–501. [CrossRef]
39. Wahid, A.; Gelani, S.; Ashraf, M.; Foolad, M.R. Heat tolerance in plants: An overview. *Environ. Exp. Bot.* **2007**, *61*, 199–223. [CrossRef]
40. Duveiller, E.; Singh, R.P.; Nicol, J.M. The challenges of maintaining wheat productivity: Pests, diseases, and potential epidemics. *Euphytica* **2007**, *157*, 417–430. [CrossRef]
41. Schussler, J.R.; Westgate, M.E. Maize Kernel Set at Low Water Potential: I. Sensitivity to Reduced Assimilates during Early Kernel Growth. *Crop Sci.* **1991**, *31*, 1189–1195. [CrossRef]
42. Abrecht, D.G.; Carberry, P.S. The influence of water deficit prior to tassel initiation on maize growth, development and yield. *Field Crops Res.* **1993**, *31*, 55–69. [CrossRef]
43. Bolaños, J.; Edmeades, G.O. Eight cycles of selection for drought tolerance in lowland tropical maize. II. Responses in reproductive behavior. *Field Crops Res.* **1993**, *31*, 253–268. [CrossRef]
44. Ribaut, J.M.; Hoisington, D.A.; Deutsch, J.A.; Jiang, C.; Gonzalez-de-Leon, D. Identification of quantitative trait loci under drought conditions in tropical maize. 1. Flowering parameters and the anthesis-silking interval. *Theor. Appl. Genet.* **1996**, *92*, 905–914. [CrossRef] [PubMed]
45. Maqbool, S.; Hassan, M.A.; Xia, X.; York, L.M.; Rasheed, A.; He, Z. Root system architecture in cereals: Progress, challenges and perspective. *Plant J.* **2022**, *110*, 23–42. [CrossRef] [PubMed]
46. Robinson, D. Optimal relations between root length and nutrient inflow rate in plant root systems. *J. Theor. Biol.* **1988**, *135*, 359–370. [CrossRef]
47. Robinson, D. The responses of plants to non-uniform supplies of nutrients. *New Phytol.* **1994**, *127*, 635–674. [CrossRef]
48. Robinson, D.A.; Campbell, C.S.; Hopmans, J.W.; Hornbuckle, B.K.; Jones, S.B.; Knight, R.; Ogden, F.; Selker, J.; Wendroth, O. Soil Moisture Measurement for Ecological and Hydrological Watershed-Scale Observatories: A Review. *Vadose Zone J.* **2008**, *7*, 358–389. [CrossRef]
49. Ding, H.; Li, X.; Zhuge, S.; Du, J.; Wu, M.; Li, W.; Li, Y.; Ma, H.; Zhang, P.; Wang, X.; et al. Genome-Wide Identification and Functional Analysis of the Genes of the ATL Family in Maize during High-Temperature Stress in Maize. *Genes* **2024**, *15*, 1106. [CrossRef]
50. Ranjan, A.; Sinha, R.; Singla-Pareek, S.L.; Pareek, A.; Singh, A.K. Shaping the root system architecture in plants for adaptation to drought stress. *Physiol. Plant.* **2022**, *174*, e13651. [CrossRef]
51. Lynch, J. Root architecture and plant productivity. *Plant Physiol.* **1995**, *109*, 7–13. [CrossRef]
52. Fitter, A.H. Functional significance of root morphology and root system architecture. *Spec. Publ. Br. Ecol. Soc.* **1985**, 87–106.
53. Fitter, A.H. Architectural approach to the comparative ecology of plant root systems. *New Phytol.* **1987**, *106*, 61–77. [CrossRef]
54. Jackson, R.B.; Caldwell, M.M. Integrating Resource Heterogeneity and Plant Plasticity: Modelling Nitrate and Phosphate Uptake in a Patchy Soil Environment. *J. Ecol.* **1996**, *84*, 891–903. [CrossRef]
55. Qiao, M.; Hong, C.; Jiao, Y.; Hou, S.; Gao, H. Impacts of Drought on Photosynthesis in Major Food Crops and the Related Mechanisms of Plant Responses to Drought. *Plants* **2024**, *13*, 1808. [CrossRef] [PubMed]
56. Voothuluru, P.; Wu, Y.; Sharp, R.E. Not so hidden anymore: Advances and challenges in understanding root growth under water deficits. *Plant Cell* **2024**, *36*, 1377–1409. [CrossRef]
57. Li, J.; Wang, T.; Wang, C.; Cui, R.; Zhang, X.; Feng, Q.; Liu, S.; Xie, Z. Effects of drought stress on tassel growth and characteristics physiological of maize varieties. *Appl. Ecol. Environ. Res.* **2022**, *20*, 685–697. [CrossRef]
58. Wood, J.M.; Bremer, E.; Csonka, L.N.; Kraemer, R.; Poolman, B.; van der Heide, T.; Smith, L.T. Osmosensing and osmoregulatory compatible solute accumulation by bacteria. *Comp. Biochem. Physiol. Part A Mol. Integr. Physiol.* **2001**, *130*, 437–460. [CrossRef]
59. Bohnert, H.J.; Shen, B. Transformation and compatible solutes. *Sci. Hortic.* **1998**, *78*, 237–260. [CrossRef]

60. Yancey, P.H. Organic osmolytes as compatible, metabolic and counteracting cytoprotectants in high osmolarity and other stresses. *J. Exp. Biol.* **2005**, *208*, 2819–2830. [CrossRef]
61. Gagneul, D.; Aïnouche, A.; Duhazé, C.; Lugan, R.; Larher, F.R.; Bouchereau, A. A reassessment of the function of the so-called compatible solutes in the halophytic plumbaginaceae *Limonium latifolium*. *Plant Physiol.* **2007**, *144*, 1598–1611. [CrossRef]
62. Ashraf, M.; Harris, P.J.C. Potential biochemical indicators of salinity tolerance in plants. *Plant Sci.* **2004**, *166*, 3–16. [CrossRef]
63. Guei, R.G.; Wassom, C.E. Genetics of osmotic adjustment in breeding maize for drought tolerance. *Heredity* **1993**, *71*, 436–441. [CrossRef]
64. Lemcoff, J.H.; Chimenti, C.A.; Davezac, T.A.E. Osmotic adjustment in maize (*Zea mays* L.): Changes with ontogeny and its relationship with phenotypic stability. *J. Agron. Crop Sci.* **1998**, *180*, 241–247. [CrossRef]
65. Ashwini, S.; Chandrakala, N.; Ravikumar, R.L. Genetic variability for osmotic adjustment in pollen grains and its association with field tolerance to moisture stress in maize inbred lines. *Curr. Sci.* **2019**, *116*, 279–285. [CrossRef]
66. Studer, C.; Hu, Y.; Schmidhalter, U. Evaluation of the differential osmotic adjustments between roots and leaves of maize seedlings with single or combined NPK-nutrient supply. *Funct. Plant Biol.* **2007**, *34*, 228–236. [CrossRef] [PubMed]
67. Bianchi, C.A.M.; Bergonci, J.I.; Bergamaschi, H.; Dalmago, G.A. Osmotic adjustment in maize cultivated in different soil tillage systems and water availability. *Pesqui. Agropecu. Bras.* **2005**, *40*, 645–651. [CrossRef]
68. Finkelstein, R. Abscisic Acid synthesis and response. *Arab. Book* **2013**, *11*, e0166. [CrossRef]
69. Sirko, A.; Wawrzyńska, A.; Brzywczy, J.; Sieńko, M. Control of ABA Signaling and Crosstalk with Other Hormones by the Selective Degradation of Pathway Components. *Int. J. Mol. Sci.* **2021**, *22*, 4638. [CrossRef]
70. Ma, Y.; Szostkiewicz, I.; Korte, A.; Moes, D.; Yang, Y.; Christmann, A.; Grill, E. Regulators of PP2C Phosphatase Activity Function as Abscisic Acid Sensors. *Science* **2009**, *324*, 1064–1068. [CrossRef]
71. Park, S.Y.; Fung, P.; Nishimura, N.; Jensen, D.R.; Fujii, H.; Zhao, Y.; Lumba, S.; Santiago, J.; Rodrigues, A.; Chow, T.F.F.; et al. Abscisic Acid Inhibits Type 2C Protein Phosphatases via the PYR/PYL Family of START Proteins. *Science* **2009**, *324*, 1068–1071. [CrossRef]
72. Bharath, P.; Gahir, S.; Raghavendra, A.S. Abscisic Acid-Induced Stomatal Closure: An Important Component of Plant Defense Against Abiotic and Biotic Stress. *Front. Plant Sci.* **2021**, *12*, 615114. [CrossRef]
73. Sussmilch, F.C.; Schultz, J.; Hedrich, R.; Roelfsema, M.R.G. Acquiring Control: The Evolution of Stomatal Signalling Pathways. *Trends Plant Sci.* **2019**, *24*, 342–351. [CrossRef]
74. Hsu, P.-K.; Takahashi, Y.; Munemasa, S.; Merilo, E.; Laanemets, K.; Waadt, R.; Pater, D.; Kollist, H.; Schroeder, J.I. Abscisic acid-independent stomatal CO_2 signal transduction pathway and convergence of CO_2 and ABA signaling downstream of OST1 kinase. *Proc. Natl. Acad. Sci. USA* **2018**, *115*, E9971–E9980. [CrossRef] [PubMed]
75. Keller, B.U.; Hedrich, R.; Raschke, K. Voltage-dependent anion channels in the plasma membrane of guard cells. *Nature* **1989**, *341*, 450–453. [CrossRef]
76. Schroeder, J.I.; Hagiwara, S. Cytosolic calcium regulates ion channels in the plasma membrane of Vicia faba guard cells. *Nature* **1989**, *338*, 427–430. [CrossRef]
77. Jiang, Z.; Li, Y.P.; Gai, P.Z.; Gao, J.; Xu, L. Exogenously applied ABA alleviates dysplasia of maize (*Zea mays* L.) ear under drought stress by altering photosynthesis and sucrose transport. *Plant Signal. Behav.* **2025**, *20*, 2462497. [CrossRef]
78. Li, D.; Wang, H.; Luo, F.; Li, M.; Wu, Z.; Liu, M.; Wang, Z.; Zang, Z.; Jiang, L. A Maize *Calmodulin-like 3* Gene Positively Regulates Drought Tolerance in Maize and *Arabidopsis*. *Int. J. Mol. Sci.* **2025**, *26*, 1329. [CrossRef]
79. Liu, L.; Gallagher, J.; Arevalo, E.D.; Chen, R.; Skopelitis, T.; Wu, Q.; Bartlett, M.; Jackson, D. Enhancing grain-yield-related traits by CRISPR-Cas9 promoter editing of maize *CLE* genes. *Nat. Plants* **2021**, *7*, 287–294. [CrossRef]
80. Filyushin, M.A.; Kochieva, E.Z.; Shchennikova, A.V. *ZmDREB2.9* Gene in Maize (*Zea mays* L.): Genome-Wide Identification, Characterization, Expression, and Stress Response. *Plants* **2022**, *11*, 3060. [CrossRef]
81. Liu, S.; Wang, X.; Wang, H.; Xin, H.; Yang, X.; Yan, J.; Li, J.; Tran, L.-S.P.; Shinozaki, K.; Yamaguchi-Shinozaki, K.; et al. Genome-Wide Analysis of *ZmDREB* Genes and Their Association with Natural Variation in Drought Tolerance at Seedling Stage of *Zea mays* L. *PLoS Genet.* **2013**, *9*, e1003790. [CrossRef]
82. Qin, F.; Kakimoto, M.; Sakuma, Y.; Maruyama, K.; Osakabe, Y.; Tran, L.S.; Shinozaki, K.; Yamaguchi-Shinozaki, K. Regulation and functional analysis of *ZmDREB2A* in response to drought and heat stresses in *Zea mays* L. *Plant J.* **2007**, *50*, 54–69. [CrossRef]
83. Sun, Y.; Mu, C.; Zheng, H.; Lu, S.; Zhang, H.; Zhang, X.; Liu, X. Exogenous Pi supplementation improved the salt tolerance of maize (*Zea mays* L.) by promoting Na^+ exclusion. *Sci. Rep.* **2018**, *8*, 16203. [CrossRef]
84. Zhao, L.; Yan, J.; Xiang, Y.; Sun, Y.; Zhang, A. *ZmWRKY104* Transcription Factor Phosphorylated by *ZmMPK6* Functioning in ABA-Induced Antioxidant Defense and Enhance Drought Tolerance in Maize. *Biology* **2021**, *10*, 893. [CrossRef] [PubMed]
85. Zhong, Y.; Yan, X.; Wang, N.; Zenda, T.; Dong, A.; Zhai, X.; Yang, Q.; Duan, H. *ZmHB53*, a Maize Homeodomain-Leucine Zipper I Transcription Factor Family Gene, Contributes to Abscisic Acid Sensitivity and Confers Seedling Drought Tolerance by Promoting the Activity of *ZmPYL4*. *Plant Cell Environ.* **2025**, *48*, 3829–3843. [CrossRef] [PubMed]

86. Tian, M.-Z.; Wang, H.-F.; Tian, Y.; Hao, J.; Guo, H.-L.; Chen, L.-M.; Wei, Y.-K.; Zhan, S.-H.; Yu, H.-T.; Chen, Y.-F. *ZmPHR1* contributes to drought resistance by modulating phosphate homeostasis in maize. *Plant Biotechnol. J.* **2024**, *22*, 3085–3098. [CrossRef] [PubMed]
87. Wang, D.; Ni, Y.; Xie, K.; Li, Y.; Wu, W.; Shan, H.; Cheng, B.; Li, X. Aquaporin ZmTIP2;3 Promotes Drought Resistance of Maize through Symbiosis with Arbuscular Mycorrhizal Fungi. *Int. J. Mol. Sci.* **2024**, *25*, 4205. [CrossRef]
88. Feng, T.; Wang, Y.; Zhang, M.; Zhuang, J.; Zhou, Y.; Duan, L. ZmSCE1a positively regulates drought tolerance by enhancing the stability of ZmGCN5. *Plant J.* **2024**, *120*, 2101–2112. [CrossRef]
89. Mao, H.; Yu, L.; Han, R.; Li, Z.; Liu, H. ZmNAC55, a maize stress-responsive NAC transcription factor, confers drought resistance in transgenic Arabidopsis. *Plant Physiol. Biochem.* **2016**, *105*, 55–66. [CrossRef]
90. Yang, Z.; Wang, C.; Zhu, T.; He, J.; Wang, Y.; Yang, S.; Liu, Y.; Zhao, B.; Zhu, C.; Ye, S.; et al. An LRR-RLK protein modulates drought- and salt-stress responses in maize. *J. Genet. Genom.* **2025**, *52*, 388–399. [CrossRef]
91. Liu, Y.; Chen, Z.; Zhang, C.; Guo, J.; Liu, Q.; Yin, Y.; Hu, Y.; Xia, H.; Li, B.; Sun, X.; et al. Gene editing of *ZmGA20ox3* improves plant architecture and drought tolerance in maize. *Plant Cell Rep.* **2024**, *43*, 18. [CrossRef]
92. Lan, Q.; He, G.; Wang, D.; Li, S.; Jiang, Y.; Guan, H.; Li, Y.; Liu, X.; Wang, T.; Li, Y.; et al. Overexpression of ZmEULD1b enhances maize seminal root elongation and drought tolerance. *Plant Sci.* **2025**, *352*, 112355. [CrossRef]
93. Ren, Z.; Zhang, P.; Su, H.; Xie, X.; Shao, J.; Ku, L.; Tian, Z.; Deng, D.; Wei, L. Regulatory mechanisms used by ZmMYB39 to enhance drought tolerance in maize (*Zea mays*) seedlings. *Plant Physiol. Biochem.* **2024**, *211*, 108696. [CrossRef]
94. Yu, T.; Dong, W.; Hou, X.; Sun, A.; Li, X.; Yu, S.; Zhang, J. The Maize Gene *ZmGLYI-8* Confers Salt and Drought Tolerance in Transgenic *Arabidopsis* Plants. *Int. J. Mol. Sci.* **2024**, *25*, 10937. [CrossRef] [PubMed]
95. Yan, Z.; Zhang, F.; Mu, C.; Ma, C.; Yao, G.; Sun, Y.; Hou, J.; Leng, B.; Liu, X. The ZmbHLH47-ZmSnRK2.9 Module Promotes Drought Tolerance in Maize. *Int. J. Mol. Sci.* **2024**, *25*, 4957. [CrossRef] [PubMed]
96. Gao, B.; Wang, Y.; Qu, J.; Miao, M.; Zhao, Y.; Liu, S.; Guan, S.; Ma, Y. The overexpression of *ascorbate peroxidase* 2 (*APX2*) gene improves drought tolerance in maize. *Mol. Breed.* **2025**, *45*, 27. [CrossRef] [PubMed]
97. Xiang, Y.; Liu, W.; Niu, Y.; Li, Q.; Zhao, C.; Pan, Y.; Li, G.; Bian, X.; Miao, Y.; Zhang, A. The maize GSK3-like kinase ZmSK1 negatively regulates drought tolerance by phosphorylating the transcription factor ZmCPP2. *Plant Cell* **2025**, *37*, koaf032. [CrossRef]
98. Chai, W.; Li, H.; Xu, H.; Zhu, Q.; Li, S.; Yuan, C.; Ji, W.; Wang, J.; Sheng, L. *ZmDST44* Gene is a Positive Regulator in Plant Drought Stress Tolerance. *Biology* **2024**, *13*, 552. [CrossRef]
99. Wang, C.; Zhou, Y.; Wang, Y.; Jiao, P.; Liu, S.; Guan, S.; Ma, Y. CRISPR-Cas9-mediated editing of *ZmPL1* gene improves tolerance to drought stress in maize. *GM Crops Food* **2025**, *16*, 1–16. [CrossRef]
100. Liu, H.; Wu, Z.; Bao, M.; Gao, F.; Yang, W.; Abou-Elwafa, S.F.; Liu, Z.; Ren, Z.; Zhu, Y.; Ku, L.; et al. *ZmC2H2-149* negatively regulates drought tolerance by repressing *ZmHSD1* in maize. *Plant Cell Environ.* **2024**, *47*, 885–899. [CrossRef]
101. Zhai, X.; Yan, X.; Zenda, T.; Wang, N.; Dong, A.; Yang, Q.; Zhong, Y.; Xing, Y.; Duan, H. Overexpression of the peroxidase gene ZmPRX1 increases maize seedling drought tolerance by promoting root development and lignification. *Crop J.* **2024**, *12*, 753–765. [CrossRef]
102. Xiao, N.; Ma, H.; Wang, W.; Sun, Z.; Li, P.; Xia, T. Overexpression of *ZmSUS1* increased drought resistance of maize (*Zea mays* L.) by regulating sucrose metabolism and soluble sugar content. *Planta* **2024**, *259*, 43. [CrossRef]
103. Wang, D.; Liu, X.; He, G.; Wang, K.; Li, Y.; Guan, H.; Wang, T.; Zhang, D.; Li, C.; Li, Y. GWAS and transcriptome analyses unravel *ZmGRAS15* regulates drought tolerance and root elongation in maize. *BMC Genom.* **2025**, *26*, 246. [CrossRef]
104. Che, R.; Tan, X.; Meng, X.; Li, H. ZmCYB5-1, a cytochrome b5 Gene, negatively regulates drought stress tolerance in maize. *Gene* **2025**, *954*, 149422. [CrossRef] [PubMed]
105. Liu, L.; Tang, C.; Zhang, Y.; Sha, X.; Tian, S.; Luo, Z.; Wei, G.; Zhu, L.; Li, Y.; Fu, J.; et al. The SnRK2.2-ZmHsf28-JAZ14/17 module regulates drought tolerance in maize. *New Phytol.* **2025**, *245*, 1985–2003. [CrossRef] [PubMed]
106. Wei, X.; Wang, C.; Wang, Y.; Zhao, Y.; Ma, Y.; Liu, S.; Guan, S.; Jiao, P. miR166e/ZmATHB14 module contributes to drought tolerance in maize root. *Int. J. Biol. Macromol.* **2025**, *297*, 139707. [CrossRef] [PubMed]
107. Wang, F.; Chen, Y.; Yang, R.; Luo, P.; Wang, H.; Zhang, R.; Li, W.; Yang, K.; Xu, X.; Hao, Z.; et al. Identification of *ZmSNAC06*, a Maize NAC Family Transcription Factor with Multiple Transcripts Conferring Drought Tolerance in *Arabidopsis*. *Plants* **2025**, *14*, 12. [CrossRef]
108. Waqas, M.A.; Wang, X.; Zafar, S.A.; Noor, M.A.; Hussain, H.A.; Azher Nawaz, M.; Farooq, M. Thermal Stresses in Maize: Effects and Management Strategies. *Plants* **2021**, *10*, 293. [CrossRef]
109. Zhang, Z.; Zhang, X.; Lin, Z.; Wang, J.; Liu, H.; Zhou, L.; Zhong, S.; Li, Y.; Zhu, C.; Lai, J.; et al. A Large Transposon Insertion in the stiff1 Promoter Increases Stalk Strength in Maize. *Plant Cell* **2020**, *32*, 152–165. [CrossRef]
110. Li, Y.; Huang, Y.; Sun, H.; Wang, T.; Ru, W.; Pan, L.; Zhao, X.; Dong, Z.; Huang, W.; Jin, W. Heat shock protein 101 contributes to the thermotolerance of male meiosis in maize. *Plant Cell* **2022**, *34*, 3702–3717. [CrossRef]

111. Shao, R.X.; Yu, K.K.; Li, H.W.; Jia, S.J.; Yang, Q.H.; Zhao, X.; Zhao, Y.L.; Liu, T.X. The effect of elevating temperature on the growth and development of reproductive organs and yield of summer maize. *J. Integr. Agric.* **2021**, *20*, 1783–1795. [CrossRef]
112. Xue, M.; You, Y.; Zhang, L.; Cao, J.; Xu, M.; Chen, S. *ZmHsp18* screened from the *ZmHsp20* gene family confers thermotolerance in maize. *BMC Plant Biol.* **2024**, *24*, 1048. [CrossRef]
113. Cao, L.; Fahim, A.M.; Liang, X.; Fan, S.; Song, Y.; Liu, H.; Ye, F.; Ma, C.; Zhang, D.; Lu, X. Melatonin Enhances Heat Tolerance via Increasing Antioxidant Enzyme Activities and Osmotic Regulatory Substances by Upregulating *zmeno1* Expression in Maize (*Zea mays* L.). *Antioxidants* **2024**, *13*, 1144. [CrossRef]
114. Li, G.; Chen, Z.; Guo, X.; Tian, D.; Li, C.; Lin, M.; Hu, C.; Yan, J. Genome-Wide Identification and Analysis of Maize *DnaJ* Family Genes in Response to Salt, Heat, and Cold at the Seedling Stage. *Plants* **2024**, *13*, 2488. [CrossRef] [PubMed]
115. Dong, A.; Wang, N.; Zenda, T.; Zhai, X.; Zhong, Y.; Yang, Q.; Xing, Y.; Duan, H.; Yan, X. *ZmDnaJ-ZmNCED6* module positively regulates drought tolerance via modulating stomatal closure in maize. *Plant Physiol. Biochem.* **2025**, *218*, 109286. [CrossRef] [PubMed]
116. Duan, P.; Kan, Y.; Zhao, D.; Yang, S.; Chen, J. Genome-Wide Identification and Characterization of CHROMO Domain Family Genes Reveal Roles of the Maize Genes in Heat Stress Response. *Biol. Bull.* **2023**, *50* (Suppl. S3), S289–S297. [CrossRef]
117. Cao, L.; Wang, G.; Fahim, A.M.; Pang, Y.; Zhang, Q.; Zhang, X.; Wang, Z.; Lu, X. Comprehensive Analysis of the DnaJ/HSP40 Gene Family in Maize (*Zea mays* L.) Reveals that *ZmDnaJ96* Enhances Abiotic Stress Tolerance. *J. Plant Growth Regul.* **2024**, *43*, 1548–1569. [CrossRef]
118. Hunter, C.T. CRISPR/Cas9 Targeted Mutagenesis for Functional Genetics in Maize. *Plants* **2021**, *10*, 723. [CrossRef]
119. Hassan, F.U.; Nawaz, A.; Rehman, M.S.; Ali, M.A.; Dilshad, S.M.R.; Yang, C. Prospects of HSP70 as a genetic marker for thermo-tolerance and immuno-modulation in animals under climate change scenario. *Anim. Nutr.* **2019**, *5*, 340–350. [CrossRef]
120. Jiang, L.; Hu, W.; Qian, Y.; Ren, Q.; Zhang, J. Genome-wide identification, classification and expression analysis of the Hsf and Hsp70 gene families in maize. *Gene* **2021**, *770*, 145348. [CrossRef]
121. Jiang, Y.; An, X.; Li, Z.; Yan, T.; Zhu, T.; Xie, K.; Liu, S.; Hou, Q.; Zhao, L.; Wu, S.; et al. CRISPR/Cas9-based discovery of maize transcription factors regulating male sterility and their functional conservation in plants. *Plant Biotechnol. J.* **2021**, *19*, 1769–1784. [CrossRef]
122. Rai, M.N.; Rhodes, B.; Jinga, S.; Kanchupati, P.; Ross, E.; Carlson, S.R.; Moose, S.P. Efficient mutagenesis and genotyping of maize inbreds using biolistics, multiplex CRISPR/Cas9 editing, and Indel-Selective PCR. *Plant Methods* **2025**, *21*, 43. [CrossRef]
123. Liu, X.; Zhai, S.; Zhao, Y.; Sun, B.; Liu, C.; Yang, A.; Zhang, J. Overexpression of the phosphatidylinositol synthase gene (*ZmPIS*) conferring drought stress tolerance by altering membrane lipid composition and increasing ABA synthesis in maize. *Plant Cell Environ.* **2013**, *36*, 1037–1055. [CrossRef]
124. Moustafa, K. CRISPR: Beyond the Excitement. *J. Bioethical Inq.* **2024**, *21*, 7–9. [CrossRef] [PubMed]
125. Yin, W.; Chen, Z.; Huang, J.; Ye, H.; Lu, T.; Lu, M.; Rao, Y. Application of CRISPR-Cas9 gene editing technology in crop breeding. *Sheng Wu Gong Cheng Xue Bao = Chin. J. Biotechnol.* **2023**, *39*, 399–424.
126. Ebrahimi, V.; Hashemi, A. CRISPR-based gene editing in plants: Focus on reagents and their delivery tools. *Bioimpacts* **2025**, *15*, 30019. [CrossRef] [PubMed]
127. Molla, K.A.; Sretenovic, S.; Bansal, K.C.; Qi, Y. Precise plant genome editing using base editors and prime editors. *Nat. Plants* **2021**, *7*, 1166–1187. [CrossRef]
128. Zong, Y.; Song, Q.; Li, C.; Jin, S.; Zhang, D.; Wang, Y.; Qiu, J.-L.; Gao, C. Efficient C-to-T base editing in plants using a fusion of nCas9 and human APOBEC3A. *Nat. Biotechnol.* **2018**, *36*, 950–953. [CrossRef]

International Journal of
Molecular Sciences

MDPI

Review

Phosphoenolpyruvate and Related Metabolic Pathways Contribute to the Regulation of Plant Growth and Development

Runzhou Hu [1,2,†], Haiyang Yu [1,†], Jing Deng [2], Shanjing Chen [2], Ronglan Yang [2], Hongjun Xie [2], Xiao Tang [2], Yaying Yu [2], Yonghong Duan [2], Meng Zhang [1], Mingdong Zhu [1,2,*] and Yinghong Yu [1,2,*]

1 Long Ping Branch, College of Biology, Hunan University, Changsha 410125, China; pamelaparrish10232@gmail.com (R.H.); yuhaiyang@hnu.edu.cn (H.Y.); zhangmeng2019@hnu.edu.cn (M.Z.)

2 State Key Laboratory of Hybrid Rice, Hunan Rice Research Institute, Hunan Academy of Agricultural Sciences, Changsha 410125, China; dengjing@hunaas.cn (J.D.); shanjing@hun.edu.cn (S.C.); yangronglan@hnu.edu.cn (R.Y.); xhj1110@126.com (H.X.); talent_tdx@hotmail.com (X.T.); yuyaying@hunaas.cm (Y.Y.); zzk126@126.com (Y.D.)

* Correspondence: uhz_uhz@hotmail.com (M.Z.); yyh30678@163.com (Y.Y.)

† These authors contributed equally to this work.

Abstract: Phosphoenolpyruvate (PEP) plays a key role in the development of plants and exists in a wide variety of species. Research on the metabolic activities of PEP in plants has received increasing attention. PEP regulates multiple processes in plant growth and development. This article provides a comprehensive summary of these pathways, including embryo formation, root development, synthesis of secondary metabolites, and the formation of lignification. We also summarize new findings, including PEP's role in nodule energy sensing and carbon allocation under the influence of ozone. This review displays the complex and differential regulatory pathways in plant growth and development and provides a reference for basic and applied research on PEP metabolism in plants.

Keywords: phosphoenolpyruvate; carbon allocation; energy metabolism; plant growth and development

1. Background

Energy metabolism is the basis of life activities and affects the growth and development of organisms. Carbon skeletons must be synthesized for organisms to complete normal biochemical metabolism. Phosphoenolpyruvate (PEP), which is an important intermediate product during the synthesis of carbon skeletons and energy metabolism, plays a key role in the growth and development of most organisms. In bacteria, PEP acts as a source of phosphoryl groups during the conversion of glucose to 6-phosphoglucose and the phosphorylation of various sugars and their derivatives through a phosphoric acid cascade reaction prior to their transport into cells. It also provides energy for life activities. In plants, it is involved in energy metabolism [1,2], the biosynthesis of various aromatic compounds, and carbon fixation during photosynthesis, while also participating in responses to external stresses (e.g., ozone stress) and root nodule energy state perception activities [3,4]. PEP ($C_3H_4O_6P$) has a relative molecular mass of 267.22. Notably, it contains the highest-energy phosphate bond in living organisms (−61.9 kJ/mol) and exists mainly in an anionic form in living organisms [5].

As Shown in Figure 1. In plants, the PEP synthesis pathway is regulated by enolase (ENO), pyruvate, phosphate dikinase (PPDK), and phosphoenolpyruvate carboxykinase (PEPCK/PCK). The PEP metabolic pathway is regulated by pyruvate kinase (PK),

https://doi.org/10.3390/ijms26010391

3-Deoxy-D-arabinoheptulosonate 7-phosphate synthase (DHAPS), and phosphoenolpyruvate carboxylase (PEPC). It is also regulated by the PEP/phosphorus translocator (PPT). Interestingly, it affects the energy state of plants as a signaling molecule. The following sections provide an overview of the overall regulatory effects of PEP involving the above-mentioned regulatory mechanisms.

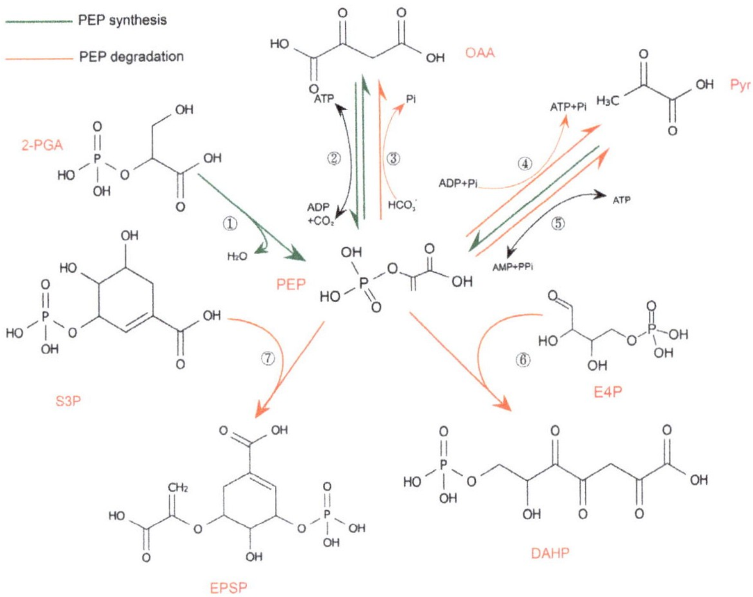

Figure 1. PEP synthesis and metabolism pathways in plants [6]: ① ENO (EC 4.2.1.11), ② PEPCK (EC 4.1.1.49), ③ PK (EC 2.7.1.40), ④ PEP (EC 4.1.1.31), ⑤ PPDK (EC 2.7.9.1), ⑥ DAHPS (EC 2.5.1.54), ⑦ 3-phosphoshikimate 1-carboxyviniltransferase (EC 2.5.1.19).

1.1. *Effects of ENO-Mediated PEP Synthesis on Plant Growth and Development*

PEP synthesis mediated by ENO is a crucial step during glycolysis, which represents the initial sugar metabolism stage in all living cells. ENO serves as the rate-limiting enzyme that catalyzes the conversion of 2-PGA to PEP in the glycolytic pathway. Accordingly, ENO is essential for normal plant growth and development as well as for plant biochemical metabolism. In *Arabidopsis thaliana* (Arabidopsis), ENO exists as three distinct isoenzymes (ENO1, ENO2, and ENO3), of which ENO1 is targeted to the plastid in most heterotrophic tissues, but not in photosynthetic tissues. In a previous study, two T-DNA insertion *eno1* mutants had distorted trichomes and fewer root hairs than the wild-type (WT) control, but the mutation did not alter plant development and metabolism at the macroscopic level [7]. In Arabidopsis, the *At2g36530* (*ENO2*, also referred to as *LOS2*) gene encodes the ENO2 protein. Mutating this gene results in mutant plants with various severe cellular and growth defects, including small leaves and short siliques (i.e., dwarfism) and increased sensitivity to abiotic stresses. Additionally, these plants accumulate more salicylic acid (SA) than WT plants [8–10]. Lignin synthesis depends on PEP, especially during late developmental stages. The reported decrease in lignin contents suggests mutating *LOS2/ENO2* affects PEP synthesis. Moreover, decreases in SA levels may be due to the collapse of improperly lignified cell walls, resulting in the induction of an endogenous defense response, thereby promoting SA synthesis [11]. These reactions reflect the importance of a regular supply of PEP for the synthesis of secondary metabolites and normal plant development and metabolism. In addition to ENO, the *LOS2/ENO2* locus also encodes the transcription factor AtMBP-1 [12], which inhibits *LOS2/ENO2* promoter activity. This dual functionality helps to maintain an appropriate ENO2 activity level [13]. However, a new hypothesis posits that LOS2 does not participate in the immune response by replacing MBP-1. Rather, LOS2 decreases the classical ENO activity, which disrupts glycolysis and consequently activates the immune response involving NOD-like receptors. This, in turn, increases the accumulation of sugars and organic acids while depleting glycolytic metabolites, which ultimately affect plant growth. This is consistent with decreases in 3-phosphate, PEP, and relative total content of 2-phosphoglyceric acid and 3-phosphoglyceric acid levels in the *los2-2* mutant (relative to the corresponding contents in the WT control). Compared with the WT control, the *los2-2* mutant has shorter siliques, less chlorophyll, and flowers earlier, underscoring the intricate interconnections between primary metabolism and plant immunity [14]. In all organs, *ENO3* is consistently expressed at lower levels than *ENO1* or *ENO2*. In Arabidopsis, ENO3 is undetectable throughout the seed development stage. Furthermore, when expressed as a soluble full-length protein in an *Escherichia coli* ENO-deficient strain, ENO3 is inactive. However, if the ENO3 amino acid sequence is modified by replacing glutamic acid at position 251 with aspartic acid and aspartic acid at position 252 with glutamic acid, the ENO-deficient *E. coli* strain can grow normally. Thus, differences in the amino acid residues at the active site of ENO3 are the underlying cause of the observed lack of activity in *E. coli* [15].

1.2. *Effects of PEPCK-Mediated PEP Synthesis and Degradation on Plant Growth and Development*

Gluconeogenesis, which is mediated by PEPCK, occurs in most plants. Glycolysis synthesizes glucose from specific non-carbohydrate carbon substrates. This is a biochemical metabolic process common among diverse organisms, including fungi, yeast, bacteria, and other microorganisms. PEPCK plays a pivotal role in the glycolytic pathway. In the initial stage of glycolysis, when ATP is present, PEPCK is decarboxylated and phosphorylates oxaloacetic acid (OAA), thereby converting it to PEP. The transfer of phosphate results in the production of an ADP molecule. PEPCK directs the carbon released from fatty acid reserves to form sugars during seed germination until the photosynthetic apparatus is fully developed [16,17]. Numerous studies confirm that PEPCK affects plant growth and development, for example, disrupting *PEPCK* expression has detrimental effects on plant growth and development. Hypocotyl lengths in 5 d old 35S-*PCK1* antisense seedlings were reduced by up to 68% [18]. In transgenic tomato plants, the suppression of *PEPCK* expression via RNA interference inhibits root growth and decreases organic acid and soluble sugar contents. The decrease in the sugar content and accumulation of malate may be attributed to the inhibition of the gluconeogenesis pathway due to PEPCK deficiency [19]. In *Capsicum annuum*, *CaPEPCK*-overexpressing plants grow faster and produce larger leaves than WT plants. Moreover, CaPEPCK contributes to plant immune processes. Homozygous T-DNA insertion Arabidopsis mutants *pck1-3* and *pck1-4* are more susceptible to a pathogenic bacterium and a pathogenic oomycete than WT plants. Considered together, these results indicate that CaPEPCK enhances the innate immunity of plants against hemibiotrophic bacteria and obligate biotrophic oomycetes. Levels of glycolytic components in glycolysis activated by CaPEPCK1 overexpression may regulate defense signaling in plants [20]. For example, There is experimental evidence that levels of glycerol-3-phosphate are associated with basal defenses of the semi-living trophic fungus *Colletotrichum higginsianum* in *Arabidopsis thaliana* [21].

In plants, PEPCK is regulated by phosphorylation. Light significantly influences the phosphorylation of PEPCK. In *Zea mays* (maize) leaves, PEPCK is phosphorylated in dark-adapted leaves and dephosphorylated in light-adapted leaves. Phosphorylation and dephosphorylation decrease and increase PEPCK activity, respectively [22–24]. Sufficient illumination regulates the dephosphorylation of PEPCK. In maize, the extent of the phosphorylation at 11 sites of ZmPEPCK1 and at all sites of ZmPEPCK2 markedly increases as the light intensity increases, reaching considerable levels [25]. Additionally, PEPCK is phosphorylated in the cotyledons and endosperm of germinated seeds. Earlier research showed the dephosphorylation of PEPCK in *Cucumis sativus* (cucumber) cotyledons is stimulated by light [26]. Four phosphorylated residues (Ser-55, Thr-58, Thr-59, and Thr-120) were identified in maize ZmPEPCK1; the phosphorylation of these four residues is positively regulated by light. Moreover, mass spectrometry has been used to investigate post-translational phosphorylations. Both AthPEPCK1 and AthPEPCK2 are phosphorylated at multiple sites. A phosphomimic mutation at the Ser-62 site leads to increased enzymatic activity, whereas a mutation at the Thr-56 site results in decreased enzymatic activity. In contrast, a phosphomimetic mutation at the Thr-66 site does not alter enzymatic activity. This further supports the hypothesis that Athpepck1 is regulated by a complex mechanism involving a variety of signals, including those mediated by effectors and post-translational modifications [27].

Furthermore, the N-terminus of PEPCK is targeted for proteolysis in a process that is influenced by pH (i.e., decreased activity at pH 9–10). Additionally, protease inhibitors have no inhibitory effect on this process, and the quaternary structure of the enzyme remains unaltered [26,28,29]. In Arabidopsis, AthPEPCK1 is targeted by the cysteine protease METACASPASE9 (AthMC9) [30], which is present in the nucleus, cytoplasm, and peroxisome [30–32]. This enzyme is involved in the regulation of cell death in various physiological contexts, including immune responses [33,34] and vascular tissue development [35,36]. Specifically, the hydrolysis of the N-terminal domain of PEPCK appears to promote PEPCK activity. This is supported by a decrease in PEPCK activity in the crude extract of the Arabidopsis *mc9* mutant, as well as an increase in PEPCK activity in a 35S:*MC9*-overexpressing strain [30]. A comparison with the WT enzyme indicated that a salient characteristic of the truncated AthPEPCK1 is the altered allosteric regulation of metabolites. Glucose-6-phosphate (Glc6P) functions as an inhibitor of the WT enzyme but a weak activator of the Δ101 truncated form, whereas malate activates the WT enzyme but has no effect on the Δ101 truncated form [29]. Furthermore, both of these earlier studies suggest that the structural characteristics of the N-terminal are associated with the regulatory properties of PEPCK [37].

PEPCK is precisely regulated by many metabolites. In Arabidopsis, AthPEPCK1 and AthPEPCK2 are primarily inhibited by Glc6P, shikimate, and inorganic pyrophosphate (PPi) but activated by malate. The allosteric regulation of AthPEPCK1 and AthPEPCK2 by key metabolic intermediates, namely succinate, fumarate, citrate, and α-ketoglutarate, has been demonstrated. Intriguingly, malate and Glc6P activate and inhibit AthPEPCK1, respectively, but have no differential effect on AthPEPCK2 [38]. The regulatory effect of malate may be critical for stimulating the entry of carbon released via lipid degradation into the glycolytic pathway during seed germination. Once the photosynthetic apparatus is fully developed, increases in PEP and hexose phosphate levels may result in the inhibition of PEPCK, thereby limiting glycolysis. Furthermore, AthPEPCK1 is inhibited by shikimate, which is the precursor of aromatic amino acids and defense compounds. This regulation occurs during the synthesis of PEP, which is the initial substrate of the shikimate pathway [38]. The regulation of PEPCK by malic acid and Glc6P is the inverse of the regulation of PEPC. This may represent an important regulatory mechanism involving two enzymes in the plant cytoplasm that catalyze opposite reactions. This regulation may be the key to preventing an inefficient carboxylation/decarboxylation cycle, which depletes cytoplasmic ATP [38–40].

1.3. Effects of PPDK-Mediated PEP Synthesis and Degradation on Plant Growth and Development

PPDK plays a crucial role in the synthesis of PEP in plants. The plant C4 pathway represents one of three recognized mechanisms mediating carbon fixation during photosynthesis. PPDK catalyzes the conversion of pyruvate to PEP in the presence of ATP and phosphate. In plants, this reaction, which is reversible, occurs in the gluconeogenesis and C4 pathways [41]. PPDK is essential for regenerating PEP in C4 plants, functioning as the primary rate-limiting enzyme in the C4 pathway [42]. In terms of its contribution to gluconeogenesis, cytosolic PPDK uses pyruvate generated by amino acid catabolism during Arabidopsis seed germination as a substrate for gluconeogenesis [16]. Additionally, PPDK affects leaf senescence, converting pyruvate derived from amino acid catabolism to PEP, which is converted to glutamine for amino acid transport. In Arabidopsis and tobacco, the overexpression of *PPDK* during the aging process increases the seed size and seed nitrogen content. Arabidopsis plants overexpressing *At4g15530(PPDK)* during the aging process are reportedly larger than control plants [43]. PPDK was revealed to influence the fruit ripening process, as well as leaf maturation and senescence, in tomato but not in peach or pepper. This suggests that PPDK plays a role in the gluconeogenesis pathway, but this role may vary across species [44]. Mutating *LOC4338750(OsPPDKB/FLO4)* in *Oryza sativa* (rice) cells decreases the starch content of mutant grains, which is accompanied by a decrease in the amylose:amylopectin ratio and an increase in the lipid content [45]. These findings imply that cytosolic PPDK is an important regulator of the formation of carbon skeletons [46], including during the biosynthesis of starch and fatty acids. It is conceivable that increased PPi levels during the conversion of PEP to pyruvate by PK and PPDK may stimulate the sucrose synthase-dependent starch biosynthesis pathway [47].

The C4 pathway uses PPDK in leaf chloroplasts to regenerate the initial CO_2 acceptor (i.e., PEP). A bifunctional regulatory protein (PDRP) controls PPDK activity. The reversible phosphorylation of the Thr residue of PPDK inactivates the enzyme. PDRP can promote the dark-induced phosphorylation and light-induced dephosphorylation of this residue, thereby controlling PPDK activation/inactivation [48,49]. Recent studies have further clarified the specific effects of PDRP on C4 crop photosynthesis. For example, the Fv/Fm value of the *PDRP* knockout *Setaria viridis* mutant gradually decreases over several days under fluctuating light conditions (i.e., repeated exposure to 1 min light phases during a 16 h photoperiod). This may be attributed to the fact that PPDK activity is not regulated, resulting in an unstable and limited supply of primary carboxylated PEP. This impairs the ability of photosynthetic organs to use excitation energy for photochemical reactions. One of the functions of PDRP in C4 photosynthesis is to maintain optimal photoreaction efficiency during changes in incident light [50].

Environmental conditions also substantially affect PPDK activity. In response to drought stress, the specific activity of tobacco PPDK increases significantly after 11 days, with increases in PPDK contents (2.7 times) also enhancing PPDK activity. The reaction catalyzed by PPDK produces ATP and Pi, resulting in a significant increase in biological energy under stress conditions. At this time, the ATP produced by mitochondria through oxidative phosphorylation may be limited, or the demand for ATP for biosynthesis reactions may increase [51].

2. PEP Is Involved in Plant Catabolic Activities and the Regulation of Plant Growth and Development

2.1. Effects of PK-Mediated PEP Degradation on the Regulation of Plant Growth and Development

Pyruvate kinase (PK) plays a significant role in PEP metabolism, as well as in the final step of the glycolytic pathway in plants. This enzyme catalyzes the transfer of a phosphate group from PEP to adenosine diphosphate, producing one molecule of pyruvate and one molecule of ATP. This reaction represents the fundamental metabolic process underlying cytosolic glycolysis and gluconeogenesis in plant cells, providing the energy necessary for sucrose synthesis, translocation, and distribution [52].

In plants, PK is divided into two distinct forms: plastidial pyruvate kinase (PKp) and cytosolic pyruvate kinase (PKc). These two isoenzymes, which are ubiquitous in plants, differ significantly in terms of physical and kinetic/regulatory properties [53], which affects crop maturation. A study of rice PKc revealed that a T-DNA insertion mutation to *OsPK1* causes dwarfism and panicle enclosure while also decreasing the seed setting rate. The sucrose content in the panicles of the *ospk1* mutant decreases by approximately 84%. These findings reflect the changes in glucose and fructose metabolism and sugar transport in the *ospk1* mutant [54]. OsPK2, which is encoded by a gene in the rice chloroplast genome, is a PKp that plays a role in the synthesis of starch in the endosperm, the formation of starch granules, and the grain filling process. The *ospk2* mutant is characterized by significant decreases in grain weight and starch content as well as altered starch physicochemical properties (relative to WT grains). The normal starch compound granules were drastically reduced and more single granules filled the endosperm cells of *ospk2*. Additionally, after a 1-year storage period, the *ospk2* seed germination rate is much lower than the WT seed germination rate [55]. OsPK3 in the rice cytoplasm is responsible for regulating the transport of sugars from the source to the sink, which in turn controls the grain filling process. A T-DNA mutant OsPK3 delays amyloplast development in endosperm cells. Moreover, the abundance of storage compounds (predominantly starch, protein, and lipids) is significantly lower in *ospk3* seeds than in WT seeds [56]. Furthermore, ABA plays a pivotal role in regulating the expression of PK-encoding genes *LOC4349774(OsPK1)*, *LOC4351493(OsPK4)*, *LOC4349454(OsPK9)*, and *LOC4350078(OsPK10)* in rice. In a study of PK-related genes in *Fragaria ananassa* (strawberry), the overexpression of *FxaC_15g00080 (FaPKc2.2)* was observed to negatively regulate strawberry seed maturation and inhibit the accumulation of organic acids [57]. Earlier research on GhPKc, which is a PKc in *Gossypium hirsutum* (cotton), showed that the rapid elongation of fiber cells is accompanied by a continuous decrease in GhPKc activity. Inhibiting *KX369413(GhPK6)* expression promotes cotton fiber elongation. One potential explanation is that during the rapid elongation of cotton fibers, GhPKc activity is relatively low, which ensures that PEP is effectively used by PEPC to synthesize malic acid, thereby increasing cell turgor pressure and facilitating cell elongation [58–60]. Therefore, PK functions are critical for crop production.

In Arabidopsis, the energy status affects PK activities. An examination of five PK isoenzymes in Arabidopsis revealed that PKc1, PKc2, and PKc3 activities are negatively regulated by ATP. Hence, these isoenzymes are inhibited when the energy supply in vivo is sufficient. In contrast, PKc4 and PKc5 activities are unaffected by ATP, indicating that the functions of these two isoenzymes are not associated with the energy state of the cell [60]. Studies on PKc from *Ricinus communis* demonstrated that glutamate is the most effective inhibitor, whereas aspartate acts as an activator [61]. Furthermore, the TCA cycle intermediates citric acid, 2-oxoglutaric acid, fumaric acid, and malic acid can decrease the activity of certain plant PKs [60], implying that the production of pyruvate in a process involving PEP is regulated by a range of metabolic intermediates.

Environmental stress significantly influences PK activities. A *Glycine max* PK-encoding gene involved in the response to salt stress has been identified. The ectopic expression of *SoyZH13_19G000600.m1(GmPK21)* adversely affects the tolerance of Arabidopsis plants to salinity, indicating that GmPK21 may be a negative regulator of plant salt tolerance [62]. In mulberry trees, the expression levels of genes encoding PKs are significantly upregulated in response to submergence stress. This may be attributed to the fact that flooding and submergence stress directly limit oxygen exchange, thereby causing energy shortages in plants and promoting glycolysis in plants [63].

2.2. PEP Degradation Mediated by PEPC Regulates Plant Growth and Development

PEPC is a cytoplasmic enzyme that is ubiquitous in higher plants and also widely distributed in bacteria, cyanobacteria, and green algae [64,65]. It catalyzes the irreversible β-carboxylation of PEP in the presence of HCO_3^- and Mg_2^+ to produce OAA and Pi. Thus, it is closely related to C4-dicarboxylic acid metabolism in plants [64]. In addition to its essential role in the initial fixation of atmospheric CO_2 during C4 photosynthesis and C4 acid metabolism (CAM), PEPC also plays a compensatory role in various non-photosynthetic processes, including C/N allocation in C3 leaves, seed formation and germination, and fruit ripening [64]. Additionally, non-photosynthetic isoforms of PEPC play a specific role in carbon metabolism in guard cells during the opening of stomata, as well as in the formation of C4 acids in plant host cells in the root nodules of nitrogen-fixing legumes [66,67]. The disparate functions of these enzymes in photosynthetic and non-photosynthetic tissues reflect their distinct contributions to PEP catabolism.

Plant *PEPC* genes are included in a small gene family that encodes several plant-type PEPC (PTPC) and distantly related bacterial-type PEPC(BTPC) [68,69]. A class 1 PEPC in developing *R. communis* seeds (COS) was revealed to be a 410 kDa homotetramer composed of identical p107 subunits that are phosphorylated in vivo [70]. Class 1 PEPC activities in diverse plant tissues under varying physiological conditions are regulated by various mechanisms (e.g., phosphorylation, monoubiquitination, alterations in intracellular pH, and allosteric effectors). In *R. communis* plants, monoubiquitination is more prevalent than phosphorylation and appears to be the dominant post-translational modification of class 1 PEPCs [69,71]. In contrast, a class 2 PEPC in developing COS was identified as a novel 910 kDa heterooctameric complex, in which the PTPC subunit, which is identical to a subunit in the class 1 PEPC homolog, is tightly associated with four BTPC subunits. Class 2 PEPC may function in a metabolic overflow mechanism, capable of maintaining a substantial flux from PEP to L-malate under physiological conditions, thereby suppressing class 1 PEPCs [69]. It has also been postulated that a COS type 2 PEPC is capable of supporting a substantial flux of PEP to malate, which is essential for fatty acid synthesis and dominates the metabolism of developing COS white bodies. The PTPC and BTPC subunits of the type 2 PEPC complex have PEPC activities during COS development [69]. However, there is a notable difference in their respective affinities and sensitivities. Specifically, the PTPC subunit has a higher affinity for PEP and is more sensitive than the BTPC subunit, which is relatively insensitive to PTPC inhibitors, including L-malate and L-aspartate [71,72].

2.3. PEP Degradation Mediated by DHAPS Regulate Plant Growth and Development

DHAPS is a crucial enzyme for the biosynthesis of secondary metabolites in plants. The shikimate pathway, which comprises seven steps, is a metabolic pathway in bacteria, archaea, fungi, algae, some protozoa, and plants for the biosynthesis of folic acid and aromatic amino acids (tryptophan, phenylalanine, and tyrosine). In addition, this pathway serves as a conduit between carbohydrate metabolism and the biosynthesis of aromatic compounds [73]. In the shikimate pathway, PEP and 4-phospho-erythrose (E4P) are involved in a condensation reaction catalyzed by DHAPS that produces 3-deoxy-D-arabino-heptulosonate-7-phosphate (DAHP) and inorganic phosphate. Accordingly, DHAPS in the shikimate pathway influences PEP flux [73].

At the transcriptional level, certain transcription factors can regulate the abundance of metabolites associated with the shikimate pathway by modulating the expression of genes involved in this pathway. The *Solyc02g088190(SlMIXTA-like)* gene, which is expressed in the hairy roots of *Solanum lycopersicum*, encodes a protein that binds directly to the promoter region of the gene encoding 3-deoxy-7-phosphoheptulonate synthase and activates expression [74]. Induced *LOC544153 (SlDAHPS)* expression enhances the activity of the shikimate pathway and leads to the production of substrates for downstream secondary metabolism [74–76]. The overexpression of *PgMyb308*-like, which encodes a transcription factor in *Punica granatum*, in pomegranate hairy roots increases the accumulation of shikimic acid, aromatic amino acids, isoferulic acid, and lignin while decreasing the contents of gallic acid and its downstream product HT as well as various flavonoids [77]. These changes are attributed to an increase in the expression of *LOC116188438 (PgSDH1)*, which encodes shikimate dehydrogenase 1, an SDH isoenzyme involved in shikimate biosynthesis, and a decrease in the expression of *OWM85405.1(PgSDH4)*, an SDH isoenzyme presumed to be responsible for the production of gallic acid [78].

Arabidopsis contains three DHAPS isoenzymes (DHAPS1, DHAPS2, and DHAPS3), among which DHAPS2 is influenced by the Tyr concentration. A Tyr treatment decreases DHAPS2 contents in chloroplasts. DHAPS2 interacts with 14-3-3 proteins in the cytoplasm. This interaction, which is enhanced by a Tyr treatment, may result in the retention of DHAPS2 in the cytoplasm, thereby preventing it from functioning in chloroplasts with elevated Tyr levels [79].

In the plant developmental cycle, *DHAPS1* is predominantly expressed in mature leaves, whereas *DHAPS2* is primarily expressed in seedlings [80]. In contrast to DHAPS1 and 3, DHAPS2 is moderately active even in the absence of a reducing agent. However, it is inhibited by aromatic amino acids [80]. Considering *DHAPS2* expression is specifically upregulated in seedlings, and *DHAPS2* is co-expressed with genes involved in plastid development rather than aromatic amino acid metabolism [80], redox-independent DHAPS2 activity may be essential for maintaining the activity of the shikimate pathway in fast-growing non-photosynthetic tissues, wherein basal levels of aromatic amino acids must be maintained to support protein synthesis for growth, even in the absence of photosynthesis.

3. PEP Transport and Distribution Affect the Regulation of Plant Growth and Development

The orderly progression of cellular metabolism is achieved through the division of metabolic pathways between organelles. However, this compartmentalization necessitates the transport of specific key metabolites across relatively impermeable membranes. The plastid inner membrane serves as the primary site for the transport of metabolic intermediates between the plastid matrix and the cytosol, which is facilitated by specific transport proteins. PEP is imported into plastids via PPT for the production of aromatic amino acids [81–84]. Arabidopsis *At5g33320(CUE1)* encodes a PPT localized in the plastid inner membrane [81]. Notable characteristics of the homozygous *cue1*/heterozygous *eno1* mutant [*cue1*/*eno1* (+/−)] include delayed vegetative growth, disrupted floral development, and considerable seed abortion (up to 80%). Other phenotypes, such as decreased oil contents in seeds, decreased flavonoid and aromatic amino acid contents in flowers, impaired lignin biosynthesis in stems, and abnormal outer wall formation in pollen, have also been observed [85]. This suggests that PEP plays a significant role in the plastid. Compared with C3 plants, C4 plants typically have a higher photosynthetic CO_2 uptake rate in their leaves. Consequently, C4 photosynthesis necessitates a greater capacity for transporting metabolites. According to a phylogenetic analysis of the two PPT homologs, PPT1 and PPT2, PPT1 was recruited for C4 photosynthesis in multiple C4 lineages. C4 evolution was accompanied by increases in *PPT1* transcription. Furthermore, *At4g23660(PPT1)* expression was transferred from the roots to the leaves and from bundle sheath cells to mesophyll cells. Additionally, the rapid and sustained response of *PPT1* to light has been demonstrated.

One potential explanation for the elevated transcription of C4 *PPT1* is the presence of a MEM1 B submodule in its promoter region [86].

The redistribution of PEP may serve as an indicator of the energy state of plant root nodules. A restricted supply of sucrose causes root nodule cells to enter a relatively low-energy state, which is accompanied by increases in adenosine monophosphate (AMP) levels. This process promotes the formation of a GmNAS1–GmNAP1 heterodimer and leads to the accumulation of GmNFYC10 in the nucleus, thereby driving glycolysis and pyruvate production. With an additional supply of sucrose when the energy state of the root tuber increases as the AMP level decreases, GmNAS1 and GmNAP1 mainly form homodimers that maintain GmNFYC10 in mitochondria, thereby preventing it from accumulating in the nucleus. This results in decreased pyruvate production, as well as the conversion of PEP to OAA. Relatively low GmNFYC10a concentrations in the nucleus result in decreased expression of glycolysis-related genes involved in pyruvate production, which alters PEP allocation in favor of nitrogen fixation. Accordingly, the ratio of PEP converted into pyruvate and OAA is largely dependent on nodule energy state in the wild type (36:64 under a low root nodule energy status and 3:97 under a high root nodule energy status) [4].

The metabolic flux of PEP is critical for the formation of downstream carbon skeletons, including the methylerythritol phosphate pathway, shikimate and phenylpropanoid pathways, and anaplerotic pathway [3]. In response to ozone stress, the downstream energy metabolism and the demand for carbon skeletons increase because of increased metabolic activities. Therefore, the source and fate of PEP must be considered. The PEP metabolic pathway in leaves is influenced by ozone stress. An imbalance between PEP production and use may disrupt homeostasis and ultimately lead to cell death [3]. Furthermore, light influences the metabolic flux of PEP. A metabolic analysis of the rapid response of Arabidopsis to light stress revealed a rapid depletion of the PEP pool in cells. In terms of direct energy production, the conversion of PEP to pyruvate produces ATP, From the standpoint of immediate energy production, the conversion of PEP to pyruvate generates ATP that could be used as an energy source to maintain the cell at a high redox state. Therefore, PEP may play an important role in plants adapting to rapid changes in light conditions [87].

4. Conclusions

As shown in Table 1, the synthesis, metabolism, and transportation related to PEP have significant impacts on the growth and development of plants. PEP is a molecule with high-energy phosphate bonds as an energy donor and can also serve as a signaling molecule to participate in the regulation of growth and development. In this article, we have elaborated on the role of PEP in plant growth and development in terms of synthesis and degradation energy allocation, as well as signaling molecules. We have not established a comprehensive and systematic regulatory network centered on PEP to elucidate the important roles of plant cell tissue organ development, from vegetative growth to reproductive growth, from abiotic stress to biotic stress, and other processes. Our review can provide a reference for the basic and applied research of PEP in plants in the future, and we look forward to more research focusing on the important role of PEP in plant growth and development in the future.

Table 1. Phenotypic of PEP-related metabolism and transport processes.

PEP-Related Metabolism and Transport Processes	Gene	Mutation Type/Stress Conditions	Common Name	Latin Name	Phenotypic	Year	Reference
ENO	*AtENO1*	T-DNA insertion	Thale cress	*Arabidopsis thaliana*	distorted trichomes and fewer root hairs	2009	[7]
ENO	*AtENO2*	T-DNA insertion	Thale cress	*Arabidopsis thaliana*	small leaves, short siliques, increased sensitivity to abiotic stresses	2021	[9]
PEPCK	*AtPEPCK*	antisense *35S-PEPCK1*	Thale cress	*Arabidopsis thaliana*	soluble sugar content, stunted seedling growth, cotyledons are unable to unfold, roots cannot elongate	2004	[17]
PEPCK	*SlPEPCK*	RNA interference	tomato	*Solanum lycopersicum*	decrease in the sugar content and accumulation of malate	2015	[19]
PEPCK	*CaPEPCK*	CaPEPCK-overexpressing	chilli	*Capsicum annuum*	grow faster and produce larger leaves	2015	[20]
PEPCK	*ZmPEPCK*	sufficient light	maize	*Zea mays*	ZmPEPCK phosphorylation increases as the light intensity increases	2023	[25]
PEPCK	*AtMC9*	MC9 overexpressing	Thale cress	*Arabidopsis thaliana*	hypocotyls grow longer and enhanced PEPCK activity	2013	[30]
PPDK	*PDRP*	PDRP knockout, fluctuating light conditions	green foxtail	*Setaria viridis*	the Fv/Fm gradually decreases over several days	2022	[50]
PPDK	*OsPPDKB*	single base conversion	rice	*Oryza sativa*	decrease in the amylose:amylopectin ratio and an increase in the lipid content	2022	[45]
PK	*OsPK2*	OsPK2 loss of function	rice	*Oryza sativa*	significant decreases in grain weight and starch	2018	[55]
PK	*OsPK3*	T-DNA insertion	rice	*Oryza sativa*	delays amyloplast development in endosperm cells; abundance of storage compounds (predominantly starch, protein, and lipids) is significantly lower in seeds	2020	[56]
PK	*FaPKc2.2*	FaPKc2.2 overexpression	strawberry	*Fragaria × ananassa*	negatively regulate strawberry seed maturation and inhibit the accumulation of organic acids	2022	[57]
PK	*GhPK6*	GhPK6 silenced	cotton	*Gossypium hirsutum*	increased fiber cell elongation and reduced reactive oxygen species (ROS) accumulation	2010	[59]
PK	*GmPK21*	Gene ectopic expression in Arabidopsis thaliana	Thale cress	*Arabidopsis thaliana*	decreased in tolerance of Arabidopsis plants to salinity	2024	[62]
DHAP	*SlMIXTA-like*	*SlMIXTA*-like overexpressing	tomato	*Solanum lycopersicum*	activates SlDHAPS expression	2020	[74]
DHAP	*PgMyb308-like*	PgMyb308-like overexpression	pomegranate	*Punica granatum*	increases the accumulation of shikimic acid, aromatic amino acids, isoferulic acid, and lignin, decreasing the contents of gallic acid and its downstream product HT, as well as various flavonoids	2022	[77]
PEP transport and distribution	*CUE1*	CUE1 knockout	Thale cress	*Arabidopsis thaliana*	delayed vegetative growth, disrupted floral development, considerable seed abortion decreased oil contents in seeds, decreased flavonoid and aromatic amino acid contents in flowers, impaired lignin biosynthesis in stems, and abnormal outer wall formation in pollen	2010	[85]
PEP transport and distribution	*GmNAS1, GmNAP1*	Sucrose,*GmNAS1 GmNAP1* double knockout	Soybean	*Glycine Soja*	2-phosphoglycerate (2-PG) and phosphoenolpyruvate (PEP) contents pyruvate lower and oxaloacetic acid (OAA) contents higher	2022	[4]

Author Contributions: Conceptualization: Y.Y. (Yinghong Yu) and M.Z. (Mingdong Zhu); writing—original draft preparation: R.H. and J.D.; writing—review and editing: H.Y., S.C., R.Y., H.X., X.T., Y.Y. (Yaying Yu), Y.D. and M.Z. (Meng Zhang). All authors have read and agreed to the published version of the manuscript.

Funding: Supported by the Open Project Program (20240201) of State Key Laboratory of Rice Biology and Breeding; Changsha Natural Science Foundation (kq2208128); Changsha Science and Technology Plan Project (kq2404013); Hunan Province Agricultural Science and Technology Innovation Fund Project (2024CX18).

Institutional Review Board Statement: There is no ethics approval and consent to participate in this manuscript.

Acknowledgments: We thank Liwen for editing the English text of a draft of this manuscript.

Conflicts of Interest: The authors declare no competing interests.

Abbreviations

abbreviations	full name
PEP	phosphoenolpyruvate
ENO	enolase
PPDK	pyruvate, phosphate dikinase
PEPCK	phosphoenolpyruvate carboxykinase
PK	pyruvate kinase
DHAPS	3-Deoxy-D-arabinoheptulosonate 7-phosphate synthase
PEPC	phosphoenolpyruvate carboxylase
PPT	the phosphoenolpyruvate/phosphate translocator
WT	wild type
SA	salicylic acid
E4P	4-phospho-erythrose
PKp	plastidial pyruvate kinase
PKc	cytosolic pyruvate kinase
Glc6P	Glucose-6-phosphate
OAA	oxaloacetic acid

References

1. Wang, H.; Chen, J.; Guo, R.; Wang, D.; Wang, T.; Sun, Y. Exogenous brassinolide treatment regulates phenolic accumulation in mung bean sprouts through the modulation of sugar and energy metabolism. *J. Sci. Food Agric.* **2024**, *104*, 1656–1667. [CrossRef]
2. Plaxton, W.C.; Podestá, F.E. The functional organization and control of plant respiration. *Crit. Rev. Plant Sci.* **2006**, *25*, 159–198. [CrossRef]
3. Dizengremel, P.; Vaultier, M.-N.; Le Thiec, D.; Cabané, M.; Bagard, M.; Gérant, D.; Gérard, J.; Dghim, A.A.; Richet, N.; Afif, D.; et al. Phosphoenolpyruvate is at the crossroads of leaf metabolic responses to ozone stress. *New Phytol.* **2012**, *195*, 512–517. [CrossRef]
4. Ke, X.; Xiao, H.; Peng, Y.; Wang, J.; Lv, Q.; Wang, X. Phosphoenolpyruvate reallocation links nitrogen fixation rates to root nodule energy state. *Science* **2022**, *378*, 971–977. [CrossRef]
5. Nelson, D.; Cox, M. *Lehninger Principles of Biochemistry*; Worth Publishers: New York, NY, USA, 2000.
6. Rojas, B.E.; Iglesias, A.A. Integrating multiple regulations on enzyme activity: The case of phosphoenolpyruvate carboxykinases. *AoB PLANTS* **2023**, *15*, plad053. [CrossRef]
7. Prabhakar, V.; Löttgert, T.; Gigolashvili, T.; Bell, K.; Flügge, U.-I.; Häusler, R.E. Molecular and functional characterization of the plastid-localized phosphoenolpyruvate enolase (ENO1) from *Arabidopsis thaliana*. *FEBS Lett.* **2009**, *583*, 983–991. [CrossRef] [PubMed]
8. Liu, Z.; Zheng, L.; Pu, L.; Ma, X.; Wang, X.; Wu, Y.; Ming, H.; Wang, Q.; Zhang, G. ENO2 affects the seed size and weight by adjusting cytokinin content and forming ENO2-bZIP75 complex in *Arabidopsis thaliana*. *Front. Plant Sci.* **2020**, *11*, 574316. [CrossRef]

9. Ma, X.; Wu, Y.; Ming, H.; Liu, H.; Liu, Z.; Li, H.; Zhang, G. AtENO2 functions in the development of male gametophytes in *Arabidopsis thaliana*. *J. Plant Physiol.* **2021**, *263*, 153417. [CrossRef]
10. Liu, W.-C.; Song, R.-F.; Qiu, Y.-M.; Zheng, S.-Q.; Li, T.-T.; Wu, Y.; Song, C.-P.; Lu, Y.-T.; Yuan, H.-M. Sulfenylation of ENOLASE2 facilitates H_2O_2-conferred freezing tolerance in Arabidopsis. *Dev. Cell* **2022**, *57*, 1883–1898.e1885. [CrossRef]
11. Gallego-Giraldo, L.; Jikumaru, Y.; Kamiya, Y.; Tang, Y.; Dixon, R.A. Selective lignin downregulation leads to constitutive defense response expression in alfalfa (*Medicago sativa* L.). *New Phytol.* **2011**, *190*, 627–639. [CrossRef]
12. Barkla, B.J.; Vera-Estrella, R.; Hernández-Coronado, M.; Pantoja, O. Quantitative proteomics of the tonoplast reveals a role for glycolytic enzymes in salt tolerance. *Plant Cell* **2009**, *21*, 4044–4058. [CrossRef]
13. Eremina, M.; Rozhon, W.; Yang, S.; Poppenberger, B. ENO2 activity is required for the development and reproductive success of plants, and is feedback-repressed by A t MBP-1. *Plant J.* **2015**, *81*, 895–906. [CrossRef] [PubMed]
14. Yang, L.; Wang, Z.; Zhang, A.; Bhawal, R.; Li, C.; Zhang, S.; Cheng, L.; Hua, J. Reduction of the canonical function of a glycolytic enzyme enolase triggers immune responses that further affect metabolism and growth in Arabidopsis. *Plant Cell* **2022**, *34*, 1745–1767. [CrossRef] [PubMed]
15. Andriotis, V.M.; Kruger, N.J.; Pike, M.J.; Smith, A.M. Plastidial glycolysis in developing Arabidopsis embryos. *New Phytol.* **2010**, *185*, 649–662. [CrossRef]
16. Eastmond, P.J.; Astley, H.M.; Parsley, K.; Aubry, S.; Williams, B.P.; Menard, G.N.; Craddock, C.P.; Nunes-Nesi, A.; Fernie, A.R.; Hibberd, J.M. Arabidopsis uses two gluconeogenic gateways for organic acids to fuel seedling establishment. *Nat. Commun.* **2015**, *6*, 6659. [CrossRef]
17. Penfield, S.; Rylott, E.L.; Gilday, A.D.; Graham, S.; Larson, T.R.; Graham, I.A. Reserve mobilization in the Arabidopsis endosperm fuels hypocotyl elongation in the dark, is independent of abscisic acid, and requires PHOSPHOENOLPYRUVATE CARBOXYKINASE1. *Plant Cell* **2004**, *16*, 2705–2718. [CrossRef] [PubMed]
18. Rylott, E.L.; Gilday, A.D.; Graham, I.A. The gluconeogenic enzyme phospho enol pyruvate carboxykinase in Arabidopsis is essential for seedling establishment. *Plant Physiol.* **2003**, *131*, 1834–1842. [CrossRef] [PubMed]
19. Huang, Y.-X.; Yin, Y.-G.; Sanuki, A.; Fukuda, N.; Ezura, H.; Matsukura, C. Phosphoenolpyruvate carboxykinase (PEPCK) deficiency affects the germination, growth and fruit sugar content in tomato (*Solanum lycopersicum* L.). *Plant Physiol. Biochem.* **2015**, *96*, 417–425. [CrossRef] [PubMed]
20. Choi, D.S.; Kim, N.H.; Hwang, B.K. The pepper phosphoenolpyruvate carboxykinase CaPEPCK1 is involved in plant immunity against bacterial and oomycete pathogens. *Plant Mol. Biol.* **2015**, *89*, 99–111. [CrossRef] [PubMed]
21. Chanda, B.; Venugopal, S.C.; Kulshrestha, S.; Navarre, D.A.; Downie, B.; Vaillancourt, L.; Kachroo, A.; Kachroo, P. Glycerol-3-phosphate levels are associated with basal resistance to the hemibiotrophic fungus Colletotrichum higginsianum in Arabidopsis. *Plant Physiol.* **2008**, *147*, 2017–2029. [CrossRef] [PubMed]
22. Walker, R.P.; LEEGOOD, R.C. Phosphorylation of phosphoenolpyruvate carboxykinase in plants. Studies in plants with C_4 photosynthesis and Crassulacean acid metabolism and in germinating seeds. *Biochem. J.* **1996**, *317*, 653–658. [CrossRef]
23. Walker, R.P.; Chen, Z.-H.; Acheson, R.M.; Leegood, R.C. Effects of phosphorylation on phospho enol pyruvate carboxykinase from the C_4 plant Guinea grass. *Plant Physiol.* **2002**, *128*, 165–172. [CrossRef]
24. Chao, Q.; Liu, X.-Y.; Mei, Y.-C.; Gao, Z.-F.; Chen, Y.-B.; Qian, C.-R.; Hao, Y.-B.; Wang, B.-C. Light-regulated phosphorylation of maize phosphoenolpyruvate carboxykinase plays a vital role in its activity. *Plant Mol. Biol.* **2014**, *85*, 95–105. [CrossRef] [PubMed]
25. Gao, Z.F.; Yang, X.; Mei, Y.; Zhang, J.; Chao, Q.; Wang, B.C. A dynamic phosphoproteomic analysis provides insight into the C4 plant maize (*Zea mays* L.) response to natural diurnal changes. *Plant J.* **2023**, *113*, 291–307. [CrossRef]
26. Walker, R.P.; Leegood, R.C. Purification, and phosphorylation In Vivo and In Vitro, of phosphoenolpyruvate carboxykinase from cucumber cotyledons. *FEBS Lett.* **1995**, *362*, 70–74. [CrossRef]
27. Shen, Z.; Dong, X.-M.; Gao, Z.-F.; Chao, Q.; Wang, B.-C. Phylogenic and phosphorylation regulation difference of phosphoenolpyruvate carboxykinase of C_3 and C_4 plants. *J. Plant Physiol.* **2017**, *213*, 16–22. [CrossRef]
28. Walker, R.P.; Trevanion, S.J.; Leegood, R.C. Phosphoenolpyruvate carboxykinase from higher plants: Purification from cucumber and evidence of rapid proteolytic cleavage in extracts from a range of plant tissues. *Planta* **1995**, *196*, 58–63. [CrossRef]
29. Rojas, B.E.; Hartman, M.D.; Figueroa, C.M.; Iglesias, A.A. Proteolytic cleavage of *Arabidopsis thaliana* phospho enol pyruvate carboxykinase-1 modifies its allosteric regulation. *J. Exp. Bot.* **2021**, *72*, 2514–2524. [CrossRef] [PubMed]
30. Tsiatsiani, L.; Timmerman, E.; De Bock, P.-J.; Vercammen, D.; Stael, S.; Van De Cotte, B.; Staes, A.; Goethals, M.; Beunens, T.; Van Damme, P.; et al. The Arabidopsis metacaspase9 degradome. *Plant Cell* **2013**, *25*, 2831–2847. [CrossRef] [PubMed]
31. Vercammen, D.; Belenghi, B.; Van De Cotte, B.; Beunens, T.; Gavigan, J.-A.; De Rycke, R.; Brackenier, A.; Inzé, D.; Harris, J.L.; Van Breusegem, F. Serpin1 of *Arabidopsis thaliana* is a suicide inhibitor for metacaspase 9. *J. Mol. Biol.* **2006**, *364*, 625–636. [CrossRef]
32. Kwon, S.I.; Hwang, D.J. Expression analysis of the metacaspase gene family in Arabidopsis. *J. Plant Biol.* **2013**, *56*, 391–398. [CrossRef]
33. Kim, S.M.; Bae, C.; Oh, S.K.; Choi, D. A pepper (*C apsicum annuum* L.) metacaspase 9 (C amc9) plays a role in pathogen-induced cell death in plants. *Mol. Plant Pathol.* **2013**, *14*, 557–566. [CrossRef] [PubMed]

34. Shen, W.; Liu, J.; Li, J.-F. Type-II metacaspases mediate the processing of plant elicitor peptides in Arabidopsis. *Mol. Plant* **2019**, *12*, 1524–1533. [CrossRef] [PubMed]
35. Escamez, S.; André, D.; Zhang, B.; Bollhöner, B.; Pesquet, E.; Tuominen, H. METACASPASE9 modulates autophagy to confine cell death to the target cells during Arabidopsis vascular xylem differentiation. *Biol. Open* **2016**, *5*, 122–129. [CrossRef] [PubMed]
36. Escamez, S.; Stael, S.; Vainonen, J.P.; Willems, P.; Jin, H.; Kimura, S.; Van Breusegem, F.; Gevaert, K.; Wrzaczek, M.; Tuominen, H. Extracellular peptide Kratos restricts cell death during vascular development and stress in Arabidopsis. *J. Exp. Bot.* **2019**, *70*, 2199–2210. [CrossRef] [PubMed]
37. Furumoto, T.; Hata, S.; Izui, K. cDNA cloning and characterization of maize phosphoenolpyruvate carboxykinase, a bundle sheath cell-specific enzyme. *Plant Mol. Biol.* **1999**, *41*, 301–311. [CrossRef] [PubMed]
38. Rojas, B.E.; Hartman, M.D.; Figueroa, C.M.; Leaden, L.; Podestá, F.E.; Iglesias, A.A. Biochemical characterization of phospho enol pyruvate carboxykinases from *Arabidopsis thaliana*. *Biochem. J.* **2019**, *476*, 2939–2952. [CrossRef] [PubMed]
39. Leegood, R.C.; Walker, R.P. Regulation and roles of phosphoenolpyruvate carboxykinase in plants. *Arch. Biochem. Biophys.* **2003**, *414*, 204–210. [CrossRef] [PubMed]
40. Martín, M.; Rius, S.P.; Podestá, F.E. Two phosphoenolpyruvate carboxykinases coexist in the Crassulacean Acid Metabolism plant Ananas comosus. Isolation and characterization of the smaller 65 kDa form. *Plant Physiol. Biochem.* **2011**, *49*, 646–653. [CrossRef]
41. Sheen, J. Molecular mechanisms underlying the differential expression of maize pyruvate, orthophosphate dikinase genes. *Plant Cell* **1991**, *3*, 225–245.
42. Wang, H.; Ham, T.-H.; Im, D.-E.; Lar, S.M.; Jang, S.-G.; Lee, J.; Mo, Y.; Jeung, J.-U.; Kim, S.T.; Kwon, S.-W. A new SNP in rice gene encoding pyruvate phosphate dikinase (PPDK) associated with floury endosperm. *Genes* **2020**, *11*, 465. [CrossRef] [PubMed]
43. Taylor, L.; Nunes-Nesi, A.; Parsley, K.; Leiss, A.; Leach, G.; Coates, S.; Wingler, A.; Fernie, A.R.; Hibberd, J.M. Cytosolic pyruvate, orthophosphate dikinase functions in nitrogen remobilization during leaf senescence and limits individual seed growth and nitrogen content. *Plant J.* **2010**, *62*, 641–652. [CrossRef] [PubMed]
44. Famiani, F.; Paoletti, A.; Battistelli, A.; Moscatello, S.; Chen, Z.-H.; Leegood, R.C.; Walker, R.P. Phosphoenolpyruvate carboxykinase, pyruvate orthophosphate dikinase and isocitrate lyase in both tomato fruits and leaves, and in the flesh of peach and some other fruits. *J. Plant Physiol.* **2016**, *202*, 34–44. [CrossRef] [PubMed]
45. Muroyama, R.; Ito, H.; Takahashi, S.; Kang, D.-J.; Hamada, S. Biochemical analysis of a novel allele of the OsPPDKB gene associated with floury endosperm. *J. Cereal Sci.* **2022**, *107*, 103529. [CrossRef]
46. Kang, M.; Abdelmageed, H.; Lee, S.; Reichert, A.; Mysore, K.S.; Allen, R.D. At MBP-1, an alternative translation product of LOS 2, affects abscisic acid responses and is modulated by the E 3 ubiquitin ligase A t SAP 5. *Plant J.* **2013**, *76*, 481–493. [CrossRef] [PubMed]
47. Farré, E.M.; Geigenberger, P.; Willmitzer, L.; Trethewey, R.N. A possible role for pyrophosphate in the coordination of cytosolic and plastidial carbon metabolism within the potato tuber. *Plant Physiol.* **2000**, *123*, 681–688. [CrossRef]
48. Burnell, J.N.; Hatch, M.D. Regulation of C_4 photosynthesis: Purification and properties of the protein catalyzing ADP-mediated inactivation and Pi-mediated activation of pyruvate,Pi dikinase. *Arch. Biochem. Biophys.* **1985**, *237*, 490–503. [CrossRef] [PubMed]
49. Raghavendra, A.S.; Sage, R.F. *C4 Photosynthesis and Related CO_2 Concentrating Mechanisms*; Springer Science & Business Media: Dordrecht, The Netherlands, 2010.
50. Om, K.; Arias, N.N.; Jambor, C.C.; MacGregor, A.; Rezachek, A.N.; Haugrud, C.; Kunz, H.-H.; Wang, Z.; Huang, P.; Zhang, Q.; et al. Pyruvate, phosphate dikinase regulatory protein impacts light response of C_4 photosynthesis in Setaria viridis. *Plant Physiol.* **2022**, *190*, 1117–1133. [CrossRef]
51. Plaxton, W.C.; Tran, H.T. Metabolic adaptations of phosphate-starved plants. *Plant Physiol.* **2011**, *156*, 1006–1015. [CrossRef]
52. Dong, N.; Chen, L.; Ahmad, S.; Cai, Y.; Duan, Y.; Li, X.; Liu, Y.; Jiao, G.; Xie, L.; Hu, S.; et al. Genome-wide analysis and functional characterization of pyruvate kinase (PK) gene family modulating rice yield and quality. *Int. J. Mol. Sci.* **2022**, *23*, 15357. [CrossRef]
53. Plaxton, W.C.; Smith, C.R.; Knowles, V.L. Molecular and regulatory properties of leucoplast pyruvate kinase from *Brassica napus* (rapeseed) suspension cells. *Arch. Biochem. Biophys.* **2002**, *400*, 54–62. [CrossRef]
54. Zhang, Y.; Xiao, W.; Luo, L.; Pang, J.; Rong, W.; He, C. Downregulation of OsPK1, a cytosolic pyruvate kinase, by T-DNA insertion causes dwarfism and panicle enclosure in rice. *Planta* **2012**, *235*, 25–38. [CrossRef]
55. Cai, Y.; Li, S.; Jiao, G.; Sheng, Z.; Wu, Y.; Shao, G.; Xie, L.; Peng, C.; Xu, J.; Tang, S. Os PK 2 encodes a plastidic pyruvate kinase involved in rice endosperm starch synthesis, compound granule formation and grain filling. *Plant Biotechnol. J.* **2018**, *16*, 1878–1891. [CrossRef] [PubMed]
56. Hu, L.; Tu, B.; Yang, W.; Yuan, H.; Li, J.; Guo, L.; Zheng, L.; Chen, W.; Zhu, X.; Wang, Y. Mitochondria-associated pyruvate kinase complexes regulate grain filling in rice. *Plant Physiol.* **2020**, *183*, 1073–1087. [CrossRef] [PubMed]
57. Chen, Q.; Long, Y.; Yang, M.; Wang, H.; Sun, Y.; Liu, X.; Wang, L.; Hou, G.; Jiang, Y.; Li, M.; et al. FaPKc2. 2 negatively modulates strawberry fruit ripening by reprograming the carbon metabolic pathway. *Sci. Hortic.* **2022**, *301*, 111114. [CrossRef]
58. Dhindsa, R.S.; Beasley, C.A.; Ting, I.P. Osmoregulation in cotton fiber: Accumulation of potassium and malate during growth. *Plant Physiol.* **1975**, *56*, 394–398. [CrossRef]

59. Li, X.-R.; Wang, L.; Ruan, Y.-L. Developmental and molecular physiological evidence for the role of phospho enol pyruvate carboxylase in rapid cotton fibre elongation. *J. Exp. Bot.* **2010**, *61*, 287–295. [CrossRef] [PubMed]
60. Wulfert, S.; Schilasky, S.; Krueger, S. Transcriptional and Biochemical Characterization of Cytosolic Pyruvate Kinases in *Arabidopsis thaliana*. *Plants* **2020**, *9*, 353. [CrossRef] [PubMed]
61. Hu, Z.; Plaxton, W.C. Purification and characterization of cytosolic pyruvate kinase from leaves of the castor oil plant. *Arch. Biochem. Biophys.* **1996**, *333*, 298–307. [CrossRef] [PubMed]
62. Liu, W.; Wang, Y.; Zhang, Y.; Li, W.; Wang, C.; Xu, R.; Dai, H.; Zhang, L. Characterization of the pyruvate kinase gene family in soybean and identification of a putative salt responsive gene GmPK21. *BMC Genom.* **2024**, *25*, 88. [CrossRef]
63. Hu, J.; Duan, Y.; Yang, J.; Gan, L.; Chen, W.; Yang, J.; Xiao, G.; Guan, L.; Chen, J. Transcriptome analysis reveals genes associated with flooding tolerance in mulberry plants. *Life* **2023**, *13*, 1087. [CrossRef]
64. Lepiniec, L.; Vidal, J.; Chollet, R.; Gadal, P.; Crétin, C. Phosphoenolpyruvate carboxylase: Structure, regulation and evolution. *Plant Sci.* **1994**, *99*, 111–124. [CrossRef]
65. Outlaw, W.H., Jr. Kinetic properties of guard-cell phosphoenolpyruvate carboxylase. *Biochem. Und Physiol. Der Pflanz.* **1990**, *186*, 317–325. [CrossRef]
66. Deroche, M.E.; Carrayol, E. Nodule phosphoenolpyruvate carboxylase: A review. *Physiol. Plant.* **1988**, *74*, 775–782. [CrossRef]
67. Vance, C.P.; Gregerson, R.G.; Robinson, D.L.; Miller, S.S.; Gantt, J.S. Primary assimilation of nitrogen in alfalfa nodules: Molecular features of the enzymes involved. *Plant Sci.* **1994**, *101*, 51–64. [CrossRef]
68. Lepiniec, L.; Keryer, E.; Philippe, H.; Gadal, P.; Crétin, C. Sorghum phosphoenolpyruvate carboxylase gene family: Structure, function and molecular evolution. *Plant Mol. Biol.* **1993**, *21*, 487–502. [CrossRef]
69. O'Leary, B.; Fedosejevs, E.T.; Hill, A.T.; Bettridge, J.; Park, J.; Rao, S.K.; Leach, C.A.; Plaxton, W.C. Tissue-specific expression and post-translational modifications of plant-and bacterial-type phosphoenolpyruvate carboxylase isozymes of the castor oil plant, *Ricinus communis* L. *J. Exp. Bot.* **2011**, *62*, 5485–5495. [CrossRef] [PubMed]
70. Uhrig, R.G.; She, Y.-M.; Leach, C.A.; Plaxton, W.C. Regulatory Monoubiquitination of Phosphoenolpyruvate Carboxylase in Germinating Castor Oil Seeds. *J. Biol. Chem.* **2008**, *283*, 29650–29657. [CrossRef] [PubMed]
71. O'Leary, B.; Rao, S.K.; Kim, J.; Plaxton, W.C. Bacterial-type phosphoenolpyruvate carboxylase (PEPC) functions as a catalytic and regulatory subunit of the novel class-2 PEPC complex of vascular plants. *J. Biol. Chem.* **2009**, *284*, 24797–24805. [CrossRef]
72. Caburatan, L.; Park, J. Differential expression, tissue-specific distribution, and posttranslational controls of phosphoenolpyruvate carboxylase. *Plants* **2021**, *10*, 1887. [CrossRef]
73. Yokoyama, R.; Kleven, B.; Gupta, A.; Wang, Y.; Maeda, H.A. 3-Deoxy-D-arabino-heptulosonate 7-phosphate synthase as the gatekeeper of plant aromatic natural product biosynthesis. *Curr. Opin. Plant Biol.* **2022**, *67*, 102219. [CrossRef]
74. Ying, S.; Su, M.; Wu, Y.; Zhou, L.; Fu, R.; Li, Y.; Guo, H.; Luo, J.; Wang, S.; Zhang, Y. Trichome regulator SlMIXTA-like directly manipulates primary metabolism in tomato fruit. *Plant Biotechnol. J.* **2020**, *18*, 354–363. [CrossRef] [PubMed]
75. Tzin, V.; Malitsky, S.; Zvi, M.M.B.; Bedair, M.; Sumner, L.; Aharoni, A.; Galili, G. Expression of a bacterial feedback-insensitive 3-deoxy-d-arabino-heptulosonate 7-phosphate synthase of the shikimate pathway in Arabidopsis elucidates potential metabolic bottlenecks between primary and secondary metabolism. *New Phytol.* **2012**, *194*, 430–439. [CrossRef]
76. Tzin, V.; Rogachev, I.; Meir, S.; Moyal Ben Zvi, M.; Masci, T.; Vainstein, A.; Aharoni, A.; Galili, G. Tomato fruits expressing a bacterial feedback-insensitive 3-deoxy-D-arabino-heptulosonate 7-phosphate synthase of the shikimate pathway possess enhanced levels of multiple specialized metabolites and upgraded aroma. *J. Exp. Bot.* **2013**, *64*, 4441–4452. [CrossRef] [PubMed]
77. Dhakarey, R.; Yaritz, U.; Tian, L.; Amir, R. A Myb transcription factor, Pg Myb308-like, enhances the level of shikimate, aromatic amino acids, and lignins, but represses the synthesis of flavonoids and hydrolyzable tannins, in pomegranate (*Punica granatum* L.). *Hortic. Res.* **2022**, *9*, uhac008. [CrossRef] [PubMed]
78. Habashi, R.; Hacham, Y.; Dhakarey, R.; Matityahu, I.; Holland, D.; Tian, L.; Amir, R. Elucidating the role of shikimate dehydrogenase in controlling the production of anthocyanins and hydrolysable tannins in the outer peels of pomegranate. *BMC Plant Biol.* **2019**, *19*, 476. [CrossRef] [PubMed]
79. Kanaris, M.; Poulin, J.; Shahinas, D.; Johnson, D.; Crowley, V.M.; Fucile, G.; Provart, N.; Christendat, D. Elevated tyrosine results in the cytosolic retention of 3-deoxy-d-arabino-heptulosonate 7-phosphate synthase in *Arabidopsis thaliana*. *Plant J.* **2022**, *109*, 789–803. [CrossRef] [PubMed]
80. Yokoyama, R.; de Oliveira, M.V.; Kleven, B.; Maeda, H.A. The entry reaction of the plant shikimate pathway is subjected to highly complex metabolite-mediated regulation. *Plant Cell* **2021**, *33*, 671–696. [CrossRef]
81. Fischer, K.; Kammerer, B.; Gutensohn, M.; Arbinger, B.; Weber, A.; Häusler, R.E.; Flügge, U.-I. A new class of plastidic phosphate translocators: A putative link between primary and secondary metabolism by the phosphoenolpyruvate/phosphate antiporter. *Plant Cell* **1997**, *9*, 453–462. [PubMed]
82. Knappe, S.; Löttgert, T.; Schneider, A.; Voll, L.; Flügge, U.I.; Fischer, K. Characterization of two functional phosphoenolpyruvate/phosphate translocator (PPT) genes in Arabidopsis–AtPPT1 may be involved in the provision of signals for correct mesophyll development. *Plant J.* **2003**, *36*, 411–420. [CrossRef] [PubMed]

83. Streatfield, S.J.; Weber, A.; Kinsman, E.A.; Häusler, R.E.; Li, J.; Post-Beittenmiller, D.; Kaiser, W.M.; Pyke, K.A.; Flügge, U.-I.; Chory, J. The phosphoenolpyruvate/phosphate translocator is required for phenolic metabolism, palisade cell development, and plastid-dependent nuclear gene expression. *Plant Cell* **1999**, *11*, 1609–1621. [CrossRef]
84. Voll, L.; Häusler, R.E.; Hecker, R.; Weber, A.; Weissenböck, G.; Fiene, G.; Waffenschmidt, S.; Flügge, U.I. The phenotype of the Arabidopsis cue1 mutant is not simply caused by a general restriction of the shikimate pathway. *Plant J.* **2003**, *36*, 301–317. [CrossRef]
85. Prabhakar, V.; Löttgert, T.; Geimer, S.; Dörmann, P.; Krüger, S.; Vijayakumar, V.; Schreiber, L.; Göbel, C.; Feussner, K.; Feussner, I.; et al. Phosphoenolpyruvate provision to plastids is essential for gametophyte and sporophyte development in *Arabidopsis thaliana*. *Plant Cell* **2010**, *22*, 2594–2617. [CrossRef]
86. Lyu, M.-J.A.; Wang, Y.; Jiang, J.; Liu, X.; Chen, G.; Zhu, X.-G. What Matters for C4 Transporters: Evolutionary Changes of Phospho enol pyruvate Transporter for C_4 Photosynthesis. *Front. Plant Sci.* **2020**, *11*, 935. [CrossRef] [PubMed]
87. Choudhury, F.K.; Devireddy, A.R.; Azad, R.K.; Shulaev, V.; Mittler, R. Rapid accumulation of glutathione during light stress in Arabidopsis. *Plant Cell Physiol.* **2018**, *59*, 1817–1826. [CrossRef] [PubMed]

International Journal of
Molecular Sciences

MDPI

Perspective

Phytochelatins and Cadmium Mitigation: Harnessing Genetic Avenues for Plant Functional Manipulation

Deyvid Novaes Marques [1,*], Cássio Carlette Thiengo [2] and Ricardo Antunes Azevedo [1]

1 Department of Genetics, Luiz de Queiroz College of Agriculture (ESALQ), University of São Paulo (USP), Piracicaba, São Paulo 13418-900, SP, Brazil
2 Luiz de Queiroz College of Agriculture (ESALQ), University of São Paulo (USP), Piracicaba, São Paulo 13418-900, SP, Brazil
* Correspondence: deyvidnovaes@gmail.com

Abstract: Among the highly toxic heavy metals, cadmium (Cd) is highlighted as a persistent environmental pollutant, posing serious threats to plants and broader ecological systems. Phytochelatins (PCs), which are synthesized by phytochelatin synthase (PCS), are peptides that play a central role in Cd mitigation through metal chelation and vacuolar sequestration upon formation of Cd-PC complexes. PC synthesis interacts with other cellular mechanisms to shape detoxification outcomes, broadening the functional scope of PCs beyond classical stress responses. Plant Cd-related processes have has been extensively investigated within this context. This perspective article presents key highlights of the panorama concerning strategies targeting the PC pathway and PC synthesis to manipulate Cd-exposed plants. It discusses multiple advances on the topic related to genetic manipulation, including the use of mutants and transgenics, which also covers gene overexpression, PCS-deficient and PCS-overexpressing plants, and synthetic PC analogs. A complementary bibliometric analysis reveals emerging trends and reinforces the need for interdisciplinary integration and precision in genetic engineering. Future directions include the design of multigene circuits and grafting-based innovations to optimize Cd sequestration and regulate its accumulation in plant tissues, supporting both phytoremediation efforts and food safety in contaminated agricultural environments.

Keywords: abiotic stress; cadmium; genetic engineering; heavy metals; phytochelatin; phytochelatin synthase; plant biotechnology; tolerance; transgenics

1. Cadmium and Phytochelatins as Strategic Targets

Cadmium (Cd) is a highly toxic pollutant for plants and other living organisms. This heavy metal poses serious risks to human health and is also one of the most extensively studied environmental factors in the field of abiotic stress within the plant research context, frequently occurring under various agricultural conditions (see articles we previously reviewed [1,2]). Here, phytochelatins (PCs), referred to as metal-binding compounds, are highlighted. They are cysteine-rich peptides regarded as key molecules in the detoxification of heavy metals in plants. In response to Cd exposure, plants synthesize PCs, which are able to chelate Cd ions in the cytosol, forming stable complexes. These complexes can then be transported into the vacuole. This mechanism helps lower the toxic concentration of Cd in sensitive cellular compartments [3–5]. As such, both the synthesis and chelation capacity of PCs are closely linked to how plants modulate their tolerance and response to Cd upon different experimental settings.

Phytochelatin synthase (PCS) in turn is the key enzyme in the biosynthesis of PCs, catalyzing their formation from the precursor glutathione (GSH) through a transpeptidation reaction involving the polymerization of GSH molecules. PCS activity can be triggered by the presence of heavy metal ions, with Cd recognized as a particularly strong inducer. Its enzymatic activity and gene expression are regulated at both transcriptional and post-transcriptional levels in response to metal exposure and other conditions (see information previously reviewed [3–5]). Given its crucial role in metal detoxification, PCS holds significant potential for use in genetic manipulation and genetic engineering applications within the context of heavy metal mitigation.

Considering the growing concern over Cd contamination in agricultural systems and its impact on food safety, there is an increasing demand for innovative and relevant strategies to modulate plant tolerance and control the metal accumulation, along with other relevant topics regarding Cd mitigation. PC and their biosynthetic enzyme, as further presented in the following section, play important roles in Cd detoxification, making them strategic targets for both basic and applied research. This article aims to provide information on the current knowledge of PC-related genetic manipulation to explore how these approaches, along with relevant genetic engineering strategies, may contribute to Cd mitigation in plants, particularly through the enzyme involved. By integrating relevant insights, this paper offers a perspective on the potential of PC-centered approaches in plant biotechnology, also highlighting future research directions in this field.

2. Advances in Phytochelatin Synthase Research and Plant Genetic Manipulation

The use of genetic manipulation and various approaches, including mutagenesis and genetic engineering, has established PCS as a relevant and promising genetic target in order to investigate plants with differential tolerance and accumulation of Cd within the context of PCS synthesis. Among such experimental approaches, the use of PCS-deficient mutants, transgenic lines with *PCS* overexpression, and studies that combine both strategies have stood out, in addition to silencing-focused techniques (which include genetic manipulation through RNA interference and virus-induced gene silencing). These approaches have been applied across different Cd-exposed plant species. This includes model organisms such as *Arabidopsis thaliana* and tobacco (*Nicotiana tabacum*), as well as crop species like rice (*Oryza sativa*) and approaches that simultaneously employ both mutants and overexpressing lines (such as the complementation of the *cad1-3* mutant with heterologous *PCS* genes) (Table 1).

Table 1. Examples of functional genetic manipulations for Cd mitigation in plants, focusing on PCS-related insights and modulation of PC synthesis.

Overall Experimental Genetic Approach	Genes; Locus; Allelic Series	Examples of Main Findings Related to PCs in Cd-Exposed Plants	Citation
Isolation and characterization of Cd-sensitive mutants	*CAD1* locus (*cad1-1*, *cad1-2*)	Mutants show reduced Cd sequestration, but BSO (PC biosynthesis inhibitor) further increases sensitivity non-additively, suggesting PCs are still produced and *cad1* affects a related detoxification aspect.	[6]
Isolation and characterization of Cd-sensitive *cad1* mutants	*CAD1* locus (*cad1-1*, *cad1-2*, *cad1-3*, *cad1-4*, *cad1-5*)	Mutants show deficient PC accumulation and PCS activity correlating with increased Cd sensitivity; *cad1-3* (most sensitive) shows no detectable PCs; BSO increases wild-type Cd sensitivity but has little effect on *cad1-3*, suggesting a primary role for PCs in Cd detoxification.	[7]

Table 1. *Cont.*

Overall Experimental Genetic Approach	Genes; Locus; Allelic Series	Examples of Main Findings Related to PCs in Cd-Exposed Plants	Citation
Positional cloning and characterization of *cad1* mutants; functional complementation; heterologous expression in *E. coli*	*CAD1* gene (MRH10.11)	*cad1* mutants are Cd-sensitive and PC-deficient with lack of PC synthase activity; *CAD1* encodes a GSH-dependent, heavy metal-activated PC synthase; *CAD1* mRNA expression is not induced by Cd.	[8]
Functional characterization and expression analysis of a second *PCS* gene in *Arabidopsis thaliana* through heterologous expression in yeast and comparison with *cad1* mutants	*AtPCS2*, *AtPCS1* (*CAD1*)	*AtPCS2* encodes a functional PC synthase in yeast, but *cad1* mutants (lacking functional *AtPCS1*) show no detectable PCs upon Cd exposure in *Arabidopsis*, indicating non-redundancy and that *AtPCS1* is likely the primary Cd-inducible PC synthase in *Arabidopsis*.	[9]
Comparison of Cd uptake and accumulation in PC-deficient *cad1-3* mutant and wild type *Arabidopsis* with and without prior Cd exposure.	*cad1-3* (deficient in PC synthase activity)	The PC-deficient mutant accumulated less Cd than the wild type at higher growth Cd concentrations, with lower Cd translocation to the shoot, suggesting PCs enhance Cd uptake and translocation.	[10]
Characterization of AtPCS1 promoter activity using GUS reporter gene in transgenic *Arabidopsis*; analysis of AtPCS1 protein accumulation using FLAG-tagged constructs; investigation of Cd's effect on AtPCS1 protein levels; comparison of expression using genomic DNA vs. cDNA of *AtPCS1*.	*AtPCS1*	*AtPCS1* promoter shows differential spatial and temporal activity, including strong expression in leaf trichomes; AtPCS1 protein levels correlate with promoter activity; AtPCS1 protein expression is not post-transcriptionally regulated by Cd; introns enhance *AtPCS1* mRNA and protein accumulation.	[11]
Overexpression of wheat *PCS* gene (*TaPCS1*) in *Nicotiana glauca*	*TaPCS1*	Transgenic *N. glauca* seedlings showed increased tolerance to Cd (longer roots and greener leaves) compared to wild-type, suggesting enhanced PC activity due to *TaPCS1* overexpression.	[12]
Expression of wheat *TaPCS1* cDNA in Arabidopsis *cad1-3* mutant (root-specific and ectopic promoters); analysis of Cd transport and PC levels	Wheat *TaPCS1*, Arabidopsis *AtPCS1* (*cad1-3* locus)	Root-specific *TaPCS1* expression leads to PC transport from roots to shoots, enhanced root-to-shoot Cd transport, and reduced Cd accumulation in roots, all of which are PC-dependent.	[13]
Generation and characterization of transgenic *Arabidopsis* overexpressing *AtPCS1* under the CaMV 35S promoter; analysis of Cd tolerance, accumulation, and PC production	*AtPCS1*	High levels of *AtPCS1* overexpression led to increased PC production but did not necessarily enhance Cd tolerance or accumulation, and in some cases caused increased Cd sensitivity; moderately overexpressing lines showed increased tolerance and accumulation.	[14]
Overexpression of *AtPCS1* in transgenic *Arabidopsis* under the control of the *AtPCS1* promoter	*AtPCS1*	Overexpression led to increased *AtPCS1* mRNA and PC production under Cd stress but paradoxically resulted in hypersensitivity to Cd and Zn (not Cu), possibly due to toxicity of PCs at supraoptimal levels compared to GSH.	[15]
Overexpression of *AtPCS1* in *Arabidopsis thaliana*	*AtPCS1*	Overexpression led to increased PC levels upon Cd exposure but surprisingly resulted in Cd hypersensitivity compared to wild-type plants.	[16]
Expression analysis and complementation studies of AtPCS2 in the AtPCS1-deficient *cad1-3* mutant using various promoters	*AtPCS2*, *AtPCS1* (*CAD1*), *cad1-3*	*AtPCS2* expression is weak compared to *AtPCS1* and cannot fully complement the *cad1-3* mutant's Cd sensitivity even when overexpressed, suggesting low expression levels limit its role in PC production in this context.	[17]

Table 1. *Cont.*

Overall Experimental Genetic Approach	Genes; Locus; Allelic Series	Examples of Main Findings Related to PCs in Cd-Exposed Plants	Citation
Transcript level analysis of AtMRPs under Cd treatment using wild-type and mutant plants	*AtMRP3, cad2 (GSH deficient), cad1-3 (PC deficient)*	AtMRP3-related induction by Cd is likely independent of GSH and PC biosynthesis/levels.	[18]
Analysis of Cd sensitivity in PC-deficient Arabidopsis mutants	*cad1-3 (PCS deficient)*	Deficiency in PC production leads to increased sensitivity to Cd.	[19]
Analysis of Cd sensitivity and root-to-shoot translocation using hma2, hma4, and cad1 single and multiple mutants	*HMA2, HMA4, CAD1; hma2-4, hma4-2, cad1-3*	PC deficiency resulted in increased shoot Cd and percentage translocation, but had a minor effect compared to hma2 and hma4 loss.	[20]
Analysis of Cd tolerance in metallothionein and PC-deficient Arabidopsis mutants	*cad1-3, mt1a-2, mt2b-1, mt1a-2, mt2b-1, cad1-3*	Combined deficiency of PCs and MTs (MT1a, MT2b) leads to increased sensitivity to Cd.	[21]
Analysis of Cd sensitivity in Arabidopsis PC-deficient mutants	*AtPCS1; cad1-3, cad1-6*	PC deficiency leads to pronounced hypersensitivity to Cd.	[22]
Characterization of knockout mutants (single and double) and overexpression lines in Arabidopsis; heterologous expression in yeast.	*AtABCC1 (At1g30400); AtABCC2 (At2g34660); atabcc1, atabcc2, atabcc1 atabcc2 (knockouts); AtABCC1 (overexpression)*	AtABCC1 and AtABCC2 are important for vacuolar sequestration of PC–Cd(II), conferring tolerance; atabcc1 atabcc2 shows reduced vacuolar Cd.	[23]
Characterization of a loss-of-function mutant	*PCS1 (CAD1); cad1-3*	Loss of PCS1 function leads to Cd hypersensitivity due to the inability to synthesize PCs under Cd stress.	[24]
Characterization of a T-DNA tagged knockout mutant in Arabidopsis; in silico 3D structure prediction and in vivo metal accumulation analysis.	*PCS1 (At5g44070); SAIL_650_C12 (homozygous T-DNA insertion mutant)*	The *pcs1* mutant (SAIL_650_C12) shows higher Cd accumulation in shoots and increased Cd sensitivity due to the inability to synthesize PCs.	[25]
Characterization of knockout mutants	*PCS (non-functional in cad1-3); cad2-1, rax1-1, cad1-3*	*cad1-3* mutant deficient in PCS shows no PC accumulation under Cd exposure, confirming its role in PC synthesis; Cd is a stronger inducer of PCs than Hg.	[26]
Characterization of knockout mutant (abcc3) and overexpression lines (AtABCC3ox) in Arabidopsis; complementation assay in atabcc1 atabcc2 mutant; pharmacological inhibition of PC synthesis (BSO); analysis of Cd localization using fluorescent dyes.	*AtABCC3 (AtMRP3); abcc3 (knockout); AtABCC3ox (overexpression); cad1-3 (PCS deficient); atabcc1 atabcc2 (double knockout)*	AtABCC3 is involved in vacuolar transport of PC–Cd complexes, contributing to Cd tolerance; abcc3 mutants are Cd-sensitive with reduced vacuolar Cd; *AtABCC3ox* lines show increased Cd tolerance and vacuolar Cd; AtABCC3 function in Cd tolerance is PC-dependent and can complement atabcc1 atabcc2 in the presence of PCs.	[27]
Directed evolution of *Arabidopsis thaliana* PCS (AtPCS1); heterologous expression of wild-type and mutant AtPCS1 in yeast (Saccharomyces cerevisiae); ectopic expression of wild-type and mutant AtPCS1 in Arabidopsis thaliana and Brassica juncea.	*AtPCS1 (GenBankTMAF085230); various mutant variants* (e.g., *Y186C*)	Specific *AtPCS1* mutants with diminished catalytic activity conferred improved Cd tolerance and increased PC accumulation in yeast, *Arabidopsis*, and *B. juncea*. This was linked to the maintenance of glutathione and γ-glutamylcysteine precursor levels and redox homeostasis during Cd exposure, unlike the overexpression of wild-type *AtPCS1*, which could deplete these metabolites.	[28]
Characterization of Arabidopsis AtPCS1 knockout mutants (*cad1-3* and *cad1-6*) grown on Cd-contaminated soil; heterologous expression of truncated AtPCS1 in yeast.	*AtPCS1; cad1-3 (null mutant), cad1-6 (T-DNA insertion)*	*cad1-3* shows Cd hypersensitivity and no detectable PC accumulation; *cad1-6* is more Cd tolerant with trace PCs in leaves; both mutants have reduced leaf Cd; truncated AtPCS1 can retain Cd-activated PC synthesis.	[29]

Table 1. *Cont.*

Overall Experimental Genetic Approach	Genes; Locus; Allelic Series	Examples of Main Findings Related to PCs in Cd-Exposed Plants	Citation
Characterization of a PC-deficient knockout mutant (*cad1-3*) in Arabidopsis; comparison with wild-type under Cd exposure (with/without supplemental Cu).	*PCS1* (*CAD1*); *cad1-3*	*cad1-3* mutant lacking PCs is more sensitive to Cd; Cd induces a Cu deficiency response dependent on PCs (reduced in *cad1-3*); PCs likely complex both Cd and Cu; *cad1-3* shows altered metal accumulation and reduced Cd translocation.	[30]
Comparison of Arabidopsis thaliana wild-type and the *cad1-3* knockout mutant (deficient in AtPCS1) under control and Cd exposure conditions.	*PCS1* (*AtPCS1*); *cad1-3*	*cad1-3* mutant shows defective callose deposition after Cd exposure and increased Cd sensitivity, suggesting AtPCS1 contributes to Cd tolerance not only through PC synthesis but also via callose deposition.	[31]
Forward genetics to isolate a Cd-hypersensitive mutant in PC-deficient (cad1-3) background; characterization of the identified PP2A-4C mutant and its interaction with ethylene production under Cd stress.	*PCS1* (*CAD1*); *cad1-3* (*PC-deficient mutant*); *PP2A-4C*; *cdsr1* (*allele isolated in cad1-3 background*); *pp2a-4c-1* (*T-DNA insertion mutant for PP2A-4C*).	The study used a PC-deficient mutant (*cad1-3*) to explore Cd toxicity responses beyond the PC pathway. The identified *PP2A-4C* mutation exacerbated Cd sensitivity in the *cad1-3* background, indicating a PC-independent mechanism of Cd tolerance modulated by PP2A via ethylene production.	[32]
Forward genetic screen for Cd-hypersensitive mutants in PC-deficient (cad1-3) background; identification and characterization of cdsr2 mutant with a mutation in IAR4; analysis of Cd accumulation and thiol compound biosynthesis (including PCs).	*PCS1* (*CAD1*); *cad1-3* (*PC-deficient mutant*); *IAR4*; *cdsr2* (*allele isolated in cad1-3 background*); *cdsr2 cad1-3* (*double mutant*); *iar4-2*, *iar4-8* (*T-DNA insertion mutants for IAR4*).	The study used the PC-deficient *cad1-3* mutant to isolate further Cd-sensitive mutants, revealing that IAR4 mutation exacerbates Cd sensitivity independently of PC synthesis. In the Col-0 background, IAR4 mutation led to increased Cys, GSH, PC2, and PC3 under Cd stress, suggesting a heightened stress response. As expected, PCs were undetectable in *cad1-3* and *cdsr2 cad1-3* under Cd stress due to a defective PCS1 gene.	[33]
Expression analysis (RT-PCR, promoter-LUC assays), protein analysis (Western blot), characterization of sbp1 null mutant and SBP1 overexpression lines in wild-type and PC-deficient (cad) Arabidopsis, and in ycf1 yeast mutant; in vitro Cd binding assays of SBP1.	*SBP1*, *SBP2*, *SBP3*; *sbp1* (*null mutant*); *SBP1* (*overexpression lines*); *PCS1* (*CAD1*) (*cad mutants lacking PCs*); *YCF1* (*Saccharomyces cerevisiae mutant*).	*SBP1* expression is induced by Cd. The sbp1 mutant showed no clear phenotype, possibly due to SBP2 upregulation. Overexpression of SBP1 enhanced Cd accumulation in roots but reduced Cd sensitivity in wild-type and, notably, in cad mutants (lacking PCs), suggesting a PC-independent Cd detoxification role potentially through direct Cd binding.	[34]
Overexpression of *AtPCS1* (cytosol and chloroplast targeting)	*AtPCS1* from *Arabidopsis thaliana*, targeted to cytosol and chloroplast.	Increased PC production upon Cd exposure; cytosolic overexpression led to Cd hypersensitivity; plastidial overexpression showed no significant effect on Cd tolerance.	[35]
Overexpression of *AtPCS1* in *Nicotiana tabacum*	*AtPCS1* from *Arabidopsis thaliana* expressed in tobacco.	Increased PC production and Cd tolerance/accumulation in roots and shoots (GSH-dependent), but no enhanced Cd translocation to the shoot.	[36]
Overexpression of *E. coli cysE*, *E. coli gshA*, and *S. pombe PCS*	Genes involved in cysteine and glutathione biosynthesis from *E. coli* and PC synthase from *Schizosaccharomyces pombe*.	Increased non-protein thiols; higher Cd concentration in roots, no significant increase in shoots after Cd exposure.	[37]
Overexpression of *CdPCS1* in tobacco	*CdPCS1* from *Cynodon dactylon* expressed in tobacco.	Increased PC levels and enhanced Cd accumulation and tolerance in transgenic plants.	[38]

Table 1. *Cont.*

Overall Experimental Genetic Approach	Genes; Locus; Allelic Series	Examples of Main Findings Related to PCs in Cd-Exposed Plants	Citation
Overexpression of *AtPCS1* in *Arabidopsis thaliana*	*AtPCS1* from *Arabidopsis thaliana* (pcs9 line).	Overexpression did not increase the maximum capacity for non-protein thiol (NPT) production upon Cd exposure; transgenic line showed Cd hypersensitivity.	[39]
Overexpression of *AtPCS1* in *N. tabacum*	*AtPCS1* from *Arabidopsis thaliana* expressed in tobacco.	Increased PCS activity, moderate PC increase, strong γ-EC accumulation, GSH depletion in leaves led to Cd hypersensitivity.	[40]
Simultaneous overexpression of *AsPCS1* and *GSH1* in *A. thaliana*	*AsPCS1* (Arsenate reductase-coupled PCS 1) and *GSH1* (Glutathione synthetase) from an unspecified source, co-expressed in *Arabidopsis*.	Dual-gene transgenic lines showed higher Cd accumulation but not the highest PC content under Cd exposure compared to single-gene overexpressing lines and wild-type.	[41]
Overexpression of *AtPCS1* in *Nicotiana tabacum* (lines PaII12)	*AtPCS1* from *Arabidopsis thaliana* expressed in specific transgenic lines of tobacco.	Decreased cytosolic and vacuolar PC levels in Cd-exposed plants led to Cd hypersensitivity despite no significant change in vacuolar Cd accumulation compared to wild-type.	[42]
Overexpression of *AtPCS1* in *Arabidopsis*	*AtPCS1* from *Arabidopsis thaliana* expressed in *Arabidopsis*.	Increased PC content in *AtPCSox* lines upon Cd exposure; lower Cd tolerance at low Cd but higher tolerance at high Cd compared to wild type; exogenous GSH did not increase Cd tolerance in *AtPCSox* lines.	[43]
Overexpression of *TcPCS1* in *N. tabacum*	*TcPCS1* from *Thlaspi caerulescens* expressed in tobacco.	Enhanced PC production and increased Cd accumulation in roots and shoots; longer root length under Cd stress.	[44]
Simultaneous overexpression of *AsPCS1* and *YCF1* in *A. thaliana*	*AsPCS1* (Arsenate reductase-coupled PCS 1) and *YCF1* (vacuolar membrane protein involved in heavy metal tolerance) from unspecified sources, co-expressed in *Arabidopsis*.	Dual-gene lines showed higher PC content after Cd exposure and higher Cd accumulation than single-gene lines and wild type, indicating increased tolerance.	[45]
Heterologous overexpression of *NnPCS1* in *A. thaliana*	*NnPCS1* from *Nelumbo nucifera* expressed in *Arabidopsis thaliana*.	Increased PC content and enhanced Cd accumulation in transgenic *Arabidopsis* plants.	[46]
Heterologous overexpression of *CdPCS1* in *Arabidopsis thaliana*	*CdPCS1* from *Cynodon dactylon* expressed in *Arabidopsis thaliana*.	Enhanced accumulation of heavy metal(loid)s including Cd in aerial parts without significant growth difference compared to wild type, suggesting potential for phytoremediation.	[47]
Overexpression of *PtPCS* in *N. tabacum*	*PtPCS* from an unspecified plant source expressed in tobacco.	Transgenic plants showed enhanced Cd tolerance and higher Cd accumulation in roots and leaves upon Cd exposure compared to wild type, but a lower transfer coefficient.	[48]
Overexpression of *NtPCS1* in *Nicotiana tabacum* (sense lines)	*NtPCS1* from *Nicotiana tabacum* overexpressed in tobacco.	Enhanced Cd tolerance but no significant change in Cd accumulation was observed compared to control plants.	[49]

Table 1. *Cont.*

Overall Experimental Genetic Approach	Genes; Locus; Allelic Series	Examples of Main Findings Related to PCs in Cd-Exposed Plants	Citation
Expression of *NtPCS1* in *Nicotiana tabacum* (antisense lines)	*NtPCS1* from *Nicotiana tabacum* underexpressed in tobacco.	Showed growth retardation in the early stage suggesting a role in plant development; PC levels under Cd exposure were not explicitly detailed.	[49]
Overexpression of *AtPCS1* in *Nicotiana tabacum* (*rolB-AtPCS1*)	*AtPCS1* from *Arabidopsis thaliana* expressed in tobacco lines with the *rolB* gene.	Transgenic plants showed higher PC levels and increased Cd accumulation in roots when exposed to Cd, but inhibited Cd extrusion from leaves.	[50]
Overexpression of *MnPCS1* and *MnPCS2* in *Arabidopsis thaliana*	*MnPCS1* and *MnPCS2* from *Mimulus guttatus* expressed in *Arabidopsis thaliana.*	Transgenic *Arabidopsis* lines showed higher Cd accumulation in shoots and roots compared to wild type when grown in Cd-containing medium, suggesting enhanced PC activity led to increased metal uptake and sequestration.	[51]
Overexpression of *AdPCS2* or *AdPCS3* in *Arabidopsis thaliana*	*AdPCS2* and *AdPCS3* from *Arundo donax* expressed in *Arabidopsis thaliana.*	Transgenic *Arabidopsis* lines exhibited significant growth reduction and chlorosis when treated with Cd, suggesting that the overexpression of these *A. donax* PCS genes can lead to enhanced Cd sensitivity in *A. thaliana.*	[52]
Overexpression of truncated *IpPCS1* in *Arabidopsis*	Truncated *IpPCS1* from *Ipomoea pes-caprae* expressed in *Arabidopsis.*	Transgenic seedlings showed slightly improved Cd tolerance, but adult plants displayed no obvious effects on growth or Cd accumulation compared to the wild type under Cd stress.	[53]
Overexpression of *BnPCS1* from *Boehmeria nivea* in *Arabidopsis thaliana*	*BnPCS1* from ramie expressed in *Arabidopsis thaliana.*	Transgenic *Arabidopsis* showed enhanced Cd tolerance, higher Cd accumulation in shoots, and an increased translocation factor of Cd from roots to shoots compared to wild type under Cd stress, suggesting improved phytoremediation potential.	[54]
Transformation of *cad1-3* mutant and wild type *Arabidopsis* with leaf-specifically expressed *AtPCS1*	*AtPCS1; cad1-3*	Leaf-specific AtPCS1 restored leaf PCS activity and PC production in *cad1-3* without increasing total leaf PC levels compared to wild type, increased Cd tolerance but did not limit Cd accumulation to leaves	[55]
Analysis of cad1-3 mutant and transgenic Arabidopsis ectopically expressing wheat *TaPCS1*	*AtPCS1;cad1-3;TaPCS1*	Transgenic plants showed higher Cd-PC2 levels in roots and shoots than wild type; more Cd-PC2 found in shoots than roots in both, indicating PC contribution to shoot Cd accumulation	[56]
Complementation of *Arabidopsis cad1-3* mutant and overexpression in transgenic tobacco with CdPCS1 from *Ceratophyllum demersum*	*CdPCS1;AtPCS1(CAD1); cad1-3*	Transgenic tobacco showed increased PC content and enhanced Cd accumulation; CdPCS1 complemented *cad1-3* Cd sensitivity.	[57]
Expression of synthetic PC (EC) genes in wild type and PC-deficient *cad1-3* mutant *Arabidopsis*	*EC14,EC16,EC20;AtPCS1 (CAD1);cad1-3*	Expression of ECs complemented *cad1-3* Cd sensitivity; transgenic plants showed enhanced heavy metal(loid) accumulation (including Cd) compared to controls.	[58]
Analysis of AtPCS2 overexpression in cad1-3 mutant to assess Cd tolerance and PC accumulation	*AtPCS2* overexpression in AtPCS1 (*CAD1*) mutant *cad1-3*	Overexpression led to constitutive PC2 production, partial rescue of *cad1-3* Cd hypersensitivity on soil, and reduced PC3 levels upon Cd exposure compared to AtPCS1 overexpression	[59]
Expression of C. elegans *CePCS* in Arabidopsis *AtPCS1*-deficient *cad1-3* mutant	*AtPCS1* (CAD1); *cad1-3*; *CePCS*	*CePCS* expression in *cad1-3* restored PC synthesis and Cd tolerance	[60]

Table 1. *Cont.*

Overall Experimental Genetic Approach	Genes; Locus; Allelic Series	Examples of Main Findings Related to PCs in Cd-Exposed Plants	Citation
Overexpression of *Vicia sativa* PCS1 homolog (*VsPCS1*) in wild-type and AtPCS1-deficient *Arabidopsis* (*atpcs1*)	*VsPCS1;AtPCS1;atpcs1*	*VsPCS1* overexpression increased PC synthesis and Cd tolerance; complemented *atpcs1* PC deficiency and Cd sensitivity; led to higher-order PCs (PC4).	[61]
Overexpression of three duplicated *BnPCS* genes from *Brassica napus* in *Arabidopsis thaliana cad1–3* mutant.	*BnPCS* (three duplicated genes); *AtPCS1* (*CAD1*); *cad1-3*	Transgenic lines showed higher PC content and enhanced Cd tolerance, accumulation, and translocation compared to *cad1-3* under Cd stress.	[62]
Overexpression of maize *ZmPCS1* in *Arabidopsis* and complementation of *Arabidopsis atpcs1* mutant	*ZmPCS1; AtPCS1; atpcs1*	Overexpression in *Arabidopsis* enhanced Cd tolerance and accumulation, along with increased GSH and PC contents; *ZmPCS1* rescued the Cd-sensitive phenotype of the *atpcs1* mutant.	[63]
Isolation and functional characterization of SepPCS from *Sedum plumbizincicola* by complementation in yeast and *Arabidopsis cad1-3* mutant	SepPCS (from *S. plumbizincicola*); AtPCS1 (*CAD1*); cad1-3	*SepPCS* expression restored PCs biosynthesis and Cd tolerance in mutants; PCs' role in *S. plumbizincicola* Cd tolerance might increase with elevated Cd.	[64]
RNAi silencing of PCS in rice seeds.	OsPCS1 (GenBank AF439787); ZMM1 (seed-specific promoter).	*OsPCS1* silencing in seeds reduced Cd accumulation by ~50%.	[65]
Overexpression of Arabidopsis PCS in Indian mustard.	*AtPCS1* cDNA under the 35S promoter.	Moderate *AtPCS1* overexpression increased PCs and Cd tolerance but decreased Cd accumulation.	[66]
Overexpression of Arabidopsis PCS in Indian mustard.	AtPCS1 gDNA under its native promoter.	Higher PC levels under Cd stress and lower Cd accumulation in shoots with enhanced Cd tolerance.	[67]
Overexpression of *Arabidopsis PCS* in rice.	*AtPCS* gene; hpt II marker gene; *Oryza sativa* genotype Azucena.	Enhanced Cd tolerance, increased PC content, reduced GSH depletion, higher root Cd.	[68]
Heteroexpression of wheat PCS in rice.	*TaPCS1* cDNA under the 35S promoter; *Oryza sativa* L., cv. Zhonghua 11.	Enhanced Cd sensitivity, increased shoot Cd, higher PC and NPT in shoots under Cd.	[69]
Heterologous expression of Phragmites australis PCS in tall fescue.	*PaPCS* cDNA under the 35S promoter.	Higher PC levels, enhanced Cd tolerance, and increased Cd accumulation.	[70]
Seed-specific RNAi to silence *OsPCS* genes.	*OsPCS1; OsPCS2*; endosperm-specific GluC promoter.	*OsPCS1* and *OsPCS2* silencing in rice grains led to ~51% reduction in Cd content.	[71]
Characterization of a PCS mutant.	*OsPCS1* (Os05g0415200); has2 (three-base deletion in catalytic domain).	*has2* mutant (impaired OsPCS1) showed similar Cd tolerance and grain Cd content to wild type.	[72]
Characterization of OsPCS1 loss-of-function mutants (T-DNA and Tos17 insertion lines) in rice.	OsPCS1 (Os06g0102300/ LOC_Os06g01260); T-DNA insertion; Tos17 insertion (NG5045).	*OsPCS1* mutants showed increased Cd sensitivity, decreased Cd in shoots/grains, reduced PCs under Cd.	[73]
RNAi-mediated silencing of *OsPCS2* expression.	*OsPCS2* gene; constitutive maize ubiquitin1 promoter.	*OsPCS2*-silenced plants showed reduced PC levels under Cd stress, but Cd tolerance was little affected.	[74]
Overexpression of two rice *PCS* genes in Arabidopsis thaliana.	*OsPCS5; OsPCS15*; under the CaMV 35S promoter.	Overexpression paradoxically increased sensitivity to Cd, possibly due to excess PC.	[75]

Table 1. *Cont.*

Overall Experimental Genetic Approach	Genes; Locus; Allelic Series	Examples of Main Findings Related to PCs in Cd-Exposed Plants	Citation
Functional characterization of the first *PCS* gene from a liverwort using yeast and Arabidopsis cad1-3 complementation.	*MpPCS* (Mapoly0046s0028.1); cad1-3 (Arabidopsis AtPCS1 null mutant).	*MpPCS* overexpression in cad1-3 complemented Cd sensitivity and restored PC2/PC3 production.	[76]
Generation and characterization of *M. polymorpha* knockout mutants and overexpressing lines for *PCS*.	*MpPCS* gene; CRISPR/Cas9 knockout (four alleles); overexpression lines (MpPCS-ox-3, MpPCS-ox-7).	*Mppcs* mutants Cd hypersensitive, no PC; overexpressors also Cd hypersensitive, reduced GSH/PCn, higher Cd accumulation.	[77]
Overexpression of a rice *PCS* gene.	OsPCS1 (Os05g0415200) cDNA under the rice OsActin1 promoter.	Increased PC, decreased grain Cd, but caused seedling Cd hypersensitivity by suppressing vacuolar Cd sequestration.	[78]
Heterologous expression of cyanobacterial PCS in *Artemisia annua* hairy roots.	*anaPCS* gene (from Anabaena PCC 7120) under CMV35S promoter.	Transformed hairy roots showed significantly higher PC content and greater Cd accumulation.	[79]
Overexpression of *Tamarix hispida PCS* gene.	*ThPCS1* gene (full length, 1581 bp ORF) from *Tamarix hispida*; pROKII-ThPCS1 vector.	ThPCS1 overexpressors showed less reactive oxygen species and Cd ions under Cd stress.	[80]
VIGS of *PCS* in tomato.	*PCS*	PCS silencing aggravated Cd phytotoxicity; melatonin-induced Cd tolerance partly depends on PCs.	[81]

The following abbreviations are used in the table: AdPCS2 = *Arundo donax* phytochelatin synthase 2; AdPCS3 = *Arundo donax* phytochelatin synthase 3; anaPCS = Anabaena PCC 7120 phytochelatin synthase; AsPCS1 = arsenate-coupled phytochelatin synthase 1; AtMRP3 = *Arabidopsis* Multidrug Resistance-associated Protein 3; AtPCS1 = *Arabidopsis* phytochelatin synthase 1; AtPCS2 = *Arabidopsis* phytochelatin synthase 2; BnPCS1 = *Boehmeria nivea* phytochelatin synthase 1; BSO = buthionine sulfoximine (PC biosynthesis inhibitor); cad1 = gene encoding AtPCS1 (*Arabidopsis* phytochelatin synthase 1); cad1-1 to cad1-6 = allelic series of cad1 mutants; cad1-3 = mutant allele of cad1 with impaired PC synthesis and enhanced Cd sensitivity; mutant containing a defective *AtPCS1* gene; CAD1 = gene encoding AtPCS1; Cd = cadmium; CdPCS1 = *Cynodon dactylon* phytochelatin synthase 1; CePCS = *Caenorhabditis elegans* phytochelatin synthase; Col-0 = Columbia-0 ecotype of Arabidopsis thaliana; EC = synthetic phytochelatin-like peptide (e.g., EC14, EC16, EC20); γ-EC = γ-glutamylcysteine; GSH = glutathione; GUS = β-glucuronidase; HMA2, HMA4 = Heavy Metal ATPases 2 and 4; IAR4 = IAA-Ala Resistant 4; IpPCS1 = *Ipomoea pes-caprae* phytochelatin synthase 1; LUC = luciferase; MnPCS1 = *Mimulus guttatus* phytochelatin synthase 1; MnPCS2 = *Mimulus guttatus* phytochelatin synthase 2; MpPCS = *Marchantia polymorpha* phytochelatin synthase; MT1a, MT2b = metallothioneins 1a and 2b; NPT = non-protein thiols; OsPCS1 = *Oryza sativa* phytochelatin synthase 1; OsPCS2 = *Oryza sativa* phytochelatin synthase 2; OsPCS5 = *Oryza sativa* phytochelatin synthase 5; OsPCS15 = *Oryza sativa* phytochelatin synthase 15; ox = overexpression line; PC = phytochelatin; PCS = phytochelatin synthase; PCs = phytochelatins; PP2A-4C = Protein Phosphatase 2A subunit 4C; PtPCS = phytochelatin synthase from an unspecified plant species; RNAi = RNA interference (gene silencing strategy); RT-PCR = reverse transcription polymerase chain reaction; SAIL_650_C12 = specific T-DNA insertion line for PCS1; SBP1 = Selenium Binding Protein 1; T-DNA = transfer DNA; TaPCS1 = *Triticum aestivum* (wheat) phytochelatin synthase 1; ThPCS1 = *Tamarix hispida* phytochelatin synthase 1; VIGS = virus-induced gene silencing; VsPCS1 = *Vicia sativa* phytochelatin synthase 1; YCF1 = Yeast Cadmium Factor 1; ZmPCS1 = *Zea mays* phytochelatin synthase 1.

The study of PCS-deficient mutants, such as the well-characterized *cad1-3* mutant of *Arabidopsis* (described as deficient in PCS1, containing a defective *AtPCS1* gene), has provided early and robust evidence for the importance of PCS in Cd detoxification. The hypersensitivity of these mutants to Cd directly demonstrated the role of PCS in protecting cells against the toxicity of this metal. Comparing the behavior of *cad1-3* mutants to wild-type plants under Cd exposure made it possible to determine the extent of PCs' contribution to Cd tolerance and accumulation. Although PCS mutants in general are more sensitive to Cd, many still exhibit some level of tolerance at low concentrations, suggesting the existence of additional metal defense mechanisms [6–8,10,13,21].

In transgenic plants, while *PCS* overexpression has, in some cases, led to increased Cd tolerance and accumulation, other studies have reported hypersensitivity (Table 1), indicating that the homeostasis of PCs is finely regulated. Furthermore, different studies have speculated and suggested various reasons for such sensitivity [15,40], which may result from PC toxicity at high concentrations, cytosolic accumulation of PC–Cd complexes exceeding vacuolar sequestration capacity, depletion of GSH, or oxidative stress; and alternatively, excess PCs may be toxic or interfere with alternative detoxification pathways [78].

To provide a clearer understanding of PC-mediated detoxification, it is crucial to distinguish between the regulation of *PCS* gene expression, the actual enzymatic activity of the PCS protein, and the subsequent localization and transport of the synthesized PCs. While PCS gene expression is inducible by heavy metals and exhibits differential spatial and temporal patterns, the activity of the PCS enzyme, which catalyzes PC synthesis from the precursor GSH, is directly triggered by metal ions. Following their synthesis in the cytosol, PCs chelate metals, and the resulting complexes must be transported into the vacuole for effective sequestration (see information previously reviewed [3–5]). Crucially, simply increasing *PCS* gene expression or enzyme activity does not guarantee effective detoxification if downstream processes, such as GSH availability or the capacity for vacuolar transport, are limiting. In fact, existing studies demonstrate that PCs play a crucial role in the plant response to Cd, influencing not only tolerance but also the accumulation and translocation of the metal. Even so, *PCS* overexpression does not always lead to consistent, desirable, or positive effects, due to aspects such as the complexity of interactions among different enzyme isoforms, metal transport, and the cellular compartmentalization of Cd-PC complexes (Table 1).

Moreover, Gong et al. [13] and Chen et al. [82], using different experimental approaches (transgenic expression and grafting, respectively) and *cad1-3* plants, provided key evidence supporting the property of PCs to be translocated within plants, their long-distance transport, and their role in the movement and potential detoxification of heavy metals such as Cd. These strategies offer valuable insight into the role of PCS in vivo, but caution is also necessary due to potential pleiotropic effects and agronomic impacts. Although increased Cd tolerance and certain integrative aspects (which includes the modulation of Cd accumulation) may occur, they do not always lead to improvements in desirable agronomic traits or meet agricultural expectations, as we previously discussed [1].

3. Research Avenues and Approaches in PC-Related Genetic Manipulation

This section outlines key research avenues and emerging perspectives on the use of genetic manipulation strategies to enhance PC-mediated Cd mitigation in plants. An overview of the main approaches investigated to date on this research theme is illustrated in Figure 1.

Other authors have also demonstrated the promising use of synthetic molecules and PC analogs in genetically modified plants as relevant biotechnological tools for research on heavy metal tolerance and remediation [83–85]. Postrigan et al. [83] designed and expressed a synthetic pseudophytochelatin gene in tobacco, observing increased resistance to Cd and enhanced accumulation of the metal in the aerial parts. In turn, Zheng et al. [84] investigated a PC-like gene (PCL) encoding a peptide with an α-Glu-Cys linkage, showing improved performance of transgenic tobacco plants under Cd^{2+} stress and greater Cd accumulation, mainly in the roots. The localization of PCL in the cytoplasm and nucleus suggests potential involvement in various physiological processes. Additionally, Vershinina et al. [85] generated transgenic tobacco plants expressing the *pph6his* gene, encoding a PC analog with an α-peptide linkage, which demonstrated stable growth in the presence of

Cd and sufficient accumulation in the aerial tissue for Cd mitigation in the given research context. Optimizing the design of these synthetic molecules and analogs with different structures and sizes, exploring regulable and reusable expression systems to enhance phytoremediation efficiency, investigating the transport and complexation mechanisms of these peptides in various plant species and their relevance for Cd mitigation in different agricultural contexts, conducting direct comparative studies between different PC analogs and PCS, evaluating their effectiveness under field conditions and with various types of contamination, as well as analyzing and comparing their short and long-term effects on ecosystems are key perspectives for this research field.

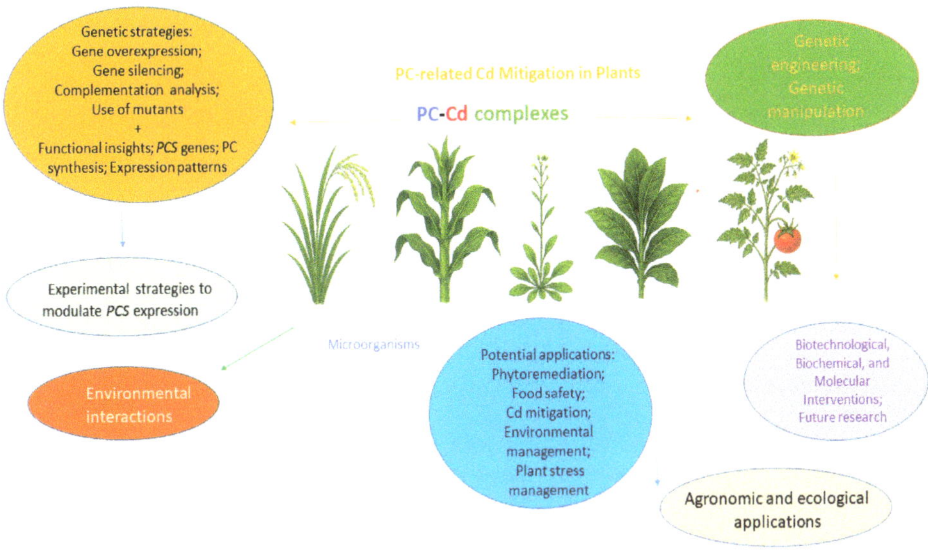

Figure 1. Overview of PC-related Cd mitigation research in plants employing genetic manipulation strategies.

Although PCS1 has historically been the primary focus in this research field, investigations have also explored the role of PCS2 in metal responses under various conditions, including its ability to complement PCS1 function in deficient mutants [17] and the regulation of PCS2 in genetically engineered rice plants (Table 1). Indeed, despite assumptions about PCS1's primary role in metal tolerance [9], other studies indicate that *Arabidopsis thaliana* PCS2 is constitutively active in vivo [59]. Additionally, the recent study by Li et al. [86] examines the use of both the cad1-3 mutant (linked to the *PCS1* gene) and the pcs2 mutant (associated with *PCS2*) to further support the involvement of these genes in Cd tolerance. Their finding that the overexpression of either *PCS1* or *PCS2* in a *wrky45* mutant background can restore the mutant's tolerance to Cd stress suggests that both genes participate in the signaling pathway downstream of WRKY45. This evidence was observed in *Arabidopsis thaliana* [86]. Also, in addition to the localization in the cytosol, AtPCS2 was shown to be detected in the nucleus, besides being tightly regulated at the transcriptional level in *Arabidopsis* and potentially playing additional roles beyond PC biosynthesis [87]. Such a regulatory landscape is complex, marked by alternative splicing of genes like *OsPCS2* [71], as well as varied expression patterns dependent on species, exposure time, and metal concentration [Table 1]. Moreover, studies involving the *cad1-3 atpcs2-1* double mutant have shown phenotypes similar to cad1-3 and a lack of detectable PCs, indicating that the low residual PCS activity in cad1-3 is attributable to AtPCS2 [82].

Such findings highlight the ongoing effort to understand the individual and combined contributions of PCS1 and PCS2 isoforms in the complex plant response to metal stress, expanding our knowledge across different plant species as well.

Indeed, PCS2 has been investigated as a functional enzyme capable of mediating PC biosynthesis in response to Cd exposure to some extent, with apparent non-redundancy with AtPCS1 [9]. In *Arabidopsis thaliana*, however, its endogenous expression was notably low due to weak promoter activity and inefficient mRNA translation, limiting its ability to compensate for the Cd hypersensitivity observed in AtPCS1-deficient mutants like *cad1-3* [17]. Despite this, AtPCS2 has shown constitutive PC-synthesizing activity and the capacity to increase Cd accumulation in *cad1-3* mutants under certain conditions [59]. Studies using *PCS2* genes from other species have yielded contrasting results; overexpression of *MnPCS2* from mulberry in Arabidopsis enhanced Cd tolerance and accumulation [51], whereas AdPCS2 from *Arundo donax* led to increased Cd sensitivity, marked by reduced shoot biomass and chlorosis [52]. In rice, the functional isoform OsPCS2a, distinct from the non-functional OsPCS2b, has been shown to contribute to Cd tolerance and accumulation when expressed in yeast, and its grain-specific silencing via RNAi—alongside OsPCS1—reduces Cd accumulation in rice grains [71]. Moreover, OsPCS2 appears to be the predominant isozyme driving PC synthesis in roots under Cd stress, with its knockdown leading to lower root Cd and PC levels but only marginally affecting whole-plant Cd tolerance [74]. Despite these advances and the contrasting outcomes across species and genetic and experimental contexts, knowledge of PCS2 function within the genetic engineering and manipulation context remains limited to a few plant species and experimental setups, highlighting the need for broader and more systematic investigations as well as including more plant species.

PCS mutants, besides their role in metal detoxification, are involved in processes like nutrient dynamics and pathogen interactions (Table 1). For example, *cad1-3* mutants show altered copper transport, revealing the link between PCs and other metal-binding proteins like metallothioneins (which also comprise other major classes of metal-binding molecules). The triple mutant *mt1a-2 mt2b-1 cad1-3* (showing metallothionein deficiency combined with PC deficiency) exhibited increased sensitivity to Cd and copper, highlighting compensatory or additive roles of different molecules and their contribution to plant metal homeostasis [21]. PC-deficient mutants also have heightened susceptibility to pathogens, linking Cd stress to weakened defense mechanisms, besides exhibiting impaired callose deposition and lignification [31]. These findings suggest that PCS and PC play a broader role in both abiotic and biotic stress responses, besides providing insights into plant survival under combined metal and biotic stress. Future studies could also focus on understanding how PCS mutations and the use of PC-focused transgenes affect plant-pathogen interactions and identify potential compensatory mechanisms in a broader agricultural context.

The study by Cahoon et al. [28] offers a compelling foundation for future biotechnological strategies targeting metal stress tolerance. By using directed evolution, they generated *PCS1* variants in *Arabidopsis thaliana* that, despite reduced catalytic efficiency, conferred enhanced Cd tolerance and accumulation in *Saccharomyces cerevisiae*, *Arabidopsis*, and *Brassica juncea*. These findings suggest that PCS activity can be fine-tuned through targeted mutations, opening new avenues for engineering crops with improved detoxification capacity and resilience to heavy metal stress. Future research may build on this approach to design custom PCS alleles tailored to specific environmental challenges.

Recent investigations have expanded our understanding of PCS and its applications in enhancing Cd tolerance and mitigation in plants. For instance, Chen et al. [33] demonstrated that the heterologous co-overexpression of *SpGSH1* and *SpPCS1* from *Spirodela polyrhiza* enhances both Cd tolerance and accumulation more effectively than individual gene overexpression. In parallel, Gui et al. [78] engineered rice plants with reduced levels of arsenic and Cd in grains through the co-overexpression of *OsPCS1*, *OsABCC1*, and *OsHMA3*, revealing a synergistic reduction in metal accumulation without compromising plant growth. Additionally, Li et al. [86] identified the transcription factor *AtWRKY45* as a positive regulator of Cd tolerance in *Arabidopsis*, acting via upregulation of *PCS1* and *PCS2* expression. Recent studies have also begun to elucidate how PC synthesis is modulated by hormonal cues and integrated with broader physiological responses to Cd stress. Xing et al. [81] for example provided compelling evidence that melatonin enhances Cd tolerance in tomato not only by mitigating oxidative damage but also by directly regulating PC biosynthesis. Using gene silencing of *PCS* and *COMT* (caffeic acid O-methyltransferase), the authors demonstrated that endogenous melatonin levels and PC content are both upregulated under Cd stress, and that melatonin-deficient plants are more susceptible to metal toxicity. These findings highlight a hormonal regulatory layer that influences the antioxidant system, redox homeostasis, nutrient absorption, and ultimately PC-mediated detoxification.

Notably, while *OsPCS1* overexpression alone has been associated with Cd hypersensitivity in some experiments based on rice research [78], combinatorial strategies involving genes encoding metal transporters have shown promise in mitigating this sensitivity and limiting Cd accumulation in edible tissues. This reinforces the notion that multigenic engineering strategies, particularly those that integrate PCS overexpression with genes involved in vacuolar sequestration and long-distance transport, are crucial for balancing detoxification efficiency with agronomic performance.

In this context, mutants such as *cad1-3* (defective in *PCS1*) and *nramp3nramp4* (defective in vacuolar metal remobilization) have provided significant mechanistic insights. As shown by Molins et al. [24], both mutants display enhanced Cd sensitivity, yet their physiological responses diverge under stress. For instance, *nramp3nramp4* exhibits severe damage to the photosynthetic apparatus under Cd exposure, a phenotype not equally observed in *cad1-3*. These findings suggest that vacuolar metal stores play a critical role in safeguarding plastid function during Cd and oxidative stress and highlight the necessity of exploring compensatory pathways when PCS function is impaired.

Taken together, these results underscore the importance of moving beyond single-gene approaches. Future research should prioritize integrative strategies that combine PCS manipulation with regulators of redox homeostasis, vacuolar sequestration, and transcriptional control. The development of synthetic circuits or inducible systems might further fine-tune PC-related responses to Cd stress, minimizing unintended physiological trade-offs and optimizing plant resilience in contaminated environments. Also, such insights can pave the way for advanced bioengineering strategies that combine PC modulation with fine-tuned control over metal transport and distribution, maximizing both tolerance and food safety outcomes in Cd-contaminated environments. Importantly, the study underscores the interconnection between PC biosynthesis and multiple transport-related processes, suggesting that future efforts should explore the cross-talk between PCs and different transport systems, including vacuolar sequestration, long-distance translocation, and efflux mechanisms.

The bibliometric analysis we conducted using the Web of Science Core Collection provided a comprehensive overview of strategies involving PC synthesis in genetically modified plants under Cd stress. Supplementary Material S1 (which also includes Tables S1–S10) provides the full set of information related to this aspect. Using the VOSviewer software (version 1.6.15), we analyzed two subsets of publications: those involving mutants and those focusing on transgenic approaches. The search and filtering process is outlined in the flowchart in Figure 2, while Figures 3 and 4 detail the collaboration networks among countries, institutions, and authors, as well as keyword co-occurrence patterns within each approach.

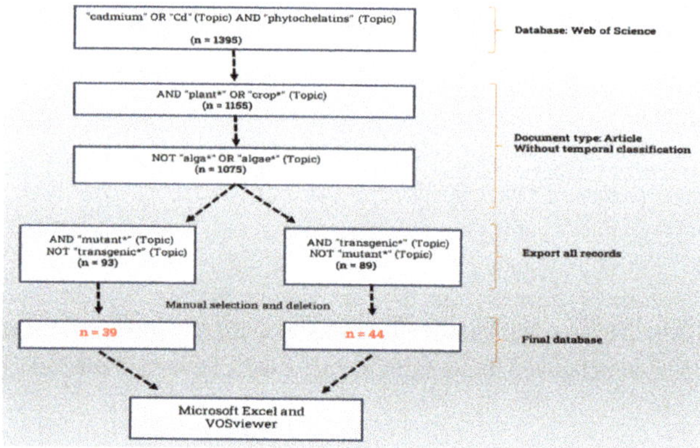

Figure 2. Flowchart for searching and filtering articles on 'phytochelatins' and 'cadmium' in plants, focusing on studies with 'mutants' and 'transgenics'. * Indicates inclusion and selection criteria used during filtering process.

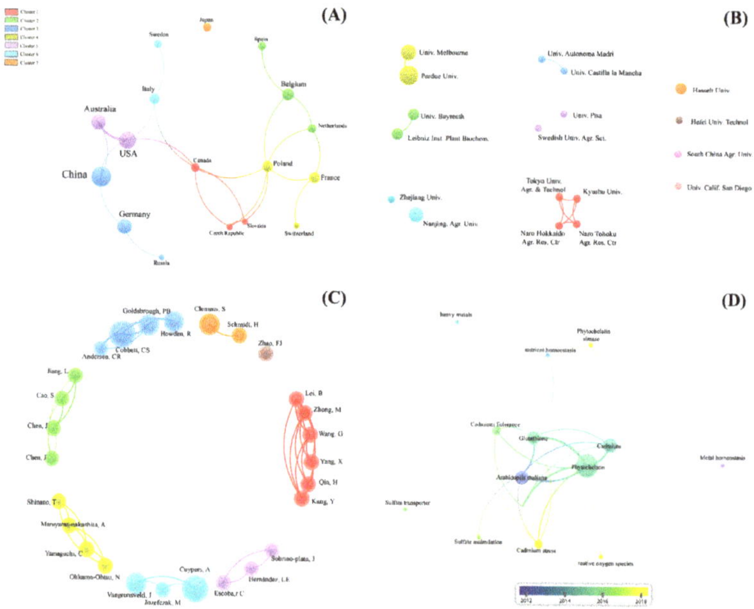

Figure 3. Bibliometric analysis of global participation and research collaboration networks by country (**A**), institutions (**B**), authors (**C**) and keyword co-occurrence network (**D**) in articles on 'phytochelatins' and 'cadmium' in plants, focusing on studies with 'mutants'. The size of the circles represents the volume of publications or the frequency of keywords, while the thickness of the lines indicates the strength of the connections or collaborations between the elements analyzed.

Our analysis revealed that research in this field extends beyond the classical phytochelatin synthase (PCS) pathway, with recurring terms such as "Glutathione", "Cadmium", and "Phytochelatin" pointing to broader strategies for responding to cadmium and other heavy metal stresses. There is a clear opportunity to deepen the integration of the PC pathway with antioxidant mechanisms (such as SOD, CAT, and POX) and with genes involved in Cd-related transport and sequestration (such as ZIP and ABC) (Supplementary Material S1), suggesting promising directions for developing plants with enhanced tolerance and accumulation capacities.

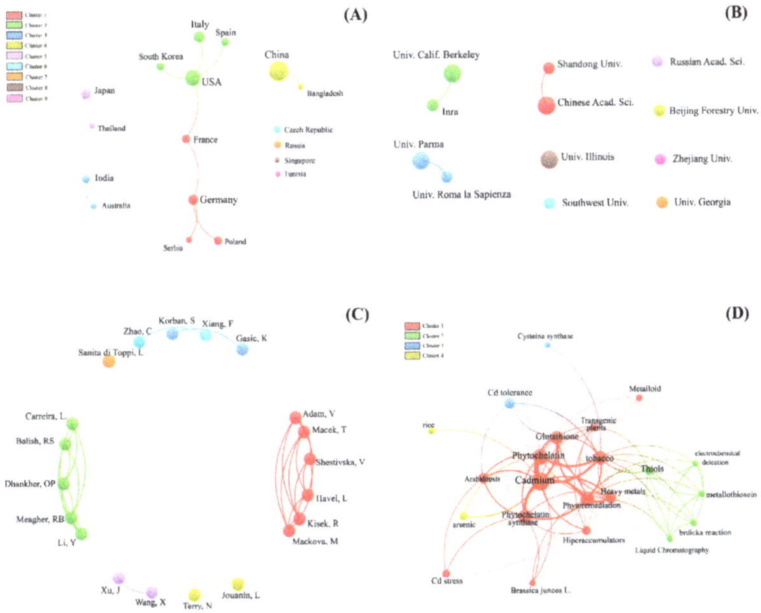

Figure 4. Bibliometric analysis of global participation and research collaboration networks by country (**A**), institutions (**B**), authors (**C**), and keyword co-occurrence network (**D**) in articles on 'phytochelatins' and 'cadmium' in plants, focusing on studies with 'transgenics'. The size of the circles represents the volume of publications or the frequency of keywords, while the thickness of the lines indicates the strength of the connections or collaborations between the elements analyzed.

The collaboration networks highlight productive and influential clusters, whose thematic analysis may uncover innovative approaches—such as the modulation of Cd bioavailability in soil or interactions with associated microbiota. These insights increase and contribute to our understanding of effective strategies in genetically modified plants and encourage interdisciplinary practices in the field. Another key point is the frequent use of model organisms like *Arabidopsis thaliana* and *Nicotiana tabacum*, highlighting the need to expand research efforts to agriculturally relevant crops or species with high phytoremediation potential, which holds substantial practical value.

Importantly, we also suggest that future bibliometric analyses take into account not only the content but also the tone and context of the publications, aiming to identify whether a balanced view is presented between the reported benefits and the potential limitations or risks of genetically manipulating the PC pathway. This critical perspective is essential for guiding more mindful, integrated, and sustainable research in plant bioengineering, particularly in the face of heavy metal contamination and its relationship with PC synthesis.

4. Concluding Remarks and Additional Future Directions

With regard to future research and ongoing investigations involving PC synthesis, mutant and transgenic plants, and Cd exposure, the integration of diverse approaches is crucial for achieving a more comprehensive and practically relevant understanding. For instance, optimizing transgenic plants through the heterologous expression of *PCS* genes from various species—while also considering factors such as subcellular localization and metabolic balance—represents a promising avenue. Transferring these genetically modified plants into agricultural systems, with an emphasis on food safety, is a critical next step. Furthermore, future studies might prioritize a fuller picture and knowledge regarding tissue-specific analyses and the use of targeted promoters to enhance the precision of transgenic and genetic engineering approaches focused on PC and PCS.

The combination of tools such as transgenic plants with omics approaches, alongside advances in gene editing technologies like CRISPR/Cas9, opens up exciting possibilities for precise modifications of PCS-related pathways. In our research group, we have also been exploring grafting as a strategy to modulate Cd tolerance. Grafting emerges as a powerful methodological innovation to dissect the specific contributions of root and shoot systems in response to Cd exposure, enabling investigations into inter-organ signaling and the modulation of metal accumulation in targeted tissues.

We have been working on the integration of grafting [1,2] with omics approaches [1,88,89] in plants, which potentially can help uncover the molecular mechanisms underlying Cd tolerance and transport related to PC synthesis. These different perspectives align with the future directions we have discussed, highlighting the importance of refining transgenic approaches, manipulating transcription factors, and applying grafting techniques to drive more robust advancements in engineering Cd-tolerant plants.

There is also a clear need for more comprehensive studies that simultaneously assess Cd accumulation, growth and productivity parameters, and the underlying mechanisms of tolerance and response. Broadening the focus to include other cellular components involved in metal homeostasis—such as ABC (ATP-binding cassette transporters) transporters and transcription factors—is equally essential.

Combining genetic manipulation of the PC pathway with agronomic strategies to modulate systemic responses holds strong potential to drive significant progress in both phytoremediation efforts and the production of safe food.

From the integrated analysis of available studies, a clear direction emerges: the development of multi-target precision engineering approaches. These would combine targeted gene editing of specific *PCS* isoforms with the simultaneous modulation of key genes involved in the transport and compartmentalization of Cd-PC complexes. The studies summarized in the Table also reveal inconsistent patterns of Cd accumulation and tolerance depending on genetic, tissue-specific, and metabolic contexts, highlighting that simple *PCS* overexpression is often insufficient to ensure consistent responses.

The given information reinforces the need for modular editing systems capable of fine-tuning not only *PCS* levels but also genes such as *ABCC1/2/3*, *HMA2/4*, and redox-related elements like *GSH1* and *YCF1*, depending on the plant species, target organ, and intended goal (phytoremediation vs. food safety). Technologies such as multiplex CRISPR/Cas9 and expression vectors with tissue-specific and/or stress-inducible promoters offer the level of precision required for such strategies.

Moreover, applying these approaches in model platforms—such as transgenic lines grafted with cultivated plant tissues—could serve as a valuable translational validation system before deploying them in economically relevant food crops. These proposals point toward a future where Cd tolerance can be modulated in a dynamic, efficient, and highly specific manner, tailored to the ecological and agricultural demands of each unique scenario.

Supplementary Materials: The supporting information can be downloaded at: https://www.mdpi.com/article/10.3390/ijms26104767/s1.

Author Contributions: D.N.M.: Conceptualization, Formal analysis, Investigation, Writing—original draft, Writing—review and editing, Visualization. C.C.T.: Formal analysis, Investigation, Visualization. R.A.A.: Supervision, Resources, Writing—review and editing. All authors have read and agreed to the published version of the manuscript.

Funding: This work was supported by the São Paulo Research Foundation—FAPESP [grant numbers 2020/12666-7 and 2022/11018–7 to DNM] (from Brazil).

Conflicts of Interest: The authors declare no conflicts of interest.

References

1. Marques, D.N.; Mason, C.; Stolze, S.C.; Harzen, A.; Nakagami, H.; Skirycz, A.; Piotto, F.A.; Azevedo, R.A. Grafting systems for plant cadmium research: Insights for basic plant physiology and applied mitigation. *Sci. Total Environ.* **2023**, *892*, 164610. [CrossRef] [PubMed]
2. Marques, D.N.; Nogueira, M.L.; Gaziola, S.A.; Batagin-Piotto, K.D.; Freitas, N.C.; Alcantara, B.K.; Paiva, L.V.; Mason, C.; Piotto, F.A.; Azevedo, R.A. New insights into cadmium tolerance and accumulation in tomato: Dissecting root and shoot responses using cross-genotype grafting. *Environ. Res.* **2023**, *216*, 114577. [CrossRef]
3. Seregin, I.V.; Kozhevnikova, A.D. Phytochelatins: Sulfur-Containing Metal(loid)-Chelating Ligands in Plants. *Int. J. Mol. Sci.* **2023**, *24*, 2430. [CrossRef]
4. Faizan, M.; Alam, P.; Hussain, A.; Karabulut, F.; Haque Tonny, S.; Cheng, S.H.; Yusuf, M.; Adil, M.F.; Sehar, S.; Alomrani, S.O.; et al. Phytochelatins: Key Regulator Against Heavy Metal Toxicity in Plants. *Plant Stress* **2024**, *11*, 10035. [CrossRef]
5. Marques, D.N.; Gaziola, S.A.; Piotto, F.A.; Azevedo, R.A. Phytochelatins: Advances in Tomato Research. *Agronomy* **2025**, *15*, 80. [CrossRef]
6. Howden, R.; Cobbett, C.S. Cadmium-sensitive mutants of *Arabidopsis thaliana*. *Plant Physiol.* **1992**, *100*, 100–107. [CrossRef] [PubMed]
7. Howden, R.; Goldsbrough, P.B.; Andersen, C.R.; Cobbett, C.S. Cadmium-Sensitive, *cad1* Mutants of *Arabidopsis thaliana* Are Phytochelatin Deficient. *Plant Physiol.* **1995**, *107*, 1059–1066. [CrossRef]
8. Ha, S.-B.; Smith, A.P.; Howden, R.; Dietrich, W.M.; Bugg, S.; O'Connell, M.J.; Goldsbrough, P.B.; Cobbett, C.S. Phytochelatin Synthase Genes from Arabidopsis and the Yeast Schizosaccharomyces pombe. *Plant Cell* **1999**, *11*, 1153–1163. [CrossRef]
9. Cazalé, A.C.; Clemens, S. *Arabidopsis thaliana* expresses a second functional phytochelatin synthase. *FEBS Lett.* **2001**, *507*, 215–219. [CrossRef]
10. Larsson, E.H.; Asp, H.; Bornman, J.F. Influence of prior Cd^{2+} exposure on the uptake of Cd^{2+} and other elements in the phytochelatin-deficient mutant, *cad1-3*, of Arabidopsis thaliana. *J. Exp. Bot.* **2002**, *53*, 447–453. [CrossRef]
11. Lee, S.; Moon, J.S.; Domier, L.L.; Korban, S.S. Molecular characterization of phytochelatin synthase expression in transgenic *Arabidopsis*. *Plant Physiol. Biochem.* **2002**, *40*, 727–733. [CrossRef]
12. Gisbert, C.; Ros, R.; De Haro, A.; Walker, D.J.; Bernal, M.P.; Serrano, R.; Navarro-Aviñó, J. A plant genetically modified that accumulates Pb is especially promising for phytoremediation. *Biochem. Biophys. Res. Commun.* **2003**, *303*, 440–445. [CrossRef] [PubMed]
13. Gong, J.-M.; Lee, D.A.; Schroeder, J.I. Long-distance root-to-shoot transport of phytochelatins and cadmium in *Arabidopsis*. *Proc. Natl. Acad. Sci. USA* **2003**, *100*, 10118–10123. [CrossRef] [PubMed]
14. Lee, S.; Petros, D.; Moon, J.S.; Ko, T.S.; Goldsbrough, P.B.; Korban, S.S. Higher levels of ectopic expression of *Arabidopsis* phytochelatin synthase do not lead to increased cadmium tolerance and accumulation. *Plant Physiol. Biochem.* **2003**, *41*, 903–910. [CrossRef]
15. Lee, S.; Moon, J.S.; Ko, T.S.; Petros, D.; Goldsbrough, P.B.; Korban, S.S. Overexpression of Arabidopsis phytochelatin synthase paradoxically leads to hypersensitivity to cadmium stress. *Plant Physiol.* **2003**, *131*, 656–663. [CrossRef]
16. Li, Y.; Dhankher, O.P.; Carreira, L.; Lee, D.; Chen, A.; Schroeder, J.I.; Balish, R.S.; Meagher, R.B. Overexpression of Phytochelatin Synthase in *Arabidopsis* Leads to Enhanced Arsenic Tolerance and Cadmium Hypersensitivity. *Plant Cell Physiol.* **2004**, *45*, 1787–1797. [CrossRef] [PubMed]
17. Lee, S.; Kang, B.S. Expression of Arabidopsis phytochelatin synthase 2 is too low to complement an *AtPCS1*-defective *Cad1-3* mutant. *Mol. Cells* **2005**, *19*, 81–87. [CrossRef]
18. Bovet, L.; Eggmann, T.; Meylan-Bettex, M.; Polier, J.; Kammer, P.; Marin, E.; Feller, U.; Martinoia, E. Transcript levels of AtMRPs after cadmium treatment: Induction of AtMRP3. *Plant Cell Environ.* **2003**, *26*, 371–381. [CrossRef]
19. Lee, S.; Kang, B.S. Phytochelatin Is Not a Primary Factor in Determining Copper Tolerance. *J. Plant Biol.* **2005**, *48*, 32–38. [CrossRef]
20. Wong, C.K.E.; Cobbett, C.S. HMA P-type ATPases are the major mechanism for root-to-shoot Cd translocation in Arabidopsis thaliana. *New Phytol.* **2009**, *181*, 71–78. [CrossRef]
21. Guo, W.J.; Meetam, M.; Goldsbrough, P.B. Examining the specific contributions of individual Arabidopsis metallothioneins to copper distribution and metal tolerance. *Plant Physiol.* **2008**, *146*, 1697–1706. [CrossRef] [PubMed]

22. Tennstedt, P.; Peisker, D.; Böttcher, C.; Trampczynska, A.; Clemens, S. Phytochelatin synthesis is essential for the detoxification of excess zinc and contributes significantly to the accumulation of zinc. *Plant Physiol.* **2009**, *149*, 938–948. [CrossRef]
23. Park, J.; Song, W.Y.; Ko, D.; Eom, Y.; Hansen, T.H.; Schiller, M.; Lee, T.G.; Martinoia, E.; Lee, Y. The phytochelatin transporters AtABCC1 and AtABCC2 mediate tolerance to cadmium and mercury. *Plant J.* **2012**, *69*, 278–288. [CrossRef] [PubMed]
24. Molins, H.; Michelet, L.; Lanquar, V.; Agorio, A.; Giraudat, J.; Roach, T.; Krieger-Liszkay, A.; Thomine, S. Mutants impaired in vacuolar metal mobilization identify chloroplasts as a target for cadmium hypersensitivity in Arabidopsis thaliana. *Plant Cell Environ.* **2013**, *36*, 804–817. [CrossRef] [PubMed]
25. Nahar, N.; Rahman, A.; Moś, M.; Warzecha, T.; Ghosh, S.; Hossain, K.; Nawani, N.N.; Mandal, A. In silico and in vivo studies of molecular structures and mechanisms of AtPCS1 protein involved in binding arsenite and/or cadmium in plant cells. *J. Mol. Model.* **2014**, *20*, 2104. [CrossRef]
26. Sobrino-Plata, J.; Carrasco-Gil, S.; Abadía, J.; Escobar, C.; Álvarez-Fernández, A.; Hernández, L.E. The role of glutathione in mercury tolerance resembles its function under cadmium stress in Arabidopsis. *Metallomics* **2014**, *6*, 356–366. [CrossRef]
27. Brunetti, P.; Zanella, L.; De Paolis, A.; Di Litta, D.; Cecchetti, V.; Falasca, G.; Barbieri, M.; Altamura, M.M.; Costantino, P.; Cardarelli, M. Cadmium-inducible expression of the ABC-type transporter AtABCC3 increases phytochelatin-mediated cadmium tolerance in Arabidopsis. *J. Exp. Bot.* **2015**, *66*, 3815–3829. [CrossRef]
28. Cahoon, R.E.; Lutke, W.K.; Cameron, J.C.; Chen, S.; Lee, S.G.; Rivard, R.S.; Rea, P.A.; Jez, J.M. Adaptive Engineering of Phytochelatin-based Heavy Metal Tolerance. *J. Biol. Chem.* **2015**, *290*, 17321–17330. [CrossRef]
29. Kühnlenz, T.; Hofmann, C.; Uraguchi, S.; Schmidt, H.; Schempp, S.; Weber, M.; Lahner, B.; Salt, D.E.; Clemens, S. Phytochelatin Synthesis Promotes Leaf Zn Accumulation of Arabidopsis thaliana Plants Grown in Soil with Adequate Zn Supply and is Essential for Survival on Zn-Contaminated Soil. *Plant Cell Physiol.* **2016**, *57*, 2342–2352. [CrossRef]
30. Gielen, H.; Vangronsveld, J.; Cuypers, A. Cd-induced Cu deficiency responses in Arabidopsis thaliana: Are phytochelatins involved? *Plant Cell Environ.* **2017**, *40*, 390–400. [CrossRef]
31. De Benedictis, M.; Brunetti, C.; Brauer, E.K.; Andreucci, A.; Popescu, S.C.; Commisso, M.; Guzzo, F.; Sofo, A.; Ruffini Castiglione, M.; Vatamaniuk, O.K.; et al. The Arabidopsis thaliana Knockout Mutant for Phytochelatin Synthase1 (cad1-3) Is Defective in Callose Deposition, Bacterial Pathogen Defense and Auxin Content, But Shows an Increased Stem Lignification. *Front. Plant Sci.* **2018**, *9*, 19. [CrossRef]
32. Chen, J.; Wang, X.; Zhang, W.; Zhang, S.; Zhao, F.J. Protein phosphatase 2A alleviates cadmium toxicity by modulating ethylene production in *Arabidopsis thaliana*. *Plant Cell Environ.* **2020**, *43*, 1008–1022. [CrossRef] [PubMed]
33. Chen, J.; Huang, S.B.; Wang, X.; Huang, L.; Gao, C.; Huang, X.Y.; Zhao, F.J. IAR4 mutation enhances cadmium toxicity by disturbing auxin homeostasis in Arabidopsis thaliana. *J. Exp. Bot.* **2024**, *75*, 438–453. [CrossRef] [PubMed]
34. Dutilleul, C.; Jourdain, A.; Bourguignon, J.; Hugouvieux, V. The Arabidopsis Putative Selenium-Binding Protein Family: Expression Study and Characterization of SBP1 as a Potential New Player in Cadmium Detoxification Processes. *Plant Physiol.* **2008**, *147*, 239–251. [CrossRef] [PubMed]
35. Picault, N.; Cazalé, A.C.; Beyly, A.; Cuiné, S.; Carrier, P.; Luu, D.T.; Forestier, C.; Peltier, G. Chloroplast targeting of phytochelatin synthase in Arabidopsis: Effects on heavy metal tolerance and accumulation. *Biochimie* **2006**, *88*, 1743–1750. [CrossRef]
36. Pomponi, M.; Censi, V.; Di Girolamo, V.; De Paolis, A.; di Toppi, L.S.; Aromolo, R.; Costantino, P.; Cardarelli, M. Overexpression of Arabidopsis phytochelatin synthase in tobacco plants enhances Cd^{2+} tolerance and accumulation but not translocation to the shoot. *Planta* **2006**, *223*, 180–190. [CrossRef]
37. Wawrzynski, A.; Kopera, E.; Wawrzynska, A.; Kaminska, J.; Bal, W.; Sirko, A. Effects of simultaneous expression of heterologous genes involved in phytochelatin biosynthesis on thiol content and cadmium accumulation in tobacco plants. *J. Exp. Bot.* **2006**, *57*, 2173–2182. [CrossRef]
38. Li, J.; Guo, J.; Xu, W.; Ma, M. Enhanced cadmium accumulation in transgenic tobacco expressing the phytochelatin synthase gene of *Cynodon dactylon* L. *J. Integr. Plant Biol.* **2006**, *48*, 928–937. [CrossRef]
39. Kim, J.H.; Lee, S. Overexpression of Arabidopsis phytochelatin synthase (atpcsi) does not change the maximum capacity for non-protein thiol production induced by Cadmium. *J. Plant Biol.* **2007**, *50*, 220–223. [CrossRef]
40. Wojas, S.; Clemens, S.; Hennig, J.; Sklodowska, A.; Kopera, E.; Schat, H.; Bal, W.; Antosiewicz, D.M. Overexpression of phytochelatin synthase in tobacco: Distinctive effects of AtPCS1 and CePCS genes on plant response to cadmium. *J. Exp. Bot.* **2008**, *59*, 2205–2219. [CrossRef]
41. Guo, J.; Dai, X.; Xu, W.; Ma, M. Overexpressing GSH1 and AsPCS1 simultaneously increases the tolerance and accumulation of cadmium and arsenic in Arabidopsis thaliana. *Chemosphere* **2008**, *72*, 1020–1026. [CrossRef] [PubMed]
42. Wojas, S.; Ruszczyńska, A.; Bulska, E.; Clemens, S.; Antosiewicz, D.M. The role of subcellular distribution of cadmium and phytochelatins in the generation of distinct phenotypes of AtPCS1- and CePCS3-expressing tobacco. *J. Plant Physiol.* **2010**, *167*, 981–988.

43. Brunetti, P.; Zanella, L.; Proia, A.; De Paolis, A.; Falasca, G.; Altamura, M.M.; Sanità di Toppi, L.; Costantino, P.; Cardarelli, M. Cadmium tolerance and phytochelatin content of Arabidopsis seedlings over-expressing the phytochelatin synthase gene AtPCS1. *J. Exp. Bot.* **2011**, *62*, 5509–5519. [CrossRef] [PubMed]
44. Liu, G.Y.; Zhang, Y.X.; Chai, T.Y. Phytochelatin synthase of Thlaspi caerulescens enhanced tolerance and accumulation of heavy metals when expressed in yeast and tobacco. *Plant Cell Rep.* **2011**, *30*, 1067–1076. [CrossRef] [PubMed]
45. Guo, J.; Xu, W.; Ma, M. The assembly of metals chelation by thiols and vacuolar compartmentalization conferred increased tolerance to and accumulation of cadmium and arsenic in transgenic Arabidopsis thaliana. *J. Hazard. Mater.* **2012**, *199*, 309–313. [CrossRef]
46. Liu, Z.; Gu, C.; Chen, F.; Yang, D.; Wu, K.; Chen, S.; Jiang, J.; Zhang, Z. Heterologous expression of a *Nelumbo nucifera* phytochelatin synthase gene enhances cadmium tolerance in *Arabidopsis thaliana. Appl. Biochem. Biotechnol.* **2012**, *166*, 722–734. [CrossRef]
47. Shukla, D.; Kesari, R.; Tiwari, M.; Dwivedi, S.; Tripathi, R.D.; Nath, P.; Trivedi, P.K. Expression of *Ceratophyllum demersum* phytochelatin synthase, CdPCS1, in *Escherichia coli* and *Arabidopsis* enhances heavy metal(loid)s accumulation. *Protoplasma* **2013**, *250*, 1263–1272. [CrossRef]
48. Chen, Y.; Liu, Y.; Ding, Y.; Wang, X.; Xu, J. Overexpression of PtPCS enhances cadmium tolerance and cadmium accumulation in tobacco. *Plant Cell Tiss. Organ Cult.* **2015**, *121*, 389–396. [CrossRef]
49. Lee, B.D.; Hwang, S. Tobacco phytochelatin synthase (NtPCS1) plays important roles in cadmium and arsenic tolerance and in early plant development in tobacco. *Plant Biotechnol. Rep.* **2015**, *9*, 107–114. [CrossRef]
50. Zanella, L.; Fattorini, L.; Brunetti, P.; Roccotiello, E.; Cornara, L.; D'Angeli, S.; Della Rovere, F.; Cardarelli, M.; Barbieri, M.; Sanità di Toppi, L.; et al. Overexpression of AtPCS1 in tobacco increases arsenic and arsenic plus cadmium accumulation and detoxification. *Planta* **2016**, *243*, 605–622. [CrossRef]
51. Fan, W.; Guo, Q.; Liu, C.; Liu, X.; Zhang, M.; Long, D.; Xiang, Z.; Zhao, A. Two mulberry phytochelatin synthase genes confer zinc/cadmium tolerance and accumulation in transgenic Arabidopsis and tobacco. *Gene* **2018**, *645*, 95–104. [CrossRef] [PubMed]
52. Li, M.; Stragliati, L.; Bellini, E.; Ricci, A.; Saba, A.; Sanità di Toppi, L.; Varotto, C. Evolution and functional differentiation of recently diverged phytochelatin synthase genes from *Arundo donax* L. *J. Exp. Bot.* **2019**, *70*, 5391–5405. [CrossRef] [PubMed]
53. Su, H.; Zou, T.; Lin, R.; Zheng, J.; Jian, S.; Zhang, M. Characterization of a phytochelatin synthase gene from Ipomoea pes-caprae involved in cadmium tolerance and accumulation in yeast and plants. *Plant Physiol. Biochem.* **2020**, *155*, 743–755. [CrossRef]
54. Zhu, S.; Shi, W.; Jie, Y. Overexpression of BnPCS1, a Novel Phytochelatin Synthase Gene From Ramie (*Boehmeria nivea*), Enhanced Cd Tolerance, Accumulation, and Translocation in Arabidopsis thaliana. *Front. Plant Sci.* **2021**, *12*, 639189. [CrossRef]
55. Peterson, A.G.; Oliver, D.J. Leaf-targeted phytochelatin synthase in *Arabidopsis thaliana. Plant Physiol. Biochem.* **2006**, *44*, 885–892. [CrossRef]
56. Sadi, B.B.; Vonderheide, A.P.; Gong, J.M.; Schroeder, J.I.; Shann, J.R.; Caruso, J.A. An HPLC-ICP-MS technique for determination of cadmium-phytochelatins in genetically modified *Arabidopsis thaliana. J. Chromatogr. B Analyt. Technol. Biomed. Life Sci.* **2008**, *861*, 123–129. [CrossRef] [PubMed]
57. Shukla, D.; Kesari, R.; Mishra, S.; Dwivedi, S.; Tripathi, R.D.; Nath, P.; Trivedi, P.K. Expression of phytochelatin synthase from aquatic macrophyte *Ceratophyllum demersum* L. enhances cadmium and arsenic accumulation in tobacco. *Plant Cell Rep.* **2012**, *31*, 1687–1699. [CrossRef]
58. Shukla, D.; Tiwari, M.; Tripathi, R.D.; Nath, P.; Trivedi, P.K. Synthetic phytochelatins complement a phytochelatin-deficient *Arabidopsis* mutant and enhance the accumulation of heavy metal(loid)s. *Biochem. Biophys. Res. Commun.* **2013**, *434*, 664–669. [CrossRef]
59. Kühnlenz, T.; Schmidt, H.; Uraguchi, S.; Clemens, S. *Arabidopsis thaliana* phytochelatin synthase 2 is constitutively active in vivo and can rescue the growth defect of the PCS1-deficient cad1-3 mutant on Cd-contaminated soil. *J. Exp. Bot.* **2014**, *65*, 4241–4253. [CrossRef]
60. Kühnlenz, T.; Westphal, L.; Schmidt, H.; Scheel, D.; Clemens, S. Expression of *Caenorhabditis elegans* PCS in the AtPCS1-deficient *Arabidopsis thaliana* cad1-3 mutant separates the metal tolerance and non-host resistance functions of phytochelatin synthases. *Plant Cell Environ.* **2015**, *38*, 2239–2247. [CrossRef]
61. Zhang, X.; Rui, H.; Zhang, F.; Hu, Z.; Xia, Y.; Shen, Z. Overexpression of a Functional *Vicia sativa* PCS1 Homolog Increases Cadmium Tolerance and Phytochelatins Synthesis in *Arabidopsis. Front Plant Sci.* **2018**, *9*, 107. [CrossRef] [PubMed]
62. Bai, J.; Wang, X.; Wang, R.; Wang, J.; Le, S.; Zhao, Y. Overexpression of Three Duplicated *BnPCS* Genes Enhanced Cd Accumulation and Translocation in *Arabidopsis thaliana* Mutant cad1-3. *Bull. Environ. Contam. Toxicol.* **2019**, *102*, 146–152. [CrossRef]
63. Jin, D.; Zhang, Q.; Liu, Y.; Liang, M.; Li, A.; Wu, J. Overexpression of the maize phytochelatin synthase gene (*ZmPCS1*) enhances Cd tolerance in plants. *Acta Physiol. Plant* **2022**, *44*, 114. [CrossRef]
64. Peng, J.S.; Ding, G.; Yi, H.Y.; Gong, J.M. Cloning and functional analysis of phytochelatin synthase gene from *Sedum plumbizincicola. Plant Physiol. J.* **2014**, *50*, 625–633.
65. Li, J.; Guo, J.; Xu, W.; Ma, M. RNA Interference-mediated Silencing of Phytochelatin Synthase Gene Reduce Cadmium Accumulation in Rice Seeds. *J. Integr. Plant Biol.* **2007**, *49*, 1032–1037. [CrossRef]
66. Gasic, K.; Korban, S.S. Expression of *Arabidopsis* phytochelatin synthase in Indian mustard (*Brassica juncea*) plants enhances tolerance for *Cd* and *Zn. Planta* **2007**, *225*, 1277–1285. [CrossRef]

67. Gasic, K.; Korban, S.S. Transgenic Indian mustard (*Brassica juncea*) plants expressing an *Arabidopsis* phytochelatin synthase (*AtPCS1*) exhibit enhanced As and Cd tolerance. *Plant Mol. Biol.* **2007**, *64*, 361–369. [CrossRef]
68. Venkataramaiah, N.; Ramakrishna, S.V.; Sreevathsa, R. Overexpression of phytochelatin synthase (*AtPCS*) in rice for tolerance to cadmium stress. *Biologia* **2011**, *66*, 1060–1073. [CrossRef]
69. Wang, F.; Wang, Z.; Zhu, C. Heteroexpression of the wheat phytochelatin synthase gene (*TaPCS1*) in rice enhances cadmium sensitivity. *Acta Biochim. Biophys. Sin.* **2012**, *44*, 886–893. [CrossRef]
70. Zhao, C.; Xu, J.; Li, Q.; Li, S.; Wang, P.; Xiang, F. Cloning and characterization of a *Phragmites australis* phytochelatin synthase (*PaPCS*) and achieving *Cd* tolerance in tall fescue. *PLoS ONE* **2014**, *9*, e103771. [CrossRef]
71. Das, N.; Bhattacharya, S.; Bhattacharyya, S.; Maiti, M.K. Identification of alternatively spliced transcripts of rice phytochelatin synthase 2 gene *OsPCS2* involved in mitigation of cadmium and arsenic stresses. *Plant Mol. Biol.* **2017**, *94*, 167–183. [CrossRef] [PubMed]
72. Hayashi, S.; Kuramata, M.; Abe, T.; Takagi, H.; Ozawa, K.; Ishikawa, S. Phytochelatin synthase OsPCS1 plays a crucial role in reducing arsenic levels in rice grains. *Plant J.* **2017**, *91*, 840–848. [CrossRef] [PubMed]
73. Uraguchi, S.; Tanaka, N.; Hofmann, C.; Abiko, K.; Ohkama-Ohtsu, N.; Weber, M.; Kamiya, T.; Sone, Y.; Nakamura, R.; Takanezawa, Y.; et al. Phytochelatin Synthase has Contrasting Effects on Cadmium and Arsenic Accumulation in Rice Grains. *Plant Cell Physiol.* **2017**, *58*, 1730–1742. [CrossRef]
74. Yamazaki, S.; Ueda, Y.; Mukai, A.; Ochiai, K.; Matoh, T. Rice *phytochelatin* synthases *OsPCS1* and *OsPCS2* make different contributions to cadmium and arsenic tolerance. *Plant Direct* **2018**, *2*, e00034. [CrossRef] [PubMed]
75. Park, H.C.; Hwang, J.E.; Jiang, Y.; Kim, Y.J.; Kim, S.H.; Nguyen, X.C.; Kim, C.Y.; Chung, W.S. Functional characterisation of two phytochelatin synthases in rice (*Oryza sativa* cv. Milyang 117) that respond to cadmium stress. *Plant Biol.* **2019**, *21*, 854–861. [CrossRef]
76. Li, M.; Barbaro, E.; Bellini, E.; Saba, A.; Sanità di Toppi, L.; Varotto, C. Ancestral function of the phytochelatin synthase C-terminal domain in inhibition of heavy metal-mediated enzyme overactivation. *J. Exp. Bot.* **2020**, *71*, 6655–6669. [CrossRef]
77. Li, M.; Leso, M.; Buti, M.; Bellini, E.; Bertoldi, D.; Saba, A.; Larcher, R.; Sanità di Toppi, L.; Varotto, C. Phytochelatin synthase de-regulation in *Marchantia polymorpha* indicates *cadmium* detoxification as its primary ancestral function in land plants and provides a novel visual bioindicator for detection of this metal. *J. Hazard. Mater.* **2022**, *440*, 129844. [CrossRef]
78. Gui, Y.; Teo, J.; Tian, D.; Yin, Z. Genetic engineering low-*arsenic* and low-cadmium rice grain. *J. Exp. Bot.* **2024**, *75*, 2143–2155. [CrossRef]
79. Pandey, N.; Rai, K.K.; Rai, S.K.; Pandey-Rai, S. Heterologous expression of cyanobacterial *PCS* confers augmented *arsenic* and *cadmium* stress tolerance and higher artemisinin in *Artemisia annua* hairy roots. *Plant Biotechnol. Rep.* **2021**, *15*, 317–334. [CrossRef]
80. Wang, P.; Yang, N.; Zhang, A.; Tangnver, S.; Li, S.; Gao, C. Cloning *ThPCS1* gene of *Tamarix hispida* to improve cadmium tolerance. *J. Nanjing For. Univ.* **2021**, *45*, 71–78.
81. Xing, Q.; Hasan, M.K.; Li, Z.; Yang, T.; Jin, W.; Qi, Z.; Yang, P.; Wang, G.; Ahammed, G.J.; Zhou, J. Melatonin-induced plant adaptation to cadmium stress involves enhanced phytochelatin synthesis and nutrient homeostasis in *Solanum lycopersicum* L. *J. Hazard. Mater.* **2023**, *456*, 131670. [CrossRef] [PubMed]
82. Chen, A.; Komives, E.A.; Schroeder, J.I. An Improved Grafting Technique for Mature *Arabidopsis* Plants Demonstrates Long-Distance Shoot-to-Root Transport of Phytochelatins in *Arabidopsis*. *Plant Physiol.* **2006**, *141*, 108–120. [CrossRef] [PubMed]
83. Postrigan, B.N.; Knyazev, A.B.; Kuluev, B.R.; Yakhin, O.I.; Chemeris, A.V. Expression of the Synthetic Phytochelatin Gene in Tobacco. *Russ. J. Plant Physiol.* **2012**, *59*, 275–280. [CrossRef]
84. Zheng, Y.; Li, M.; Liu, B.; Qin, Y.; Li, J.; Pan, Y.; Zhang, X. Effects of Phytochelatin-like Gene on the Resistance and Enrichment of Cd^{2+} in Tobacco. *Int. J. Mol. Sci.* **2022**, *23*, 16167. [CrossRef]
85. Vershinina, Z.R.; Maslennikova, D.R.; Chubukova, O.V.; Khakimova, L.R.; Mikhaylova, E.V. Synthetic *Pph6his* Gene Confers Resistance to Cadmium Stress in Transgenic Tobacco. *Russ. J. Plant Physiol.* **2023**, *70*, 83. [CrossRef]
86. Li, F.; Deng, Y.; Liu, Y.; Mai, C.; Xu, Y.; Wu, J.; Zheng, X.; Liang, C.; Wang, J. Arabidopsis transcription factor WRKY45 confers cadmium tolerance via activating *PCS1* and *PCS2* expression. *J. Hazard. Mater.* **2023**, *460*, 132496. [CrossRef]
87. Ding, G.; Peng, J.; Zhang, G.; Yi, H.; Fu, Y.; Gong, J. Regulation of the Phytochelatin Synthase Gene AtPCS2 in Arabidopsis thaliana. *Sci. Sin. Vitae* **2013**, *43*, 1112–1118. [CrossRef]
88. Marques, D.N.; Piotto, F.A.; Azevedo, R.A. Phosphoproteomics: Advances in Research on Cadmium-Exposed Plants. *Int. J. Mol. Sci.* **2024**, *25*, 12431. [CrossRef]
89. Marques, D.N.; Gaziola, S.A.; Azevedo, R.A. Phytochelatins and their relationship with modulation of cadmium tolerance in plants. In *Handbook of Bioremediation: Physiological, Molecular and Biotechnological Interventions*; Hasanuzzaman, M., Prasad, M.N.V., Eds.; Elsevier: Cambridge, MA, USA, 2021; pp. 91–113.

International Journal of
Molecular Sciences

MDPI

Perspective

Phosphoproteomics: Advances in Research on Cadmium-Exposed Plants

Deyvid Novaes Marques [1,2,*], Fernando Angelo Piotto [1] and Ricardo Antunes Azevedo [1]

1 Department of Genetics, Luiz de Queiroz College of Agriculture (ESALQ), University of São Paulo (USP), Piracicaba 13418-900, São Paulo (SP), Brazil
2 Boyce Thompson Institute, Cornell University, Ithaca, NY 14853, USA
* Correspondence: deyvidnovaes@gmail.com

Abstract: With the increasing concern on heavy metal contamination in agriculture and other environmental settings, unraveling the mechanisms of cadmium (Cd) tolerance and response in plants has become highly important. Ongoing plant Cd research over the years has focused on strategic and relevant aspects, including molecular, biochemical, and physiological processes. From this perspective, phosphoproteomics appears to be an innovative and powerful approach to investigating plant responses to Cd stress. Here, we summarize progress in plant Cd research across different plant species regarding large-scale phosphoproteomic investigations. Some studies revealed major proteins participating in detoxification, stress signaling, and metabolism, along with their regulation through phosphorylation, which modulates the plant's defense against Cd. However, many pathways remain unexplored. Expanding these studies will help our ability to alleviate Cd stress and provide further information concerning involved mechanisms. Our purpose is to inspire researchers to further explore the use of phosphoproteomics in unraveling such complex mechanisms of Cd tolerance and response across various plant species, with the ultimate aim of enhancing strategies for mitigating Cd stress in agriculture and polluted environments.

Keywords: abiotic stress response; cadmium toxicity; crops; detoxification; heavy metal; phosphoproteomics; protein phosphorylation; plant tolerance; proteomic approaches

1. Cadmium in Plants and the Phosphoproteomic Context

Cadmium (Cd), a significant abiotic stress factor, is among the most toxic heavy metals, which poses significant challenges to plant metabolism and agricultural safety. Cd is non-essential in plants and animals, and even at low concentrations it may severely affect physiological processes, thus being highly hazardous when entering the food chain (reviewed by Marques et al. [1,2]). It is released into the environment both through natural and anthropogenic sources, including mining activities, industrial releases into the air, and phosphate-based fertilizers. According to the existing literature [3,4], Cd has been reported to exist in several forms of chemical compounds; cadmium sulfide (CdS) is a prevalent form of this metal in nature. The formation of $Cd(CN)_4^{2-}$ and $Cd(NH_3)_6^{4-}$ has been known to derive from the combination of CdS with cyanide and ammonia, respectively [3]. This diversity indicates a profound influence on the behavior of Cd in the environment and its interaction with biological systems. After plants uptake Cd, it interferes with important physiological processes within the plants (some of them reviewed by Clemens et al. [5] and Clemens and Ma [6]). Thus, the understanding of how plants respond to Cd stress and elucidating related tolerance mechanisms are important prerequisites in basic plant physiology regarding Cd research and developing effective strategies for Cd mitigation.

Understanding the dynamics of cell signaling in response to Cd exposure is relevant to delineate the molecular mechanisms that allow plants to cope with and adapt to Cd-induced stress. According to the review by Vitelli et al. [7], analysis of how such signaling pathways work under Cd stress provides critical insight into the capability of the plant to temper the toxic effects of Cd and maintain cellular function.

Protein phosphorylation represents one of the major and crucial signaling regulatory mechanisms in plant responses to various types of stress, including Cd exposure. Such a process includes the addition of phosphate groups to proteins, thus changing their activity, stability, or ability to interact with other cellular components. This post-translational modification serves as a kind of molecular switch that turns on or off multiple signal pathways that are crucial for plant development, growth, and adaptation against environmental challenges such as toxicity caused by heavy metals. Importantly, protein phosphorylation turns on the detoxification mechanisms and stress signaling pathways of plants; this in turn ameliorates the detrimental action of stress [8].

Proteomics—the field of research focused on all proteins present in a biological sample, particularly with regard to various relevant aspects, including their functions, structures, and interactions in biological systems—has gained the limelight as an effective tool in unraveling the molecular basis of plants' responses to stressors. The information on the set of proteins expressed under stress conditions could enable the identification of proteins important in the defense and adaptation processes of the plant. Proteomic approaches have greatly improved our knowledge on how plants cope with Cd stress through the identification of proteins taking part in signaling, metabolic, and detoxification processes [9]. We investigated Cd-exposed plants in several studies, which included a suite of proteomic approaches to point out molecular mechanisms of tolerance to Cd [1,10] and our pioneer phosphoproteomic investigation using tomato (*Solanum lycopersicum*) plants [11].

Phosphoproteomics, a specialized proteomics field, has involved insights on the identification and quantification of phosphopeptides, as well as the phosphorylation status of proteins, which is one critical aspect in the regulation of cellular processes under stress. Therefore, it enabled the investigation into how phosphorylation dynamically regulates proteins during stress responses. This makes phosphoproteomics of particular importance for studying abiotic stresses, considering that phosphorylation controls major signal transduction pathways and metabolic adjustments that enable plants to tolerate stress [12]. Identifying the specific phosphorylation events that occur in Cd allows the use of phosphoproteomics to investigate the molecular mechanism involving tolerance and detoxification.

Although some studies using proteomic and phosphoproteomic approaches have been conducted on Cd stress responses among plant species, phosphoregulation is still in the early stages of study. Much remains to be explored in terms of specific proteins and pathways regulated by phosphorylation under Cd exposure. Thus, detailed studies of phosphorylated proteins in plants exposed to Cd will reveal more about how plants use strategies for Cd detoxification and mitigation. Clearly, there is a need for a better understanding of how Cd exposure modulates the function of proteins through phosphorylation. Herein, we summarize information on these phosphoproteomic investigations and also provide perspectives on the current state and future directions of this field.

2. Insights into Protein Phosphorylation in Plants Under Cadmium Exposure

Some of the mechanisms involved in Cd-induced protein phosphorylation without using large-scale phosphoproteomic approaches have been at least partially explained by various studies across different plant species. For instance, Reddy and Prasad [13] determined that changes occur in protein phosphorylation in rice (*Oryza sativa*), where it was found that Cd preferentially increased the phosphorylation of a 68 kDa protein (p68) while decreasing the phosphorylation level of another heat shock-related cognate protein, hsc70. In another investigation, a heat shock associated with Cd treatment inducing protein phosphorylation in soybean (*Glycine max*) seedlings has been shown [14]. Furthermore, the influence of Cd on nitrate assimilation in bean (*Phaseolus vulgaris*) through modifications of the activation states of nitrate reductase, as well as the related role of protein phosphorylation in metabolic regulation, has been investigated [15]. Some authors have analyzed protein phosphorylation events and reported the activation of some kinases (including mitogen-activated protein kinases (MAPKs) and myelin basic protein kinases) in response to Cd in alfalfa (*Medicago sativa*) [16] and rice plants [17] in a heavy metal dose-dependent manner.

Further information touches upon the capability of brassinosteroids to increase the phosphorylation level of plasma membrane H^+-ATPase under Cd stress, as reported by Jakubowska and Janicka [18]. The investigation conducted by Ma et al. [19] revealed a protein kinase SOS2L1 (a calcineurin B-Like-interacting protein kinase) in apple (*Malus domestica*), which controls malate excretion and interacts with the malate transporter MdALMT14. The phosphorylation-induced stability of MdALMT14 under Cd exposure was also noted [19].

Zhang et al. [20] pointed out the role of calcium-dependent protein kinases in regulating Cd tolerance in *Arabidopsis*; Du et al. [21] identified and demonstrated a receptor-like kinase (TaWAK20) in wheat (*Triticum aestivum*), which is induced by Cd treatment. Taken together, these studies have highlighted the importance of phosphorylation events as a post-translational change in mediating plant responses to Cd stress and the relevance of non-omic molecular and biochemical approaches for the improvement of this field of research. Phosphoproteomics in combination with other approaches will provide more detailed information about complex signaling networks regulating plant responses to Cd stress. These integrative approaches may point to key phosphorylation events associated with the modulation of Cd tolerance, thus identifying genetic targets for use in biotechnological applications. Additionally, the interaction between different signaling pathways involving kinases and plant hormones will be understood and lead to better resilience and improved plants to Cd exposure.

3. Phosphoproteomics in Cadmium-Exposed Plants

3.1. Investigations of Different Plant Species

Our research group carried out a pioneering comparative phosphoproteomic analysis of tomato genotypes with contrasting tolerance to Cd, identifying thousands of phosphopeptides [11]. This was the first large-scale study of its kind on tomato. Cd treatment led to an increased difference in phosphorylation status between the Cd-tolerant and Cd-sensitive tomato genotypes in relation to proteins associated with major functions. This included phytohormones-related proteins, kinases and phosphatases, and transcription factors, and heat shock proteins (HSPs). Other protein groups found in this tomato research included ABC (ATP-Binding Cassette) transporters and HSPs, which are proposed as important contributors to the Cd detoxification process, potentially facilitating Cd sequestration into vacuoles by working synergistically or in parallel with phytochelatins (PCs) [22] and preventing Cd-induced protein misfolding [23], respectively. These observations suggest that detoxification, stress signaling, and transcriptional pathways—essential for maintaining normal plant homeostasis under Cd stress—are regulated to some extent by phosphorylation in tomato.

Regarding rice plants, large-scale phosphoproteomic studies have reported that proteins with altered phosphorylation status under Cd treatment were involved in a wide range of cellular processes (e.g., signal transduction and transcriptional regulation), as well as ABC transporters and HSPs [24]. Protein phosphorylation has also been reported as a contributing factor in the modulation of rice's oxidative stress response and protection against Cd-induced cell damage [24]. Another comparative study by Fang et al. [25] examined the differential accumulation of phosphorylated proteins between Cd-tolerant (D28) and Cd-sensitive (D69) rice lines and identified proteins associated with key metabolic processes—such as carbon metabolism, proteolysis, DNA repair, RNA helicases, and transcription—in the tolerant one. These findings suggest that protein phosphorylation-related events also contribute to the fine-tuning of metabolism to detoxify Cd related to the proteins from these pathways in rice [25].

Other authors also conducted a phosphoproteomic investigation and observed that fructose 1,6-bisphosphate aldolase (PvFBA1), as well as proteins related to various metabolic pathways (e.g., carbohydrate and energy metabolism), were differentially phosphorylated in Seashore paspalum (*Paspalum vaginatum*) plants under Cd exposure [26]. Considering the potential of this perennial turfgrass for use in phytoremediation, these findings provide additional information on the applicability of phosphoproteomics. Altogether, these findings complement the findings from key food crop species mentioned above and indicate the potential of phosphoproteomics research beyond Cd mitigation mechanisms within a crop safety context. Within this context, strategic tools to increase the knowledge of Cd-tolerant plants that could eventually be used in the future with enhanced metal hyperaccumulation to advance environmental cleanup are also relevant aspects to be further explored.

3.2. Nitrogen Supplementation Modulating Protein Phosphorylation in Cadmium-Exposed Plants

Some authors studied the phosphoproteomic responses of poplar plants to Cd stress and the consequences of supplementation with N, respectively. The supplementation of nitrogen (N) was found to enhance plant resistance to Cd stress by modifying key signaling pathways and enhancing stress response mechanisms in such previous studies [27,28]. He et al. [27] identified significantly enriched phosphorylated protein sites and differentially expressed kinases and underlined the role of N application in increasing protein phosphorylation and boosting key proteins in sugar synthesis pathways, such as UDP-glucose dehydrogenase and sucrose-phosphate synthase.

Huang et al. [28] found that, due to Cd stress, some proteins with phosphosites were differentially regulated (e.g., proteins involved in the cellular response to abscisic acid (ABA) stimulus and those with calmodulin protein kinase activity), with some being upregulated and others downregulated. However, N + Cd treatment induced changes in such phosphorylation patterns, generally counteracting the downregulation caused by Cd stress. This shift suggests that N plays a role in modulating protein phosphorylation status under Cd stress, likely contributing to nitrogen's protective effects [28].

Potential opportunities lie in a more comprehensive phosphoproteomic view along with N-involved responses in more plant species. This includes investigations focused on N availability [29] and nitric oxide (NO) [30], which have been involved in Cd-induced responses and Cd mitigation in plants. The combination of the effect of N on plant metabolism and stress signaling with phosphoproteomic aspects and N metabolism-related molecules in multiple plant species might help to unravel more mechanisms underlying the wide array of N implications under various experimental conditions.

In Figure 1, we present a graphical overview of plant species studied, highlighting exemplary differentially regulated molecules identified in phosphoproteomic analyses under Cd exposure.

Figure 1. Graphical overview of different plant species under Cd exposure, highlighting exemplary differentially regulated molecules and mechanisms identified in phosphoproteomic studies.

4. Physiological and Biochemical Insights Complement Results of Phosphoproteomic Analyses in Cadmium-Exposed Plants

A holistic understanding of how plants respond to Cd stress requires the integration of proteomic and phosphoproteomic data with physiological and biochemical analyses. Indeed, the integration of phosphoproteomic and physiological data points to a future research direction where a more complete understanding of plant responses to Cd stress will emerge. Omics-based studies provide essential information on molecular changes that lead to the activation of physiological mechanisms in response to Cd stress across different experimental conditions (Table 1).

Table 1. Experimental conditions along with exemplary findings in phosphoproteomic studies on cadmium exposure in plants.

Plant Species	Cd Treatment Used for Phosphoproteomic Analyses	Growth Medium	Duration of Cd Treatment	Plant Age	Plant Organs Used for Phosphoproteomic Analyses	Examplary Findings in Investigation	Reference
Solanum lycopersicum	35 μM $CdCl_2$	Hydroponic solution	4 days	20-day-old seedlings	Leaves	Cd tolerance-related proteins (e.g., ABC transporters, HSPs), auxin signaling changes	[11]

Table 1. *Cont.*

Plant Species	Cd Treatment Used for Phosphoproteomic Analyses	Growth Medium	Duration of Cd Treatment	Plant Age	Plant Organs Used for Phosphoproteomic Analyses	Examplary Findings in Investigation	Reference
Oryza sativa	10 μM and 100 μM ($CdCl_2.2.5H_2O$)	Hydroponic solution (Hoagland's solution)	12 days	Seedlings after first leaf fully expanded	Shoots	Differential regulation of proteins related to signaling, stress tolerance, oxidative stress tolerance, and transcription factors	[24]
Oryza sativa	0.1 mmol/L $CdCl_2$	Hydroponic solution	7 days	Third-leaf stage	Leaves	Differential accumulation of proteins related to carbon metabolism, signal transduction, and gene regulation	[25]
Paspalum vaginatum	500 μM $CdCl_2$	1/2 Hoagland's nutrient solution	72 h	30-day-old plants	Roots	Increased phosphorylation of PvFBA1, upregulated activity of pentose phosphate pathway (PPP) pathway, enhanced soluble sugar content	[26]
Populus yunnanensis	40 mg Cd/kg soil	Soil in pots	60 days	1-year-old seedlings	Leaves	Increased sucrose and soluble sugar; reduced ROS and membrane damage	[27]
Populus yunnanensis	1.22 g $CdCl_2.2.5H_2O$ (40 mg Cd^{2+}. Kg^{-1})—added to each pot every 5 days	Soil in pots	60 days	20 cm height (with 10–15 leaves)	Leaves	Increased Cd uptake and translocation; enhanced GSH and PCs levels; reduced H_2O_2 and MDA	[28]

The different aspects of this will now be discussed in light of various perspectives as follows:

1. Our research group has been undertaking the challenge of unraveling the physiological responses of the tomato genotypes used in the phosphoproteomic approaches mentioned above [11]. Thus, this encompasses the studies covering different approaches including those involving oxidative stress parameters, other proteomic approaches, antioxidant enzyme activities, and the accumulation of Cd and nutrients in plant tissues (see insights and information reviewed in Marques et al. [1,2]). In addition, these physiological data are well-positioned to provide an overview of the broad context of mechanisms underlying Cd tolerance, pointing out that several levels of analysis have to be considered to fully understand tomato responses to Cd.
2. He et al. [27] demonstrated that N supplementation in poplar plants mitigated the Cd-induced loss of chlorophyll and increased soluble sugar content. Such physiological improvements are associated with membrane stability and ion balance during heavy metal stress. Moreover, N augmented the activities of key antioxidant enzymes that

mitigate oxidative stress imposed by Cd [27]. This demonstrates that N participates not only in supporting photosynthesis but also in enhancing cellular defense mechanisms. Further investigation showed that N supplementation in Cd-exposed poplar plants increases the synthesis and activity of stress-responsive proteins, including HSP70 and peroxidase (POD). These proteins support plants in protein stability maintenance and detoxifying reactive oxygen species (ROS), thereby reducing oxidative damage. It is interesting that N supplementation enhanced Cd uptake, suggesting that poplar plants could use N to cope better with Cd accumulation through an integrated action of more effective detoxification and sequestration mechanisms [28].

3. Studies have demonstrated that applying N to poplar plants under Cd stress leads to an increase in the levels of proteins which play key roles in sucrose and starch metabolism [27,28]. Considering that other authors have reported that different forms of N additions can alleviate stress induced by other heavy metals besides Cd [31], the exploration and phosphoproteomic observation in a multi-stress context is also relevant. Indeed, the potential of phosphoproteomic studies to identify key regulatory proteins and phosphorylation sites involved in plant responses to combined stresses or combined environmental settings, such as heavy metal toxicity and nutrient availability, is thus a potential focus in further investigations. Furthermore, future research could focus on further determining how these changes in sucrose metabolism contribute to Cd tolerance in poplar and other crops.
4. Huang et al. [28] found that exogenous N enhances the efficiency of poplar plants in responding to Cd stress by identifying differentially accumulated proteins in parallel with phosphoproteomic findings, which included translationally controlled tumor proteins (TCTPs). The relevance of TCTPs have been highlighted for its potential and application in plant growth improvement and abiotic stress tolerance [32]; however, the detailed mechanisms are yet to be determined. Even though some phosphoproteomic studies have not confirmed such a differential regulation of TCTP under Cd stress, these results suggest that a study of TCTPs within a phosphoproteomic framework may reveal their specific functions regarding their role in the development of stress tolerance. This is particularly important because TCTP-related mechanisms may differ in different plant species and under specific treatments [32], suggesting diverse phosphorylation patterns may appear together with such responses. More investigations will be required to understand the phosphorylation dynamics in parallel with TCTPs function and other candidates, as well as interactions with other stress-related proteins that could enrich our understanding of the molecular network constituting plant resilience to diverse stresses.
5. In addition to N, the investigation of other elemental treatments is equally relevant in the context of phosphoproteomics and Cd exposure in plants. For example, Cu homeostasis is important for normal plant growth under Cd toxicity [33]. Importantly, the complex interplay of other factors, such as K, Ca, and Mg accumulation, has been reviewed as playing a critical role in modulating signaling pathways under Cd stress [2]. Such elements may therefore alter the physiological and biochemical processes of the plant and provide tolerance or sensitivity mechanisms. Consequently, studies need to be expanded from N to other essential elements so that a comprehensive understanding of the responses of plants to Cd may be made at the phosphoproteomic level.
6. Using a phosphoproteome approach, Fang et al. [25] showed that a Cd-tolerant rice genotype (D28) exhibited increased expression of several phosphorylated proteins involved in key processes potentially related to Cd tolerance. These molecules included proteolytic enzymes, enzymes involved in carbon metabolism, and stress-related proteins such as F-box transcription factors and a DEAD-box ATP-dependent RNA helicase. Either way, further research, incorporating phosphoproteomic analysis to identify phosphorylation changes and status in detoxification-related proteins, is needed to definitively establish the functional connections between these metabolic changes and detoxification mechanisms in Cd-tolerant rice and across different rice

genotypes. Other findings indicated that phosphorylated proteins involved in ROS detoxification play an important role in protecting cellular structures against Cd-induced oxidative stress in rice [24]. This adds an important layer of understanding to how rice plants integrate signaling pathways and enzymatic defenses to maintain homeostasis under Cd exposure.

7. Another important approach for understanding Cd stress responses in plants is to investigate protein interactions among differentially phosphorylated proteins, as shown by Zhong et al. [24]. The protein–protein interaction network generated in their study has revealed important kinases and phosphatases and specifically showed that PP2C family phosphatases are central players in response to Cd. These findings suggest ABA signaling pathways may play an important role in Cd detoxification and tolerance. Furthermore, the identification of proteins involved in osmotic stress tolerance, such as SAPK6, along with other pathways, including MAPK and CaMK [24], point to the complex molecular responses to Cd stress that should be explored in further studies by combining protein–protein interaction analysis with phosphoproteomics. Further details on signaling networks responsible for regulating plant resilience against this environmental stress will be valuable.
8. Furthermore, protein phosphorylation plays a key role in hormone-mediated signal transduction pathways. Among them, auxins, cytokinins, brassinosteroids, gibberellin, ABA, and ethylene control such processes with phosphorylation being crucial in modulating the activity of hormone receptors and downstream signaling proteins [8]. Various phosphoproteomic analyses discussed above show that the phenomenology of phosphorylation lies at the core of cellular responses to stress, including those driven by hormones. For example, we identified auxin and ethylene signaling pathways by studying the phosphorylation of proteins in Cd-tolerant tomato genotypes [11]. Proteins in the ethylene biosynthesis pathway of Cd-treated rice plants were differentially phosphorylated as well [25]. Thus, future phosphoproteomic work, in concert with hormonal analyses, will elucidate how Cd stress specifically modifies signaling networks through phosphorylation.
9. Zhong et al. [24] and Marques et al. [11] reported that ABC transporters were induced after Cd stress. Huang et al. [28] reported that exogenous N significantly decreased Cd-induced hydrogen peroxide (H_2O_2) and malondialdehyde (MDA) generation, whereas it enhanced the accumulation of glutathione and PCs. Interestingly, the authors identified that ABC transporter proteins accounted for the transporter proteins upregulated by the N + Cd treatment [28]. As shown and previously reviewed [9,34–36], ABC proteins have been implicated in the action of PCs in the important modification of Cd tolerance by binding and sequestering heavy metals. Further still, this interplay between phosphorylation dynamics, ABC transporters, and PCs points to an important area of future research.
10. Transcription factors are important in orchestrating gene expression regarding Cd stress tolerance, and their targets include genes responsible for antioxidant defense and metal chelation [37,38]. The contrasting Cd tolerance in plant crop phosphoproteome investigations [11,25] and rice Cd-induced phosphoproteomic response [24] was shown to be related to transcription factors. It would therefore be of interest to determine how phosphorylation changes their activity in response to Cd; hence, future research could focus on how phosphoregulation modulates transcription factor networks.
11. The proteomic investigations and insights achieved so far highlighted the significant impact of Cd exposure on the differential regulation of kinases and/or phosphatases [11,24,25,27,28]. This interplay suggests a crucial regulatory mechanism for maintaining cellular homeostasis under the given heavy metal stress involving such enzymes. This topic warrants further investigation in order to further explore how targeted modulation of these molecules could modulate stress tolerance.

12. Cd-induced changes in the phosphorylation level of PvFBA1 and other potential molecules related to Cd tolerance in *Seashore Paspalum* were reported. In parallel, the ability of Cd to bind to PvFBA1, modulating its phosphorylation, in microscale thermophoresis experiments was also observed [26]. In such investigations, the authors reported the genetic transformation of *Arabidopsis thaliana* using the overexpression of the *PvFBA1* gene and related increases in soluble sugar content levels in transgenic plants [26]. These findings indicate the relevance of further exploring the combination of approaches related to plant physiology and biochemistry and genetic engineering tools (including the use of transgenic plants) within the Cd research context. These integrated strategies and applications in other plant species and experimental settings, along with phosphoproteomic insights, will also be relevant in future studies.
13. Overall, such physiological and biochemical findings broaden our knowledge of Cd tolerance by underlining the intricate interrelationship underlying the complexity of phosphoproteomic-related signaling pathways and molecular strategies involved in plant responses to Cd. Further studies need to be conducted to clarify specific phosphorylation patterns and phosphoproteomic profiling across multifaceted cellular and molecular strategies that have so far been put forward for mitigating Cd toxicity. This involves vacuolar chelation and sequestration, the dynamics of signaling and physiological pathways, investigations corresponding to transcription regulation, transport, accumulation studies, HSPs and chaperones, and antioxidant involvement. The involvement of potential molecules also needs further investigation in these aspects since their functions relating to stress response are not yet established.

5. Future Multi-Omics Perspectives

In some investigations involving Cd-exposed plants, proteomics and/or differential gene expression were investigated in parallel with phosphoproteomic parameters [24,25,27,28]. Transcriptomic experiments have shed light on the molecular mechanisms behind the plant's response to Cd stress, but many of these experiments have been performed without considering accompanying proteomic/phosphoproteomic analysis to evaluate how changes in transcriptional profile relate to protein expression and phosphorylation patterns. Gu et al. [39] conducted a comparative transcriptome analysis in *Medicago sativa* and found that melatonin pretreatment to Cd stress induced the expression of a large number of genes involved in protein folding, sorting, and degradation, as well as those involved in protein phosphorylation in sugar metabolism, along with genes encoding kinases (e.g., hexokinase-1). In the context of response to Cd stress and transcriptomic research, Yu et al. [40] identified the involvement of protein phosphorylation in maize seedlings, with key pathways including those of protein modification and signal transduction and differential expression of genes encoding ATP-dependent 6-phosphofructokinase 3 and MAP3K epsilon protein kinase 1. These studies have underlined how the transcriptional regulation related to protein phosphorylation plays an important role in plant resilience to heavy metal stress and have emphasized that future studies should consider multi-omics approaches to understand such complex interactions.

6. Exemplary Studies for Broader Implications of Phosphoproteomics

Exemplary studies, not directly addressing plant Cd exposure but giving valuable information relevant to Cd response, are discussed below. Kohorn et al. [41] examined the phosphoproteomic response of *Arabidopsis thaliana* to oligo-galacturonides, pectin fragments released in response to pathogen attacks. The study that follows highlighted that proteins undergo rapid phosphorylation changes in the WAKs (Wall-Associated Kinases) and MAPK signaling pathways, which also overlap with those activated under Cd stress. It suggests that common phosphorylation mechanisms are used by the plant to respond to the biotic and abiotic stressors, including Cd, supporting its versatility as a general defense strategy.

In an isotopic dimethyl labeling-based quantitative phosphoproteomic study of soybeans, Moradi et al. [42] defined the phosphorylation events that included proteins implicated in calcium signaling and ROS detoxification, both critical components for the management of oxidative stress induced by heavy metals. Though the focus of this study was not exactly on Cd, it highlights phosphorylation as a critical regulator of core stress response pathways that are also activated under Cd stress. These model studies illustrate how phosphoproteomic research, even in non-Cd contexts, stands to offer valuable insights into Cd mitigation mechanisms. These studies extend our knowledge of the phosphorylation networks utilized by plants during their response to a wide array of stressors. They illustrate that some of the phosphorylation events identified in the non-Cd context may also apply in the context of Cd stress responses and provide a basis for further researching Cd tolerance mechanisms in environments under different or multiple stressors.

7. Concluding Remarks and Additional Future Directions

1. Phosphoproteomic studies have significantly expanded our knowledge on plant responses to Cd stress through the identification of key phosphorylation sites and events along with regulated proteins that participate in modulating the signaling, metabolic, and detoxification cascades. The identification of conserved phosphorylation events across more plant species opens promising perspectives for breeding Cd-tolerant varieties. These studies not only reveal ways in which Cd tolerance and response pathways operate at the molecular and biochemical level but also provide a firm biological basis upon which crops that can tolerate Cd-contaminated environments could be developed. The knowledge sought and gained from all these aspects will become critical in enhancing the resilience of crops and will go a long way toward fulfilling agricultural production on contaminated soils in a sustainable manner.
2. To fully exploit this information, these individual protein groups would have to be studied in more detail, including their interactions with other proteins. Functional validation of such proteins can demonstrate more specific roles they play in Cd tolerance and response, and hence provide additional and novel targets for breeding or engineering Cd-resistant crops.
3. Furthermore, there is an increasing demand for integrative studies of physiological and biochemical analyses complementary to omics data. Using such diverse approaches will enrich the information obtained from phosphoproteomics as Cd stress induces multiple parallel responses in plants. More comprehensive studies are needed to correlate the molecular responses with the observable physiological characteristics. This may create more encompassing methods for developing Cd-tolerant crops and plant genotypes.
4. Besides the studies involving a large number of plant species and experimental settings in phosphoproteomics, Cd tolerance and sensitivity mechanisms still lack research across different plant species considering genotypes or plants with contrasting Cd tolerance. Increasing the pool will afford a look at species-specific and shared patterns of stress responses. Such a wider comparison may give indications for more focused breeding approaches and strategies and identify key regulatory pathways operating across plant species.
5. Overall, the integration of plant physiology, nutrient accumulation, and molecular responses has provided a panoramic view—from cellular level to phenotype—of how plants respond to Cd stress. With such insight, future studies will appropriately follow through on the translation gap from basic findings to field-level applications that will potentially lead to the development of crops with improved Cd tolerance, as well as to other environmental stressors.
6. The identification of specific phosphorylation sites in stress-related proteins opened new perspectives on plant defense mechanisms and furthered the impulse on research that should concentrate on the identification of more phosphorylation sites and their

functional significance in stress adaptation as a means of further exploiting these mechanisms to enhance Cd tolerance.

7. Identifying potential solutions to Cd stress through integrated multi-omics involving, for instance, phosphoproteomics, transcriptomics, metabolomics, along with other omic approaches, will provide more functional details on how plants cope with Cd toxicity; all these data will be useful in breeding programs oriented to obtain tolerant plants to Cd contamination conditions.
8. Field-level considerations, phosphoproteomic comparisons involving both short- and long-term Cd exposure settings, and interactions among the various types of stresses represent targeted approaches. It must be underlined that, in the field, stresses due to Cd are not usually incurred alone but, rather, together with other abiotic and biotic stresses, including drought and pathogen attacks. Future studies will be required to understand how co-occurring stresses interact with Cd responses at the phosphoproteomic level. A proper understanding of such complex interactions among multiple stresses will help in fine-tuning strategies for Cd mitigation, covering both phytoremediation and environmental safety purposes. Among challenges to date is our ability to translate fundamental knowledge on phosphoproteomic response to Cd stress into practical field performance. What will be needed is to strike a balance between understanding phosphoproteomics and creating plant genotypes that are pleasing to agronomists. Further confirmation through field trials will be necessary to ascertain whether this indeed translates the findings in the laboratory into improved performance of crops in the field. Full implementation of phosphoproteomic findings will necessitate a serious commitment to closing the gap between laboratory and field applications. This will involve the validation of phosphoproteomic insights under varied environmental conditions, which would ensure that stress responses identified under the complex realities of agricultural systems are effective. Indeed, this will be an important way to develop robust crops with the capability to withstand Cd stress in different environments. By addressing these directions, future research would provide further insight into Cd stress responses, integrate omics with physiological understanding, and offer practical applications to enhance crop resilience in Cd-contaminated environments.
9. Different phosphoproteomic approaches and strategies have been used in the studies on the response of plants to Cd stress. We utilized a liquid chromatography–tandem mass spectrometry (LC–MS/MS)-based high-throughput label-free quantitative approach to identify, quantify, and compare phosphopeptides between Cd-tolerant and sensitive tomato genotypes [11]. Labeling with tandem mass tags (TMT), enrichment through high-pH reverse-phase high-performance liquid chromatography (HPLC), and analysis by LC-MS/MS were employed for seashore paspalum plants under Cd treatment [26]. Labeling with TMT, IMAC (immobilized metal affinity chromatography), fractionation using HPLC, and analysis by LC-MS/MS have been employed for poplar plants treated with Cd and N [27,28]. Fang et al. [25] used 2D gel electrophoresis followed by Matrix-Assisted Laser Desorption/Ionization Time-of-Flight/Time-of-Flight Tandem Mass Spectrometry (MALDI-TOF-TOF MS/MS) for analyzing differential accumulation of phosphorylated proteins in rice lines with contrasting Cd tolerance. Zhong et al. [24] adopted an iTRAQ (isobaric tag for relative and absolute quantitation)-based quantitative phosphoproteomics approach for the investigation of rice seedling responses under Cd exposure. Further improvement of phosphoproteomic technologies in the future would go towards enhancing the sensitivity to detect low abundance phosphopeptides, improving the enrichment methods of phosphopeptides, and/or developing bioinformatics tools for analyses and interpretation of the complex phosphoproteomic data. Future improvements will allow for an in-depth understanding of the role of protein phosphorylation in plant Cd stress responses and related biological processes. In addition, some potential approaches could be incorporated into such omics perspectives, such as Nuclear

Magnetic Resonance (NMR) spectroscopy techniques, which can contribute to different stages of protein analysis, including characterizing the structure, dynamics, and interactions of proteins [43].

10. Our research group is currently working on a grafting technique—a practice with numerous basic plant physiology and real field applications, as performed and reviewed by our research group [1,2,44–46]—by combining phosphoproteomic analysis with physiological approaches. We expect that the grafted plants will provide new insights regarding differential Cd tolerance, probably by enhancing root-to-shoot communication and stress signaling. Preliminary results are encouraging, and we hope to present the results about phosphoproteome changes in grafted plants in forthcoming communications.

Author Contributions: D.N.M.: Conceptualization, Formal Analysis, Investigation, Writing—Original Draft, Writing—Review and Editing, Visualization. F.A.P.: Writing—Review and Editing. R.A.A.: Supervision, Resources, Writing—Review and Editing. All authors have read and agreed to the published version of the manuscript.

Funding: This work was supported by the São Paulo Research Foundation—FAPESP [grant numbers 2020/12666-7 and 2022/11018-7 to DNM] (from Brazil).

Institutional Review Board Statement: Not applicable.

Data Availability Statement: No new data were created or analyzed in this study. Data sharing is not applicable to this article.

Acknowledgments: D.N.M. thanks Olena K. Vatamaniuk (Cornell University, NY, USA) for her suggestions and comments on the manuscript. He also expresses gratitude to Aleksandra Skirycz and Boyce Thompson Institute, Cornell University (USA), for the support during the multi-omics journey and plant cadmium research in the United States. D.N.M. acknowledges Hirofumi Nakagami and the Protein Mass Spectrometry Group at the Max Planck Institute for Plant Breeding Research, Max Planck Society (Germany), for their collaboration, support, and guidance during the scientific phosphoproteomics journey and plant cadmium research conducted in Germany.

Conflicts of Interest: The authors declare no conflicts of interest.

References

1. Marques, D.N.; Mason, C.; Stolze, S.C.; Harzen, A.; Nakagami, H.; Skirycz, A.; Piotto, F.A.; Azevedo, R.A. Grafting systems for plant cadmium research: Insights for basic plant physiology and applied mitigation. *Sci. Total Environ.* **2023**, *892*, 164610. [CrossRef] [PubMed]
2. Marques, D.N.; Nogueira, M.L.; Gaziola, S.A.; Batagin-Piotto, K.D.; Freitas, N.C.; Alcantara, B.K.; Paiva, L.V.; Mason, C.; Piotto, F.A.; Azevedo, R.A. New insights into cadmium tolerance and accumulation in tomato: Dissecting root and shoot responses using cross-genotype grafting. *Environ. Res.* **2023**, *216*, 114577. [CrossRef] [PubMed]
3. Khan, Z.; Elahi, A.; Bukhari, D.A.; Rehman, A. Cadmium sources, toxicity, resistance, and removal by microorganisms—A potential strategy for cadmium eradication. *J. Saudi Chem. Soc.* **2022**, *26*, 101569. [CrossRef]
4. Ullah, H.; Zheng, W.; Sheng, Y. Translocation of CdS nanoparticles in maize (*Zea mays* L.) plant and its effect on metabolic response. *Chemosphere* **2023**, *343*, 140189. [CrossRef] [PubMed]
5. Clemens, S.; Aarts, M.G.; Thomine, S.; Verbruggen, N. Plant science: The key to preventing slow cadmium poisoning. *Trends Plant Sci.* **2013**, *18*, 92–99. [CrossRef]
6. Clemens, S.; Ma, J.F. Toxic heavy metal and metalloid accumulation in crop plants and foods. *Annu. Rev. Plant Biol.* **2016**, *67*, 489–512. [CrossRef]
7. Vitelli, V.; Giamborino, A.; Bertolini, A.; Saba, A.; Andreucci, A. Cadmium stress signaling pathways in plants: Molecular responses and mechanisms. *Curr. Issues Mol. Biol.* **2024**, *46*, 6052–6068. [CrossRef]
8. Zhang, W.J.; Zhou, Y.; Zhang, Y.; Su, Y.H.; Xu, T. Protein phosphorylation: A molecular switch in plant signaling. *Cell Rep.* **2023**, *42*, 112729. [CrossRef]
9. Marques, D.N.; Gaziola, S.A.; Azevedo, R.A. Phytochelatins and their relationship with modulation of cadmium tolerance in plants. In *Handbook of Bioremediation: Physiological, Molecular and Biotechnological Interventions*; Hasanuzzaman, M., Prasad, M.N.V., Eds.; Elsevier: Cambridge, UK, 2021; pp. 91–113.
10. Borges, K.L.R.; Salvato, F.; Loziuk, P.L.; Muddiman, D.C.; Azevedo, R.A. Quantitative proteomic analysis of tomato genotypes with differential cadmium tolerance. *Environ. Sci. Pollut. Res. Int.* **2019**, *26*, 26039–26051. [CrossRef]

11. Marques, D.N.; Stolze, S.C.; Harzen, A.; Nogueira, M.L.; Batagin-Piotto, K.D.; Piotto, F.A.; Mason, C.; Azevedo, R.A.; Nakagami, H. Comparative phosphoproteomic analysis of tomato genotypes with contrasting cadmium tolerance. *Plant Cell Rep.* **2021**, *40*, 2001–2008. [CrossRef]
12. Chong, L.; Hsu, C.C.; Zhu, Y. Advances in mass spectrometry-based phosphoproteomics for elucidating abscisic acid signaling and plant responses to abiotic stress. *J. Exp. Bot.* **2022**, *73*, 6547–6557. [CrossRef] [PubMed]
13. Reddy, N.G.; Prasad, M.N.V. Cadmium-induced protein phosphorylation changes in rice (*Oryza sativa* L.) seedlings. *J. Plant Physiol.* **1995**, *145*, 67–70. [CrossRef]
14. Krishnan, H.B.; Pueppke, S.G. Heat shock triggers rapid protein phosphorylation in soybean seedlings. *Biochem. Biophys. Res. Commun.* **1987**, *148*, 762–767. [CrossRef] [PubMed]
15. Gouia, H.; Ghorbal, M.H.; Meyer, C. Effects of cadmium on activity of nitrate reductase and on other enzymes of the nitrate assimilation pathway in bean. *Plant Physiol. Biochem.* **2000**, *38*, 629–638. [CrossRef]
16. Jonak, C.; Nakagami, H.; Hirt, H. Heavy metal stress: Activation of distinct mitogen-activated protein kinase pathways by copper and cadmium. *Plant Physiol.* **2004**, *136*, 3276–3283. [CrossRef]
17. Yeh, C.M.; Hsiao, L.J.; Huang, H.J. Cadmium activates a mitogen-activated protein kinase gene and MBP kinases in rice. *Plant Cell Physiol.* **2004**, *45*, 1306–1312. [CrossRef]
18. Jakubowska, D.; Janicka, M. The role of brassinosteroids in the regulation of the plasma membrane H^+-ATPase and NADPH oxidase under cadmium stress. *Plant Sci.* **2017**, *264*, 37–47. [CrossRef]
19. Ma, Q.J.; Sun, M.H.; Lu, J.; Hu, D.G.; Kang, H.; You, C.X.; Hao, Y.J. Phosphorylation of a malate transporter promotes malate excretion and reduces cadmium uptake in apple. *J. Exp. Bot.* **2020**, *71*, 3437–3449. [CrossRef]
20. Zhang, Y.; Wang, Z.; Liu, Y.; Zhang, T.; Liu, J.; You, Z.; Huang, P.; Zhang, Z.; Wang, C. Plasma membrane-associated calcium signaling modulates cadmium transport. *New Phytol.* **2023**, *238*, 313–331. [CrossRef]
21. Du, X.; Zhou, L.; Zhu, B.; Gu, L.; Yin, H.; Wang, H. The TabHLH35-TaWAK20-TaSPL5 pathway positively regulates Cd stress in wheat. *Theor. Appl. Genet.* **2023**, *136*, 153. [CrossRef]
22. Moravčíková, D.; Žiarovská, J. The Effect of Cadmium on Plants in Terms of the Response of Gene Expression Level and Activity. *Plants* **2023**, *12*, 1848. [CrossRef] [PubMed]
23. Ul Haq, S.; Khan, A.; Ali, M.; Khattak, A.M.; Gai, W.X.; Zhang, H.X.; Wei, A.M.; Gong, Z.H. Heat Shock Proteins: Dynamic Biomolecules to Counter Plant Biotic and Abiotic Stresses. *Int. J. Mol. Sci.* **2019**, *20*, 5321. [CrossRef] [PubMed]
24. Zhong, M.; Li, S.; Huang, F.; Qiu, J.; Zhang, J.; Sheng, Z.; Tang, S.; Wei, X.; Hu, P. The phosphoproteomic response of rice seedlings to cadmium stress. *Int. J. Mol. Sci.* **2017**, *18*, 2055. [CrossRef] [PubMed]
25. Fang, Y.; Deng, X.; Lu, X.; Zheng, J.; Jiang, H.; Rao, Y.; Zeng, D.; Hu, J.; Zhang, X.; Xue, D. Differential phosphoproteome study of the response to cadmium stress in rice. *Ecotoxicol. Environ. Saf.* **2019**, *180*, 780–788. [CrossRef] [PubMed]
26. Zheng, Y.; Liu, Y.; Jiang, Y.; Li, Z.; Zhang, Q.; Yu, Q.; Liu, Y.; Liu, J.; Yang, Z.; Chen, Y. Phosphoproteomics profiling reveals the regulatory role of a phosphorylated protein PvFBA1 in cadmium tolerance in seashore paspalum. *Ecotoxicol. Environ. Saf.* **2024**, *286*, 117220. [CrossRef]
27. He, F.; Zhao, Q.; Huang, J.L.; Niu, M.X.; Feng, H.C.; Shi, Y.J.; Zhao, K.J.; Cui, X.L.; Wu, X.L.; Mi, J.X.; et al. External application of nitrogen alleviates toxicity of cadmium on poplars via starch and sucrose metabolism. *Tree Physiol.* **2021**, *41*, 2126–2141. [CrossRef]
28. Huang, J.; Wu, X.; Tian, F.; Chen, Q.; Luo, P.; Zhang, F.; Wan, X.; Zhong, Y.; Liu, Q.; Lin, T. Changes in proteome and protein phosphorylation reveal the protective roles of exogenous nitrogen in alleviating cadmium toxicity in poplar plants. *Int. J. Mol. Sci.* **2019**, *21*, 278. [CrossRef]
29. Zhang, S.; Yang, C.; Chen, M.; Chen, J.; Pan, Y.; Chen, Y.; Rahman, S.U.; Fan, J.; Zhang, Y. Influence of nitrogen availability on Cd accumulation and acclimation strategy of Populus leaves under Cd exposure. *Ecotoxicol. Environ. Saf.* **2019**, *180*, 439–448. [CrossRef]
30. Wang, X.; Du, H.; Ma, M.; Rennenberg, H. The dual role of nitric oxide (NO) in plant responses to cadmium exposure. *Sci. Total Environ.* **2023**, *892*, 164597. [CrossRef]
31. Bai, T.; Liu, Y.-Y.; Muhammad, I.; Yang, X.; Yin, X.-J.; Bai, L.; Wang, Y.-J. Mixed nitrogen form addition facilitates the growth adaptation of legume plants to heavy metal contamination in degraded mining areas. *Glob. Ecol. Conserv.* **2020**, *24*, e01387. [CrossRef]
32. Marques, D.N.; Barros, N.L.F.; de Souza, C.R.B. Does translationally controlled tumor protein (TCTP) have the potential to produce crops with increased growth and tolerance to environmental stresses? *Plant Cell Rep.* **2023**, *42*, 821–824. [CrossRef] [PubMed]
33. Gayomba, S.R.; Jung, H.I.; Yan, J.; Danku, J.; Rutzke, M.A.; Bernal, M.; Krämer, U.; Kochian, L.V.; Salt, D.E.; Vatamaniuk, O.K. The CTR/COPT-Dependent Copper Uptake and SPL7-Dependent Copper Deficiency Responses Are Required for Basal Cadmium Tolerance in *A. thaliana*. *Metallomics* **2013**, *5*, 1262–1275. [CrossRef] [PubMed]
34. Kim, D.Y.; Bovet, L.; Maeshima, M.; Martinoia, E.; Lee, Y. The ABC Transporter AtPDR8 Is a Cadmium Extrusion Pump Conferring Heavy Metal Resistance. *Plant J.* **2007**, *50*, 207–218. [CrossRef]
35. Rea, P.A. Plant ATP-Binding Cassette Transporters. *Annu. Rev. Plant Biol.* **2007**, *58*, 347–375. [CrossRef]
36. Mendoza-Cózatl, D.G.; Zhai, Z.; Jobe, T.O.; Akmakjian, G.Z.; Song, W.Y.; Limbo, O.; Russell, M.R.; Kozlovskyy, V.I.; Martinoia, E.; Vatamaniuk, O.K.; et al. Tonoplast-Localized Abc2 Transporter Mediates Phytochelatin Accumulation in Vacuoles and Confers Cadmium Tolerance. *J. Biol. Chem.* **2010**, *285*, 40416–40426. [CrossRef]

37. Li, Y.; Ding, L.; Zhou, M.; Chen, Z.; Ding, Y.; Zhu, C. Transcriptional Regulatory Network of Plant Cadmium Stress Response. *Int. J. Mol. Sci.* **2023**, *24*, 4378. [CrossRef] [PubMed]
38. Zhang, H.; Lu, L. Transcription Factors Involved in Plant Responses to Cadmium-Induced Oxidative Stress. *Front. Plant Sci.* **2024**, *15*, 1397289. [CrossRef]
39. Gu, Q.; Xie, C.; Zhang, S.; Zhou, T.; Li, N.; Xu, C.; Zhou, Z.; Wang, C.; Chen, Z. Transcriptomic analysis provides insights into the molecular mechanism of melatonin-mediated cadmium tolerance in *Medicago sativa* L. *Ecotoxicol. Environ. Saf.* **2024**, *278*, 116411. [CrossRef]
40. Yu, S.; Zhu, J.; Yin, Y.; Zhang, X.; Dai, Y.; Xing, Y.; Cheng, X.; Zhang, A.; Li, C.; Zhu, Y.; et al. Dynamic transcriptome profiling revealed a key gene *ZmJMJ20* and pathways associated with cadmium stress in maize. *Ecotoxicol. Environ. Saf.* **2024**, *277*, 116352. [CrossRef]
41. Kohorn, B.D.; Hoon, D.; Minkoff, B.B.; Sussman, M.R.; Kohorn, S.L. Rapid oligo-galacturonide induced changes in protein phosphorylation in *Arabidopsis*. *Mol. Cell. Proteom.* **2016**, *15*, 1351–1359. [CrossRef]
42. Moradi, A.; Dai, S.; Wong, E.O.Y.; Zhu, G.; Yu, F.; Lam, H.M.; Wang, Z.; Burlingame, A.; Lin, C.; Afsharifar, A.; et al. Isotopically dimethyl labeling-based quantitative proteomic analysis of phosphoproteomes of soybean cultivars. *Biomolecules* **2021**, *11*, 1218. [CrossRef] [PubMed]
43. Kaas, Q.; Craik, D.J. NMR of plant proteins. *Prog. Nucl. Magn. Reson. Spectrosc.* **2013**, *71*, 1–34. [CrossRef] [PubMed]
44. Gratão, P.L.; Monteiro, C.C.; Tezotto, T.; Carvalho, R.F.; Alves, L.R.; Peters, L.P.; Azevedo, R.A. Cadmium Stress Antioxidant Responses and Root-to-Shoot Communication in Grafted Tomato Plants. *Biometals* **2015**, *28*, 803–816. [CrossRef]
45. Alves, L.R.; Monteiro, C.C.; Carvalho, R.F.; Ribeiro, P.C.; Tezotto, T.; Azevedo, R.A.; Gratão, P.L. Cadmium Stress Related to Root-to-Shoot Communication Depends on Ethylene and Auxin in Tomato Plants. *Environ. Exp. Bot.* **2017**, *13*, 102–115. [CrossRef]
46. Lupp, R.M.; Marques, D.N.; Lima Nogueira, M.; Carvalho, M.E.A.; Azevedo, R.A.; Piotto, F.A. Cadmium tolerance in tomato: Determination of organ-specific contribution by diallel analysis using reciprocal grafts. *Environ. Sci. Pollut. Res. Int.* **2024**, *31*, 215–227. [CrossRef] [PubMed]

MDPI AG
Grosspeteranlage 5
4052 Basel
Switzerland
Tel.: +41 61 683 77 34

International Journal of Molecular Sciences Editorial Office
E-mail: ijms@mdpi.com
www.mdpi.com/journal/ijms

www.ingramcontent.com/pod-product-compliance
Lightning Source LLC
Chambersburg PA
CBHW061547210926
51850CB00009B/2029
* 9 7 8 3 7 2 5 8 7 5 6 1 0 *